Evolutionary Biology – New Perspectives on Its Development

Volume 2

Evolutionary biology has been a remarkably dynamic area since its foundation. Its true complexity, however, has been concealed in the last 50 years under an assumed opposition between the "Extended Evolutionary Synthesis" and an "Alternative to the Evolutionary Synthesis". This multidisciplinary book series aims to move beyond the notion that the development of evolutionary biology is structured around a lasting tension between a Darwinian tradition and a non-Darwinian tradition, once dominated by categories like Darwinian Revolution, Eclipse of Darwinism, Evolutionary Synthesis, and Post-Synthetic Developments.

The monographs and edited volumes of the series propose an alternative to this traditional outlook with the explicit aim of fostering new thinking habits about evolutionary biology, a multifaceted area composed of changing and interacting research entities and explanatory levels. Contributions by biologists and historians/philosophers are welcomed. Topics covered in the series span from (among many other possibilities):

- An Overview of Neutralist Theories in Evolutionary Biology
- Developmental Biology: From Reductionism to Holism and Back
- Selection Theories Beyond Hard and Soft Inheritance
- Divergent, Parallel, and Reticulate Evolution: Competing or Complementary Research Programs?
- The Rise of Molecular Biology: Between Darwinian and Non-Darwinian
- Biologizing Paleontology: A Tradition with Deep Historical Roots
- The Darwinian Revolution and the Eclipse of Darwinism: Blurring the Historiographical Lines
- Darwinism, Lamarckism, Orthogenesis: Can We Really Define Them by Their Hard Explanatory Cores?
- The Evolutionary Synthesis: A Fabricated Concept?
- The Opposition to the Evolutionary Synthesis: Criticizing a Phantom?
- A Reversed Perspective: Approaching Charles Darwin from the Pre-1859 Period
- The Long Development of the Multilevel Paradigm in Evolutionary Biology
- Self-Organization: A Research Tradition from Morphology to Cosmology
- Human Evolution: Sociobiological or Sociocultural?

More information about this series at http://www.springer.com/series/16175

Salvatore J. Agosta • Daniel R. Brooks

The Major Metaphors of Evolution

Darwinism Then and Now

Springer

Salvatore J. Agosta
VCU Life Sciences, Center
for Environmental Studies
Virginia Commonwealth University
Richmond, VA, USA

Daniel R. Brooks
Department of Ecology and Evolutionary
Biology
University of Toronto
Toronto, ON, Canada

ISSN 2524-7751 ISSN 2524-776X (electronic)
Evolutionary Biology – New Perspectives on Its Development
ISBN 978-3-030-52085-4 ISBN 978-3-030-52086-1 (eBook)
https://doi.org/10.1007/978-3-030-52086-1

This Springer imprint is published by the registered company Springer Nature Switzerland AG.
The registered company address is: Gewerbestrasse 11, 6330 Cham, Switzerland

This book is dedicated to the memory of John Collier, who not only worked through and resolved many of the conceptual quagmires we wandered into, but who also set a standard of intellectual pursuit always in the context of an ethical life. We hope he would have largely approved of our efforts, and we think he would have been one of the first to begin the job of improving upon the platform we present herein.

Acknowledgements

Our thanks to Prof. Richard Delisle for inviting us to present our ideas in this series; to Eric Hoberg, Walter Boeger and Alicia Juarrero for detailed and exceptionally helpful critiques of the manuscript; and to Eloy Ortiz Hernandez, Deborah McLennan and Walter Boeger, who prepared the illustrations.

Salvatore J. Agosta and Daniel R. Brooks

I began devouring books about evolution when I was a teenager. This book represents the culmination of a 30-year journey that started then, marked by the influence of too many people and places to name. Special thanks to Mike Steele (Wilkes University), Joe Bernardo (Texas A&M) and Art Dunham (University of Pennsylvania) for their outstanding mentorship and personal friendship along the way. Thanks also to Dan Janzen (University of Pennsylvania) for introducing me to the Area de Conservación Guanacaste, Costa Rica, and the wonders of tropical field biology and for fostering key ideas that led to the evolution of this book. And most of all, special thanks to my co-author Dan Brooks for being an outstanding mentor, colleague and friend. Quite simply, his impact on my career has been enormous, and I am extremely proud and grateful of the opportunity to write this book together.

Special thanks also to Jeff Klemens (Philadelphia University), a great friend and co-founder of the concept of "ecological fitting in sloppy fitness". Jeff and I spent countless hours together in Philadelphia and Costa Rica as graduate students talking about science and just about everything else—suffice it to say, we had more than a good time doing it.

And special thanks to my family. Mom, Dad, Peter, Jill, Alexis and Devin—my foundation with which none of this could have happened. Sam and Sofia—my furry non-human writing companions who put a smile on my face even in the most stressful times.

And finally, I thank my wife and best friend, Jessica, who has given me nothing but her love and support over the years. She is my strength and the best companion I could ever hope for.

Salvatore J. Agosta

This book represents a journey of nearly a half century, and explaining why each person is acknowledged would take too much space. Special thanks to Ed Wiley, who suggested we think about trying to make some kind of connection between information theory, phylogenetic analysis and nonequilibrium thermodynamics; James D. Smith, a faculty member at California State University, Fullerton (CSUF) and editor of *Systematic Zoology*, who had the courage to publish the second article Ed and I wrote when *Nature*, which accepted the first ms, never published it or replied to inquiries about its fate. Jim also tirelessly pursued funding for a number of later events at CSUF which explored many aspects of the concepts discussed in this book; Deborah McLennan, who co-developed historical ecology in three books and numerous articles with me; and Eörs Szathmáry, who invited me to be a Senior Visiting Fellow at the Collegium Budapest in 2010–2011 to work on the idea that ultimately became this book and continued to inspire and support me in pursuing this project at the Institute for Advanced Study Köszeg (iASK) in 2016, 2018 and 2019 and at the Institute for Evolutionary Biology, Centre for Ecology, Tihany, Hungary, in 2017, 2018 and 2019. My thanks to Ferenc Miszlevitz, Director of iASK, and the institute's kind and generous staff. Thanks also to friends and colleagues who provided foundations and insights: Walter Boeger, Universidade Federal do Parana, Curitiba, Brazil; Niles Eldredge, American Museum of Natural History; Terry Erwin, National Museum of Natural History, Smithsonian Institution; Paul Leblond, Jack Maze and Geoffrey GE Scudder, University of British Columbia; Menachem Fisch, University of Tel Aviv; Michael Hammond, University of Toronto; Eric Hoberg, Museum of Southwestern Biology; Alicia Juarrero, University of Miami; John D. Lynch, University of Nebraska-Lincoln; Sven Jakobsson, Niklas Janz, Sören Nylin, Hans-Erik Wanntorp and Christer Wiklund, Stockholm University; Koichiro Matsuno, Nagaoka University of Technology; Alvaro Moreno, University of the Basque Country; Stan Salthe, CUNY; Mike Singer, Plymouth University; Jonathan Smith, Iowa State University; Alycia Stigall, Ohio University; Peter Watts, Toronto; David Sloan Wilson, SUNY Binghamton; and Rick Winterbottom, Royal Ontario Museum.

Sadly, John Collier, University of Kwa Zulu Natal, Cyril V Finnegan, University of British Columbia, Terry Erwin and Vicki Funk, US Museum of Natural History, Smithsonian Institution; David Hull, Northwestern University; Brian Maurer, Michigan State University; and Birgitta Sillen-Tullberg, Stockholm University, are not here to see this book to which their friendship and insights and support contributed so much.

My strongest thanks to my co-author Sal Agosta, who has been a personal friend and invaluable colleague since a memorable breakfast introduction in the *comedor* of the Area de Conservación Guanacaste in the summer of 2003.

Finally, I thank my wife, Zsuzsanna, who has managed a decade of disruption and chaos with far more grace and equanimity than anyone could humanly expect.

Daniel R. Brooks

Contents

About the Authors

Salvatore J. Agosta is an Associate Professor at Virginia Commonwealth University. He received a PhD in biology from the University of Pennsylvania in 2007, after which he was awarded a Postdoctoral Fellowship in Ecology and Evolutionary Biology from the University of Toronto followed by a Howard Hughes Medical Institute Postdoctoral Fellowship from Wilkes University. Sal is an ecologist and evolutionary biologist whose work ranges from field and laboratory studies of animal–plant interactions in both temperate and tropical habitats to theoretical studies of foundational concepts in ecology and evolution. His current research focuses on the physiological ecology of plant-feeding insects, particularly in the context of biological invasions and climate change.

Daniel R. Brooks is Professor Emeritus, University of Toronto. He is a Fellow of the Royal Society of Canada (Academy of Science) and Fellow of the Linnaean Society of London and has been awarded honorary doctorates from Stockholm University and the University of Nebraska. He has been a Senior Visiting Fellow of the Collegium Budapest, Stellenbosch Institute for Advanced Study, Institute of Advanced Studies Köszeg and the Hungarian National Institute of Ecology. Dan is an evolutionary biologist whose more than 375 scientific publications, including half a dozen books, range from field studies of the evolution of host–pathogen systems in tropical wildlands to foundational studies of evolutionary theory. His current focus is integrating evolutionary principles into developing proactive public policy for coping with global climate change, with an emphasis on the emerging disease crisis.

Chapter 1
Context

Abstract The unifying principle of Biology is that all life is evolved life. As part of science's social contract, the main goal of Biology should be to "put evolution to work" for humanity, which is currently facing existential threats stemming from the combination of overpopulation, globalized trade and travel, urbanization, and widespread environmental change. These are all complex phenomena involving living systems, putting Biology and its central theory at the forefront of efforts to deal with the problems they cause. Now more than ever, we need a coherent operational evolutionary theory that works for humanity. To accomplish this, we argue, requires a return to Darwinism to recover useful but discarded ideas and to extend it forward while removing barriers between specialized areas of research to integrate new ones.

This book is a contribution to the series *Evolutionary Biology: New Perspectives on its Development*. The editor of the series founded it to encourage a forum for biologists, historians, and philosophers to express ideas in an effort to form a coherent narrative for evolutionary biology. We are pleased to be able to contribute to the series. We believe humanity faces an existential threat stemming from the complex and intertwined phenomena of global climate change, overpopulation, global trade and travel, and increasing urbanization. These phenomena are all produced or influenced by living systems and affect the lives and futures of living systems, and Biology is increasingly the focus of efforts to cope with them. The unifying principle of Biology is that all life is evolved life, yet too few proposals for safeguarding the future of humanity invoke evolutionary principles, and most of those that do emphasize stasis, not change and diversification.

Humanity seems to have a love–hate relationship with evolution. Many scientists study evolution in a manner that would allow them to control, to engineer, the process. Conservationists want evolution to be a matter of equilibrium-based cycles because they think such cycles are perpetual motion machines that can be controlled, despite evidence of massive episodic disassembly and reassembly of ecosystems throughout history. Genetic engineers, especially those in the agricultural sciences, want infinite food production for an infinitely growing human population. Evolution, however, seems to place limits on our ability to accomplish whatever we wish, and

S. J. Agosta, D. R. Brooks, *The Major Metaphors of Evolution*, Evolutionary Biology – New Perspectives on Its Development 2,
https://doi.org/10.1007/978-3-030-52086-1_1

many scientists chafe at this. Finally, some think that our technological development has removed us from the arena of evolution.

Cosmologists are often in error but never in doubt—attributed to Lev Landau

The Enlightenment created a social contract in which scientists were supposed to understand nature and its laws in order to control them for the betterment of humanity. But such humanitarian impulses, religious and secular, did not result in our attempting to learn to live within our means. Rather, the Enlightenment gave our species permission to be a species of cheaters. We want to cheat on the laws of nature; we want endless energy for free; we want perpetual motion machines; we want to go faster than the speed of light; we want to live forever; we want beautiful children; and we want all of it for free. Physics has promised all that and more.

Biologists think differently than physicists—different kinds of people are attracted to the different disciplines and their training reinforces those interests and propensities for seeing the world. Many things that biologists take for granted as fundamental truths have preoccupied physicists and philosophers for centuries. Biologists tend to understand and accept complexity more easily than physicists because their subject matter is inherently more complex. Physicists operate in a world of relative simplicity in which the fundamental units are generally homogeneous (if you have met one electron, neutron, proton, hydrogen atom, water molecule, etc., you have met them all) and where a comparatively small number of principles can be used to explain the things that happen to them. Biologists operate in a world in which the fundamental units (genes, organisms, populations, species, ecosystems, etc.) are highly heterogeneous due to unique inheritances and histories of interacting with the environment. For biology, the "average individual" does not exist—even clones are not identical (Cepelewicz 2020). This makes it less fruitful to describe things using the "law of large numbers" that makes the elegant difference and differential equation approach of describing physics so successful. Biology is an emergent property of physics and chemistry and therefore exceedingly more difficult to explain. All of the patterns and processes in biology are the result of many simultaneous and potentially interacting factors most of which are scale-dependent (Quinn and Dunham 1983), so the domain of generality in describing any given part of biological systems becomes extremely limited (Dunham and Beaupre 1998). Biologists are preconditioned to *expect* that everything will be complex. And yet, one product of the Enlightenment is that certain topics are considered the exclusive property of physics, while no topic is outside its scope. In 1827, for example, the British Botanist Robert Brown described an interesting phenomenon about the movement of pollen of *Clarkia pulchella* immersed in water (Brown 1828). Almost 80 years later, Albert Einstein published an article in which he translated Brown's almost poetic prose into the cabalistic incantations of physics (Einstein 1906). Einstein's explanation of *Brownian motion* served as convincing evidence that atoms and molecules exist and helped build his career. In 1908, Jean Perrin provided experimental confirmation of what was by then considered to be Einstein's theory (Perrin 1909), for which he was awarded a Nobel Prize in 1926.

We know of one notable exception to the directional flow of epistemic trespassing from Physics and Chemistry into Biology. In our experience, nothing makes physicists more apoplectic than for biologists to mention the word *entropy*. And yet, in a public presentation in Leipzig in 1905, Ludwig Boltzmann said that Darwin had established the initial framework for a statistical mechanics of biology. Boltzmann also said that the one place he disagreed with Darwin was that he felt the struggle for survival was actually a struggle for entropy (Broda 1983). It was Boltzmann, therefore, who gave biologists permission to discuss evolution as an entropic phenomenon.

We believe it is important to rediscover the richness of the past in evolutionary biology, not just in terms of insights gained, lost, and discarded, but also more literally. If all life on this planet is evolved life, then the past is the prologue for all its narratives. Providing a common basis for a coherent narrative demands that we remember (and in some cases rediscover) the reservoir of useful ideas proposed during the last two centuries while removing barriers between specialized areas of research to effectively integrate new findings and insights. As part of science's social contract, the ultimate goal of evolutionary biology should be to create an operational framework to "put evolution to work" for humanity, to make it a theory that matters. We cannot accomplish this task by rehashing old ground, by arguing about semantics, or by leaping from one shiny object to another as the next "new idea" appears and someone decides it is the monolithic answer to all real and perceived shortcomings in existing views about evolution.

We have a single goal in mind—making the conceptual framework of evolutionary biology stronger and more useful to humanity in a time of existential crisis. At the same time, we understand that the conceptual realm for evolutionary biology is extensive and complex, comprising all areas of biological research as well as elements of physics, chemistry, and anthropology. As a consequence, we realized that a useful framework for beginning multiple conversations leading to that hoped-for common narrative needed a highly abstract metalanguage. That abstraction will lead us to veer outside the accepted academic boundaries within which we were trained. And this will inevitably lead some to accuse us of that sin of *epistemic trespassing* (Ballantyne 2019; Bristol and Rossano 2020). We think that would be wrong. We will borrow liberally from outside our training, but we are not trying to tell anyone "out there" how to do their jobs. Rather, we think we can learn from each other and generalize our knowledge as a result of peering into someone else's backyard to see if they have developed ideas that can help us. And when that happens, why should we not extend the range of those ideas and acknowledge where we got them?

> Origin of man now proved—Metaphysic must flourish—He who understands baboon would do more towards metaphysics than Locke—Charles Darwin (1838)

David Hull (1988) was the first person we encountered who actively embraced and combined history, philosophy, and sociology, weaving stories about a range of people involved in scientific controversies into his narrative account of the dynamics of science. Luckily for us, he focused on evolutionary biology rather than physics or chemistry. Hull's findings differed from the wildly popular notion of paradigm shifts espoused by Thomas Kuhn (1962), which treated change in scientific theories as exemplars of intergenerational conflict. If you do not like the Old Guard's ideas, just wait, they will die and you can take over (until you die). But Hull et al. (1978) found that one's age—*Planck's Principle* (Planck 1950)—had little to do with acceptance or rejection of Darwin's theory. Agreement or disagreement with the basic principles mattered a lot. But so did something else, something very human.

Hull discovered that the one thing of paramount importance to scientists was the one thing they themselves often failed to do. And that is a matter of respect—time and again, the scientists Hull interviewed said the most important thing to them was to be acknowledged for their original contributions—no matter how small. And yet, time and again, Hull discovered that in the published research of those same scientists, they often failed to give credit where credit was due. First and foremost, they gave credit where credit would help their careers the most, even if others had published the ideas previously. And feeling that you have been dissed is age-independent. Hull believed that this created the tension within scientific communities that Kuhn mistook for intergenerational conflict. Hull then suggested a positive aspect of such an unpleasant system—if you live in an environment full of people out to get you or looking for a reason to ignore your work, you will be extra careful with your own work. So, Hull reasoned, science is full of nasty people because the need to be factually correct outweighs the need to be polite. Needless to say, when his book appeared, Hull was attacked bitterly for painting a negative picture about how bitterly scientists deal with each other on a personal level. At some point, the attacks stopped, perhaps because the attackers realized that the more aggressive they were, the more Hull's basic thesis seemed sound.

No doubt some studied Hull's book for clues on how to be successfully aggressive, and no doubt many were helped in their career aspirations. At the end of the twentieth century, much—though not all—of the history and philosophy of Biology was geared toward describing, justifying, or celebrating the inevitability of a paradigm that many who extolled its virtues also recognized as flawed and incomplete. The historians and philosophers of biology during the past 50 years have tended to describe the discipline as passing through a few key phases: Darwinian Revolution; Eclipse of Darwinism; Evolutionary Synthesis; Expanded Evolutionary Synthesis. More recent analyses take the perspective that those phases are not so much objective mileposts as social constructs created to frame an intellectual and institutional agenda emphasizing key people rather than key contributions (e.g., Delisle 2008, 2011, 2019 and references therein).

This led a new generation of historians and philosophers to begin asking different questions. Interdisciplinary conferences sprang up, examining evolution within the context of topics never previously associated with evolutionary theory: closure, complexity, entropy, information, irreversibility, self-organization, systems theory (e.g., Weber et al. 1988; Van de Vijver et al. 1998; Chandler and Van de Vijver 2000; Taborsky 2000; Capra et al. 2010), all participants having a wonderful time in a merry ferment of epistemic trespassing. There was no doubt that the end of the twentieth and beginning of the twenty-first century were periods of enormous empirical and conceptual output. Nonetheless, some began to ask pointed questions about the official pantheon of the twentieth century evolutionary biology, specifically questions about what colleagues they might have trampled on their way to the top, and what interesting ideas they might have ignored and suppressed along the way. Everything was being questioned, sides were being taken and battle lines drawn, leading to a fragmentation of the biological sciences at precisely the time unification was needed most to confront the anthropogenic alteration of the biosphere, most notably the threat of global climate change, as well as the profound ethical issues associated with biotechnology. John Collier (2000) spoke for an emerging sensibility within history, philosophy, and sociology of science at that time when he posed the question, "Is there any virtue in modern science?" Less than a decade later, Alicia Juarrero (2008) suggested that professional philosophy needed a "To Re-think" list, focusing on causality, explanation, and ethics. We hope our contribution to this series will help advance that agenda.

References

Ballantyne N (2019) Epistemic trespassing. Mind 128:367–395

Bristol R, Rossano F (2020) Epistemic trespassing and disagreement. J Mem Lang 110. https://doi.org/10.1016/j.jml.2019.104067

Broda E (1983) Darwin and Boltzmann. In: Geissler E, Scheler W (eds) Darwin today: the 8th Kühlungsborn colloquium on philosophical and ethical problems of biosciences. Abhandlungen der Akademien der Wissenschaften der DDR. Akademie, Berlin, pp 61–70

Brown R (1828) A brief account of microscopical observations made in the months of June, July and August, 1827, on the particles contained in the pollen of plants; and on the general existence of active molecules in organic and inorganic bodies. Philos Mag 4:161–173

Capra F, Juarrero A, Sotolongo P, van Uden J (eds) (2010) Reframing complexity: reflections from the north and south. Emergent Publications, Litchfield Park, AZ

Cepelewicz J (2020) Nature versus nurture? Add "noise" to the debate. Quanta. https://quantamagazine.org/nature-versus-nurture-add-noise-to-the-debate-20200323

Chandler J, van de Vijver G (eds) (2000) Closure: emergent organizations and their dynamics, Annals of the New York Academy of Science Series, vol 901. New York Academy of Sciences, New York

Collier J (2000) Is there any virtue in modern science. Biol Philos 15:773–784

Darwin CR (1838) Notebook M: Metaphysics on morals and speculations on expression. CUL-DAR 125. Rookmaaker K (transcriber), Barrett P (ed). Darwin Online. http://darwin-online.org.uk/

Delisle RG (2008) Expanding the framework of the holism/reductionism debate in neo-Darwnism: the case of Theodosius Dobzhansky and Bernhard Rensch. Hist Philos Life Sci 30:207–226

Delisle RG (2011) What was really synthesized during the evolutionary synthesis? A historiographic proposal. Stud Hist Phil Biol Biomed Sci 42:50–59

Delisle RG (2019) Charles Darwin's incomplete revolution: the origin of species and the static worldview. Springer Nature, Cham

Dunham AE, Beaupre SJ (1998) Ecological experiments: scale, phenomenology, mechanism, and the illusion of generality. In: Bernardo J, Resetarits W (eds) Experimental ecology: issues and perspectives. Oxford University Press, New York, pp 27–49

Einstein A (1906) On the movement of small particles suspended in stationary liquids required by the molecular-kinetic theory of heat. Annalen der Physik (in German) 322:549–560
Hull DL (1988) Science as a process. University of Chicago Press, Chicago, IL
Hull DL, Tessner P, Diamond A (1978) Planck's principle. Science 202:717–723
Juarrero A (2008) On Philosophy's "To rethink" list: causality, explanation, and ethics. Ecol Psychol 20:278–282
Kuhn T (1962) The structure of scientific revolutions. University of Chicago Press, Chicago, IL
Perrin J (1909) Mouvement brownien et réalité moléculaire [Brownian movement and molecular reality]. Annales de chimie et de physique (in French) 18:5–114
Planck MK (1950) Scientific autobiography and other papers, vol 18. Philosophical Library, New York, p 117
Quinn JF, Dunham AE (1983) On hypothesis testing in ecology and evolution. Am Nat 122:602–617
Taborsky E (ed) (2000) Semiotics, evolution, energy. Shaker Verlag, Aachen
Van de Vijver G, Salthe SN, Delpos M (eds) (1998) Evolutionary systems: biological and epistemological perspectives on selection and self-organization. Kluwer Academic, Dordrecht
Weber BH, Depew DJ, Smith JD (eds) (1988) Information, entropy and evolution: new perspectives on physical and biological evolution. MIT Press, Cambridge, MA

Chapter 2
A Talking Book

Abstract Evolutionary biology is in conceptual disarray at a time when humanity cannot afford it. Never have we had more information and less integration. As well, the current consensus framework has failed substantially to provide guidance in the face of accelerating global climate change. We hope to provide an arena within which all interested people can contribute to unifying the disparate perspectives into a comprehensive theoretical framework leading to a more robust approach to ensuring humanity's future. We are by nature fearful, storytelling, dreaming beings. For the arena to be maximally inclusive, we rely on narrative approaches grounded in metaphors.

> There is another form of temptation, even more fraught with danger. This is the disease of curiosity... It is this which drives us to try and discover the secrets of nature, those secrets that are beyond our understanding, which can avail us nothing, and which man should not wish to learn—St. Augustine

Let's talk.

All living systems on the planet are linked together by a single evolutionary history. This is the unifying principle of biology, and it has guided an enormous amount of basic research that enhanced our understanding of the origins of biological diversity and the way it has coped with adversity since the origin of life more than 3 billion years ago. That understanding has led, in turn, to substantial contributions to human civilization, notably in the areas of food production and health, and more recently in advancing our awareness of the existential threat created by global climate change.

Ironically, evolutionary biology in the first quarter of the twenty-first century has never had more data and less integration. There is a consensus framework, but it has been buffeted by criticism broadly and consistently for half a century. Among the critics are many people with interesting perspectives, but to date, there is no platform that allows all of them to have a voice. The result is a fragmented and fractious group of scientists, each composed of highly intelligent people well acquainted with the empirical findings of their narrow specialty, all arguing that the rest of the biological world should be explained in terms of what they know. This is due, in large part, to

S. J. Agosta, D. R. Brooks, *The Major Metaphors of Evolution*, Evolutionary Biology – New Perspectives on Its Development 2,
https://doi.org/10.1007/978-3-030-52086-1_2

the early twentieth century emergence of specialized and canalized research programs focusing on pieces of a puzzle. As in the parable of the blind men and the elephant, people study only their particular discipline and do not ask how it might connect with data from another.

It is generally in the best interests of scientific development to have a common platform, even if it is not a universally agreed-upon framework (though most would agree this is a worthy goal). We thus think it is mildly embarrassing that the unifying theory of biology is so disunified. But we are not surprised; this is academic business as usual when an important conceptual framework is concerned. The natural history of humans in general is complex and messy. Conflicts of interest occur among humans easily and often, and scientists never escape their basic humanity. As a result, scientists are notoriously ill-mannered and intolerant when it comes to new ideas, except their own.

> A scientist will never show any kindness for a theory which he did not start himself—Mark Twain (1880)

More importantly, we find it unconscionable that the unifying theory of biology has been so poor at anticipating, coping with, and producing proactive recommendations relative to global climate change. The global climate change that is rapidly accelerating and bearing down on a largely unprepared humanity has a substantial biological, and thus evolutionary, basis. We strongly believe that the consensus framework that dominated the twentieth century largely failed to anticipate the various evolutionary implications of global climate change or offer effective solutions. Evolution has been the only process to regenerate the biosphere following a mass extinction event, and evolution has never failed in that regard. Despite this, most policies in conservation biology consist of programs designed to stop evolution at all costs and promote a nostalgia for a static past that never existed. As humanity faces an existential threat coming at an accelerating pace, we must expand and enrich evolutionary theory. We think many recent perspectives on evolutionary theory have something to offer toward that goal. The advocates of those perspectives, however, must accept that they are like the blind men examining an elephant. It will require many to make this new framework coherent, to truly understand what an elephant is.

We think a unified, coherent, and above all useful evolutionary framework is within our grasp, but we have not had a metalanguage to allow helpful communication among specialized research programs. That is the primary purpose of this book. Three basic aspects of human biology underly our effort to produce such a metalanguage.

2.1 We Are a Fearful Species

> I must not fear. Fear is the mind-killer. Fear is the little-death that brings total obliteration. I will face my fear. I will permit it to pass over me and through me. And when it has gone past I will turn the inner eye to see its path. Where the fear has gone there will be nothing. Only I will remain—Frank Herbert (1965).

In order to understand how we think about a concept like evolution, and why there is such a diversity of aggressively held views, we have to understand some critical things about our own evolutionary legacy. We are a cautious, fearful species, and we come by this fearful nature honestly, that is, evolutionarily. We are descended from prey and the life of prey is founded on fear (Brown et al. 1999; Laundre et al. 2014; Bleicher 2017). Anyone who has seen primates in the wild understands this. Faced with known predators, or animals of unknown capabilities and intent (like biologists), primates freeze and assess, then flee if the potential threat does not quickly move along. We began as Man the Hunted, descendants of primates whose cleverness, combined with caution and suspicion, allowed us to survive long enough to begin making the weapons evolution neglected to provide. Once we successfully tested those weapons on species that recently had been fellow prey, we styled ourselves Man the Hunter. But there is one thing that still scares us all.

2.1.1 The Complexity Paradox

Paradoxes offend reason and are therefore a reason to laugh—Umberto Eco (1983).

How wonderful that we have met with a paradox. Now we have some hope of making progress—Niels Bohr (quoted in Moore 1966)

The fragility of early hominids placed a premium on accurately perceiving and generalizing the complexity of their surroundings—underestimate it and you are lunch, overestimate it and you starve. Evolution worked well for us in that regard; self-awareness brought with it an excellent ability to perceive complex patterns and to accurately assess the complexity of our surroundings. But self-awareness also produced a paradox: if we are good at accurately assessing complex patterns in our surroundings, why are we so afraid of complexity? The answer is that successful generalizing of our surroundings gave us a sense of security, so we trusted it—good became synonymous with generalization, evil with contingency, the complexity that cannot be generalized, that frightens us and does us harm. We distrust complexity because that is where the saber tooths, cave bears, and boogeymen are.

Men always fear things which move by themselves—Frank Herbert (1969)

Given the recognition of complexity in the world and our fear of it, self-awareness had to be linked to psychological denial in order for us to continue to exist in a world made dangerous by that complexity. But, like all evolutionary innovations, denial has costs as well as benefits. Denial can keep us from responding assertively to threats in our surroundings that we perceive (Dor-Ziderman et al. 2019). We are so afraid of the complexities of an uncertain tomorrow and the certainty of death that we invented a universe that we can control as soon as we have accumulated enough knowledge about it. Western philosophy of science has been based entirely on the belief that nature is fundamentally simple, and any appearance of complexity is our fault—the result of incomplete information.

And yet, all of our experience tells us that we live in a complex world, not a poorly known simple one, and the more we learn about it, the more complex it becomes. We know it is a universe structured in such a way that we can control only limited amounts of our existence, and that makes us unhappy and resentful. This is why proximal explanations have always been more comfortable and familiar to humans, while ultimate explanations have tended to be a little frightening. This sense of fear is incorporated into our equating "ultimate explanations" with "final causes," truly a scary thought. Following this tradition, the Nobel laureate chemist Ilya Prigogine associated our understanding of the open universe as the "end of certainty" (Prigogine 1996).

The realities of the twentieth century have forced scientists to confront the natural complexity and to produce theories that are consistent with observations. We are so afraid of complexity that we keep wanting to (over)simplify, even if it is not helping. Like our remote ancestors, we think complexity is dangerous and frightening.

> . . .Rulag was an engineer, and he had found in her the engineer's clarity and pragmatism of mind, plus the mechanist's hatred of complexity and irregularity—Ursula LeGuin (1974)

We evolved in a world of patterned complexity, and all of our empirical experience of the world as we grow up reinforces that. The more we learn the more complex the world becomes, and yet we are still able to function; in fact, we function quite well in complex situations. The world is neither chaotic nor simple.

Depending on our personalities, our surroundings either scare us into denial or intrigue us to investigate their true nature. What do we hide from?—our fears. What do we fear the most?—the unknown. And yet, we claim to be explorers of the unknown. Most often, however, we want the unknown to be only a little unknown and ultimately controllable. We do not like big surprises.

Darwin was not afraid—the complexity excited him. And that is what allowed him to formulate the first scientific theory based on principles of complexity. What separated Darwin from most of his contemporaries, and from most of those who have followed, was his ability to accurately assess the complexity of evolution and not be afraid. He did not try to rationalize reality so it would no longer be fearful. Darwin saw great beauty as well as great tragedy in evolution, and he produced a pragmatic theory. He showed us that we do not need certainty to survive and to thrive. Despite his example, we are still largely in denial about his fundamental insight, that life on this planet is evolvable and that evolution is a complex phenomenon. We have begun to accept that this is true, but we are uncertain about what this means and what we ought to do in the face of the greatest imminent threat to humanity, global climate change. A tennis adage is "never change a winning game, always change a losing game." We are playing a losing game with respect to global climate change, and we cannot change that unless we admit it, but we do not want to. Complexity makes us afraid, and fear catalyzes denial, leading us to continue doing what we have been doing, even though it is not working.

We need to be able to overcome our fears if we are to work together. Fortunately for us, the origin of human language provided us with a means of doing both. We believe that sharing information through storytelling can help us achieve a common platform where everyone can learn, and humanity can extend itself into the future. Complexity defies prediction and so some of it will always appear mysterious. Parts can be expressed in the form of a narrative, because narratives simulate the flow of time, carrying the lessons of the past and projecting aspirations for the future. There have been many unpleasant events in evolutionary history, but there is an unbroken record of survival. Life is resilient and persistent.

2.2 We Are a Storytelling Species

> It is shocking to find out how many people do not believe they can learn, and how many more believe learning to be difficult... every experience carries its lesson—Frank Herbert (1965)

Our evolutionary legacy as social primates produced the complexity paradox and a means of coping with our fears of the complex unknowns of the world. One of the benefits of the origin of language was the ability to share information and to learn from each other. Huddled in small groups around a sputtering flame, gathered boisterously in a royal hall at banquet time, assembled solemnly in a religious convocation, or simply in pairs of parents and children, humans have long known that important lessons are best taught through narratives (Sugiyama 2001). The most important of those lessons have always been cast as narratives anchoring the great lessons in past times and places, even if mythical. These *historical narratives* can be legends or sagas. Legends, stemming from the Latin *legere*—"to read, gather, select"—are written compositions meant to be repeated in storytelling without change. Sagas, from the Old Norse *saga* and Old English *sagu*—"a saying"—are narrative compositions meant to be adapted to changing audiences over time and space. The intentionally flexible nature of sagas is the reason we tend to believe that written histories are more accurate than oral histories, even though this is often not the case. An incorrect written record is incorrect forever; a saga, no matter how often it is modified, may retain essential truths.

Storytelling is a form of social cooperation, so we can tell a story from different perspectives and still see that it is the same story. By blurring the distinctions between subjectivity and objectivity, narratives allow us to learn more easily, feeling less coerced. They play an essential role in science (Norris et al. 2005). Good narratives attempt to explain as well as describe the world, to give insights—*we never anticipated this particular thing, but we can explain it.*

> As a philosopher living in a post-modern world, I have come to accept that narratives have a power that despite all the literature on the subject I cannot quite put my finger on, a power that manages to accomplish what nonfiction explanations and accounts simply do not: a kind of integration between seemingly irreconcilable and incommensurate voices. I have grudgingly come to believe that stories might even be a better tool than nonfiction at conveying the incompressible complexity of the human element. Plato, of course, knew this early on: When the going gets tough in any of his dialogues, the old Greek always resorts to a story to get his point across—Alicia Juarrero (1999)

We are trying to create an environment for discussion that allows people to wander among different frameworks in a way that allows them to discover common perspectives that may be obscured by the use of diverse nomenclature. We are not trying to create an echo chamber, telling the story as a legend word for word to maintain group cohesion. Rather than telling a story about how things were, are, or should be, this will be more like telling a story about how to tell the story. This is the beginning of a saga, one that we hope will persist, grow, and change through time as it is told by a growing number of people.

> . . .during our whole journey, I have been teaching you to recognize the evidence through which the world speaks to us like a great book. Alanus de Insulis said that *omnis mundi creatura quasi liber et pictura nobis est in speculum*. . .But the universe is even more talkative than Alanus thought, and it speaks not only of the ultimate things (which it does always in an obscure fashion) but also of closer things, and then it speaks quite clearly—Umberto Eco (1983).

2.2.1 A Story Within a Story

> In the beginning was the Big Bang, and that was a very long time ago. This is just a reminder of this evening's extra performance. . .In brief, the encore revolves around the creation of the performance's audience. . .Seats are still available. . .The applause for the Big Bang was heard only fifteen billion years after the explosion. . .—Jostein Gaarder (1999)

We believe that setting the stage for productive discussions among many different biologists by telling stories is particularly apt for biology. Living systems are capable of acting on their own behalf but, more importantly, they regularly take the initiative—life has a life of its own. And they do this primarily through capacities to cope with their environments that they have inherited. Also, the nature of inheritance is so conservative that most explanations for how organisms look and how they function today are rooted in persistent history. In the fifteenth century, poets and natural philosophers were content with the idea that history is a dead record of the past, having nothing to do with the present or future (Huizinga 1996). Darwin changed all that.

Evolution, therefore, is a journey, not a destination, a game not a victory; it is a never-ending story in which the participants not only play the game, they change the rules and the dimensions of the playing field from time to time. It is an interwoven collection of many stories involving common times and common places, each with one narrator and many commentators. Our narrative approach to talking about evolution, therefore, is telling the story within a story.

2.3 We Are a Dreaming Species

> Life isn't a problem to solve, but a reality to experience. . .A process cannot be understood by stopping it. Understanding must move with the flow of the process, must join it and flow with it. . .What senses do we lack that we cannot see and cannot hear another world all around us?—Frank Herbert (1965)

We are not just a fearful storytelling species, cowering in the dark. Sleeping and awake, we are a dreaming species. Despite our fear of the unknown, we simply cannot stop dreaming. We do not seem to be able to resist the urge to embellish our stories of what happened with ideas about why it happened and what might happen in the future. It is in our dreams that we seek insights. We think about our dreams and worry over them. When we try to incorporate our dreams into our conscious lives, we become symbolic storytellers and generalize using metaphor. When some of those metaphorical narratives point the way to living truths, we call them scientific theories.

> The question, in fact, was whether metaphors and puns and riddles, which also seem conceived by poets for sheer pleasure, do not lead us to speculate on things in a new and surprising way, and I said this was also a virtue demanded of the wise man—Umberto Eco (1983)

Our best stories come from our dreams and are inspirational and aspirational, not merely operational. When we share our dreams, we speak metaphorically, whether we are aware of that or not. So, when we resort to metaphors, we are talking about our aspirations and fears, the dream world where our understandings and our beliefs come together (Coward and Gamble 2010).

> You see things; and you say "Why?" But I dream things that never were; and say "Why not?"—George Bernard Shaw (1921)

Scientists use two kinds of language: nomenclature and metaphor. The technical language of nomenclature attempts to eliminate ambiguity in concepts and entities within an established area of science. An excellent example in Biology is the use of scientific names for species. If a North American and a European call a bird a "robin," they are referring to two different species that are not closely related. Common names invite speculation and investigation; they are metaphorical. If, however, both say *Turdus migratorius*, there is no confusion. Choosing a "dead language" for formal nomenclature helps preclude additional meanings creeping into the nomenclatural designations.

Nomenclature can mean we know what we are talking about and we want no ambiguity. It can also mean we have no idea how to solve a problem so we will simply make up terms and hope somehow an answer will come to us. Making up new terms, giving new names to old phenomena, is a way for scientists to try to quell their fears of failure. But terminology never solves problems. Professional nomenclature is used to formalize objects and to provide internal cohesion among research groups. This is an exercise in mapping of static objects and relations. Biology is full of labeling language and much of it is in Latin because a dead language creates no new connotations or metaphors. Nomenclature is also used to eliminate metaphors because it wants to eliminate connotations and ambiguity (e.g., choose a dead language to name species rather than common names in various living languages). So, none of this can tell a story. Labeling does not allow you to judge, assess, or learn about yourself. It leads inevitably to iconotropy, pointless battling over the proper definition of a term.

Scientists who are comfortable with a given normative framework tend to embrace nomenclature as the solution to problems, which they see as mostly a matter of reducing ambiguity within the framework. Appeals to nomenclature will never lead to new insights, no matter how much basic information is obtained.

Scientists defending a conceptual framework tend to mistrust metaphors because they allow too many possibilities, thus introducing ambiguity into the framework. Metaphors may also be emotion-laden, and scientists who do not know how to integrate reason and emotions distrust them.

The essence of science is change, however, and scientific change is creative. When there is a need, or desire, to make a change in a normative framework, therefore, metaphor becomes the language of choice. Perhaps the most important function of metaphor in science is in extending existing nomenclature to accommodate new concepts and empirical findings. This may seem contrary to the normative use of nomenclature, but it is not. If we created new nomenclature for each new proposal, there would be no way to show connections between the old and the new frameworks. This means that when there is an explosion of nomenclatural proposals for a given topic, there is likely a more fundamental conflict that needs to be resolved. And metaphors set the stage for that resolution.

Theoretical advances rarely emerge from ecclesiastical dialogues among different entrenched viewpoints in which each side expects the other to convert at some point until an eventual winner emerges—what Menachem Fisch calls *inter-faith dialogue*. He argues that what we really need is *inter-faith learning*. Inter-faith learning allows the possibility of cooperation for mutual understanding. We think everyone with a perspective on evolutionary theory today has something to offer, including those who do not wish to part with the old ways. This is why the academic food fight conducted publicly in the pages of *Nature* in 2014 (Laland et al. 2014; Wray et al. 2014) resolved nothing. The disputants on each side tried to convince the other to convert to their nomenclature. They sought no common ground for turning their interdisciplinary dialogue into interdisciplinary learning.

Metaphors allow us to unify different systems of nomenclature. They are the roots of creativity in science, a common metalanguage for people to come together and find common ground to achieve needed change. Metaphors keep us from getting bogged down in trivial disputes. They are not useful for winning an argument.

> Metaphors were never made for keeping score—Jimmy Buffet (1996)

Rather, metaphor is the search engine for interdisciplinary studies.

> If you want to change the world, you have to change the metaphor—Joseph Campbell (interviewed by Bill Moyers 1987)

We intend to use metaphorical language to make readers feel ambivalent about at least some aspects of whatever explanatory framework they have adopted. This is not to provoke either the fight or the flight professional response; we want people to talk to each other and learn about how they think by learning how others see their paradigm. In fact, metaphors are a way to show the common ground between frameworks whose advocates might feel are in competition. For example, we will use the term *inheritance system* for cases in which different readers will use terms like organism, population, deme, or species, depending on their nomenclatural preferences.

Fundamentally, we seek a common framework within which all good ideas and insights reinforce each other comfortably, much as a good teacher seeks to sharpen and liberate the critical faculties of students.

> It is my opinion the purpose of a teacher is to warp young minds, and then show them how to resolve the resulting conflict. One of the problems with modern biology is there is too little warping and too much worshipping—Jack Maze, personal communication

"Telling a story" is not "naming a thing." For telling a story you need a narrative, for a narrative you need a natural language, and for telling a story of change that unites different perspectives, that language will be metaphorical. Different metaphors can tell different stories, and we can choose among them based on what, and how much, each one tells us.

The way forward lies in elaborating a framework of *suitable* metaphors for evolutionary biology. The poet Robert Graves (1948) wrote that natural metaphors can lead you to the truth, but magical metaphors, no matter how beautiful they are, will not. We believe life is astonishing, but not magical. And therefore, we strive to use natural metaphors and to be alert to metaphors that invoke magic. We will try to always distinguish metaphors that present natural truths using language that allows easy understanding by non-specialists from those that present true-sounding statements that are not actually known to be true. This second class of metaphors may ultimately be shown to be representations of natural truths, but until that is demonstrated, they are magical statements, asking for a suspension of disbelief by readers. Natural truths transcend the religious, social, and economic beliefs of any particular group of people; they are natural, therefore, because they are beyond belief. Anyone can pick up a stone and release it and see evidence of gravity. Poetry that depicts people and objects falling from a height is based on a natural truth as a natural metaphor. Poetry that depicts people spontaneously rising in the air or to the heights, to heaven, is based on a magical metaphor. Nonetheless, even natural metaphors can appear mysterious. Metaphors cannot disappear—forgotten or set aside, they can always be recovered and reexamined in light of new information or challenges to an existing framework. This means metaphors allow us to suspend the rules of time and space; sometimes the way forward requires us to recover old metaphors set aside.

Kurt Vonnegut (1963) used metaphor in a didactic sense when he characterized science as "magic that works." As scientists, we do not believe that is a useful characterization, because it invokes a magical metaphor to explain efforts to understand and use natural truths. In this book, we will attempt to use Graves' distinction between natural and magical metaphors to help us recognize aspects of scientific explanations that rely on magical metaphors and eliminate them. But to do that, we must be willing to confront our human propensity, as a cautious species, for wanting to believe in magic.

Most people, including most scientists, equate nomenclature with *fact* and metaphor with *conjecture*. As long ago as 1883, Twain recognized the essential dualism in scientific activities:

> There is something fascinating about science. One gets such wholesome returns of conjecture out of such a trifling investment of fact—Mark Twain (1883)

There is much truth to this, but it is not entirely accurate. By extending existing nomenclature to accommodate new concepts with the bank of old and new "facts," metaphors give everyone a reason and a means to learn new ideas (Cowan et al. 1994). Most importantly, natural metaphors are the poetics of science, presenting natural truths using language that allows understanding by non-specialists. Science has nothing to offer, and thus no reason to be supported, if it does not communicate its findings in ways that are intelligible and relevant to society at large. Metaphors do, indeed, give a rich return in speculation but they are paid for by a massive, not trifling, investment in fact.

Darwin's literary skills were key to the great success of *Origin of Species* among non-specialists. Darwin was a master at drawing together common everyday experiences and knowledge with technical observations by biologists, linking them metaphorically to produce his panoramic synthesis of the history of life. As Neo-Darwinism emerged as the consensus framework in the twentieth century, the language of evolution shifted rapidly from metaphor to nomenclature as it transformed from inspirational, to aspirational, then to operational. The demise of metaphor led to a predictable narrowing of vision that has been recognized by some for more than half a century.

> Our most recent, still continuing, period has been dominated by reaction against an earlier perspective considered too sweeping, too ambitious in scope, too weak in data and method. In outline caricature, the devolution from generalizations of bold scope has been first to drop the generalizations, and then the scope—Dell Hymes (quoted in George W Stocking 1968)

We need to couch our story in metaphorical terms because rallying evolutionary biology to help humanity cope with an existential threat requires a network of teams, not just a network of people within teams. The language we use to begin those discussions must be abstract enough that all who participate can see that they have a proper place around the campfire. We want every reader to be ambivalent about their own views of evolution by learning how we see those views. We do not want to make them afraid, and thus too defensive to think and imagine.

Telling stories from a metaphorical perspective also allows us to take an empathetic, or internalist view of the nature of evolvable life. Narratives are intensely personal and stem from the ability to take the viewpoint of the subject of the narrative—the worm's-eye view rather than the bird's-eye view. Koichiro Matsuno has said that the internalist perspective does not ask the system to describe itself but asks the observer to see the system as it imposes itself on its surroundings. Field biologists actively use an internalist perspective. Go into these areas with the right attitude, and they will show you things few others have ever seen. This is the reason explorations of remote and isolated places have always led to amazing epiphanies in those who visit but do not live there. And one of those epiphanies is that those who live in those places understand the same mystery and majesty. The internalist perspective invites you to go deeper, to think about connections and context, about origins and history (how did this come to be rather than simply, what does this do). Understanding the biosphere from the perspective of how it imposes itself on the surroundings is the way to learn from it how we might survive global environmental changes, rather than engaging in futile attempts to bludgeon it into controlled submission. The internalist perspective helps us remember that we should be "talking to nature" rather than "talking to ourselves."

2.4 Summary

> Truth is born into this world only with pangs and tribulations, and every fresh truth is received unwillingly. To expect the world to receive a new truth, or even an old truth, without challenging it, is to look for one of those miracles which do not occur—Alfred Russel Wallace (interview by S. Herren 1913)

Science is the very human process of changing magic into prose, the supernatural into the expected. This does not mean science is soulless—metaphor is the poetic growth medium for new conceptual frameworks in science.

We are not attempting to argue anyone into submission by advocating one particular view. Rather, we want each reader to feel ambivalent about his or her own views of evolution. Our choice of epigrams is meant to show that much of what we will discuss is common knowledge among thoughtful and observant people. The reader's ambivalence will come, therefore, from recognizing a discrepancy between what they have experienced and what they have been taught was reasonable. We are not trying to make readers afraid and thus too angry to think rationally; we are trying to encourage them to talk with those who have different views and to learn. We understand we will not succeed in all cases.

We encourage you to think of the major metaphors of evolution as friendly campfires around which people can share their stories and try to make sense of the universe. They are neutral ground to begin negotiations and conflict resolution, for which metaphor is essential. We seek mutual learning and a narrative framework based on natural metaphors that connect to everyday experience. We want that framework to be large enough to encompass what we have accumulated, resolve existing paradoxes and disagreements, and set the stage for future research. Good science should not be boring; it should be fun, and its stories should be interesting without resorting to magic. The authors of this book are naturalists by temperament and training, the academic descendants of von Humboldt, Darwin, and Wallace. Like them, we want an evolutionary theory full of wonder but without magic or fear. Magical thinking does not make science interesting; it makes it boring and authoritarian, and acceptance of authority is merely a way to shield us from fear. The framework we strive for must be useful to humanity and yet interesting enough to capture the imagination of bright creative minds in each new generation. It is okay to show the audience how you did the trick.

References

Bleicher SS (2017) The landscape of fear conceptual framework: definition and review of current applications and misuses. PeerJ 5:e3772. https://doi.org/10.7717/peerj.3772
Brown JS, Laundre JW, Gurung M (1999) The ecology of fear: optimal foraging, game theory and tropic interactions. J Mammal 80:385–399
Buffett J (1996) Only time will tell. Pan, London
Cowan G, Pines D, Melzner D (eds) (1994) Complexity: metaphors, models and reality. Addison-Wesley, Reading, MA
Coward F, Gamble C (2010) Metaphor and materiality in earliest prehistory. In: Malafouris L, Renfrew C (eds) The cognitive life of things: recasting the boundaries of the mind. McDonald Institute for Archaeological Research, Cambridge, pp 47–58
Dor-Ziderman Y, Lutz A, Goldstein A (2019) Prediction-based neural mechanisms for shielding the self from existential threat. NeuroImage 202. https://doi.org/10.1016/j.neuroimage.2019.116080
Eco U (1983) The name of the rose. Harcourt, New York
Gaarder J (1999) Maya. H. Aschehoug & Co. (W. Nygaard), Oslo
Graves R (1948) The white goddess: a historical grammar of poetic myth. Faber & Faber, London
Herbert F (1965) Dune. Hodder and Stoughton, New York
Herbert F (1969) Dune messiah. G.P Putnam's Sons, New York
Huizinga J (1996) The autumn of the middle ages. English translation of the 1921 edition. (trans: Payton RJ, Mammitzsch U). University of Chicago Press, Chicago
Juarrero A (1999) Dynamics in action. MIT Press, Boston, MA
Laland K, Uller T, Feldman M, Sterelny K, Müller GB, Moczek A, Jablonka E, Odling-Smee J (2014) Does evolutionary theory need a re-think? Yes, urgently. Nature 514:161–164
Laundre JW, Hernandez L, Lopez Medina P, Campanella A, Lopez-Portillo J, Gonzalez-Romero A, Grajales-Tam KM, Burke AM, Gronemeyer P, Browning DM (2014) The landscape of fear: the missing link to understand top-down and bottom-up controls of prey abundance. Ecology 95:1141–1152
LeGuin U (1974) The dispossessed. Harper and Row, New York
Moore RE (1966) Niels Bohr: the man, his science, & the world they changed. MIT Press, Boston
Moyers B (1987) The power of myth. Public broadcasting systems television series
Norris SP, Guilbert SM, Smith ML, Hakimelahi S, Phillips LM (2005) A theoretical framework for narrative explanation in science. Sci Educ 89:535–563
Prigogine I (1996) The end of certainty. Simon and Schuster, New York
Shaw GB (1921) Back to Methuselah. Constable, London
Sherren W (1913) Third-person interview with Alfred Russell Wallace. T.P.'s Weekly 10 January 1913:49
Stocking GW Jr (1968) Race, culture and evolution. University of Chicago Press, Chicago, IL
Sugiyama MS (2001) Food, foragers, and folklore: the role of narrative in human subsistence. Evol Hum Behav 22:221–240
Twain M (1880) A tramp abroad. American Publishing Company, New York
Twain M (1883) Life on the Mississippi. Dover, New York
Vonnegut K (1963) Cat's cradle. Holt, Rinehart and Winston, New York
Wray GA, Hoekstra HE, Futuyma DJ, Lenski RE, Mackay TFC, Schluter D, Strassmann JE (2014) Does evolutionary theory need a re-think? No, all is well. Nature 514:161–164

Chapter 3
Setting the Stage

Abstract The nineteenth century saw a major transition in thinking from a static universe made perfect by a divine creator to one full of change. The transition was led by biologists, most notably Charles Darwin. Darwin's panoramic vision of evolution involved the interaction between two major factors, the *Nature of the Organism* and *Nature of the Conditions*, with the former predominating over the latter due to the historically conservative and autonomous nature of organisms and their inheritance systems. From this emerges a "struggle for existence" producing Natural Selection—an outcome of the interaction favoring any organism adequate enough to cope with the conditions by surviving and reproducing. In today's terms, Darwinism was a theory of complex systems, which he attempted to communicate using two great visual metaphors, the Tree of Life and Entangled Bank. In Darwin's day, complexity was not in style; good theories were simple with deterministic lawlike behavior. Furthermore, social styles that originated in the eighteenth century including Naturalism, Modernism, and Romanticism, sought progress and perfection in explanations of the natural world. Within the scientific community, the mix of these social preferences and general acceptance of the notion of biological evolution led to several research programs aimed at "fixing" or replacing Darwinism. By the end of the nineteenth century, four distinct theoretical frameworks—Geographic differentiation, Orthogenesis, neo-Lamarckism, and neo-Darwinism—had emerged as rivals in the race to replace Darwin.

For almost 2000 years, explanations of biological origins were founded upon the concept of stasis. To most medieval European naturalists, adaptations were evidence of God's design, the perfection of God's mind reflected in the perfection of each organism's fit to its environment. The theory of divine creation was coupled with the idea of organic stasis: in effect, a creator would not make mistakes, so the diversity we see around us today, a reflection of the creator's perfection, is the way things have always been.

At the beginning of the nineteenth century, European natural philosophers firmly believed two things. First, the universe was a gigantic clockwork mechanism that functioned perfectly according to the dictates of a few powerful laws (specifically,

S. J. Agosta, D. R. Brooks, *The Major Metaphors of Evolution*, Evolutionary Biology – New Perspectives on Its Development 2,
https://doi.org/10.1007/978-3-030-52086-1_3

what are now called Newton's Laws). And second, every living organism was perfectly fit to its surroundings. Nature provided a place for every species and every species occupied its particular place.

For these thinkers, it seemed self-evident that the key to understanding the nature of living things lay in that perfect fit to their surroundings. The Newtonian clock-work universe must apply to living as well as nonliving components of the universe. We need to only figure out how the Newtonian laws produced the perfect fit between organisms and their surroundings, and the rest would be merely detail to occupy the minds of natural philosophers when they needed a respite from the rigors of Physics and Chemistry. Others sought a union of theology and Newton. William Paley's (1802) "argument from design" was perhaps the most influential of these proposals. Suppose you are a natural philosopher out walking one day, and you find a watch lying on the ground. You conclude that such an intricate and complex thing could not have appeared by a random conjunction of various metals, so you conclude that the existence of the watch implies the existence of a watchmaker. So far, so good. Now substitute "cat" for "watch": just as the intricate watch requires a designer, so to the complex cat seems to require a creator.

Regardless of debates about the ultimate meaning of life, the belief that organisms were perfectly fit to their environments guided the expectation of most Europeans in the great voyages of discovery of the sixteenth and seventeenth centuries. And because such perfect fit was what explorers expected to find, it is what they tended to report. Such wonderfully optimistic reports from the far corners of the earth increasingly attracted well-trained naturalists to these voyages of discovery. As their findings began to appear, the idea of a perfect (and easily understood) fit between organisms and their surroundings began to erode. One of the most remark-able of those naturalists was Alexander von Humboldt. Von Humboldt believed that a theoretical understanding of the natural world should emerge from exacting observations of it. His extensive studies in South America contributed greatly to the recognition that the more diversity we discovered, the more complexity in the ways in which organisms coped with their surroundings we encountered. The fit of organisms to their surroundings might not be a simple matter.

At the same time, discoveries of fossilized remains of creatures no longer alive added to the debate. Naturalists found themselves in a logical paradox. If the fossils were once life forms, and if the theory of stasis was to be upheld, then there had to be living representatives of those forms still on earth. In other words, fossils had to represent dead individuals, not dead species (Ruse 1979). If fossils were once life forms, and if they had been created perfectly fit to their surroundings, the surround-ings must have changed, leading to the extinction of the forms perfectly fitted to the original surroundings. Also, because there were living species seemingly perfected fitted to their surroundings today, there must be a means by which such well-fitted biodiversity was perpetuated. One explanation was that both the environmental conditions and the organisms perfectly fitted to them were evidence of a process of ongoing divine Creation. This was an explanation that very few found satisfac-tory. If God, or natural processes, had produced both changing environments and a consistent supply of living species perfectly fitted to the conditions at any particular place and time, without creating new life, then there must be a mechanism whereby some preexisting life must be capable of changing in response to the changes in the surroundings.

These lines of thinking emerged in the early nineteenth century and marked the transition of our understanding from a static universe to one full of change. And that transition was led by biologists. One of the first, and most influential of those biologists was Jean-Baptiste de Lamarck. Although Lamarck never ventured forth on voyages of discovery, his appointment as a professor at the Natural History Museum in Paris allowed him to immerse himself in the museum's extensive invertebrate collection assembled by those who had experienced such field studies. Lamarck spent years describing the ways in which organisms defended themselves from a hostile world. From this starting point, he developed his theory of evolution (Lamarck 1809). He argued that (1) Living organisms have many capacities to survive in a harsh world. Given this, we must reject the theory of extinction. (2) Many fossils represent previously living forms that are no longer present on the earth. Given this, we must reject the theory of stasis. These two points begged the following question: If species do not go extinct, how do we explain fossils? Answer: There are no living representatives of many fossil species *because those species evolved into something else*. Having argued for the existence of biological evolution, Lamarck provided a detailed mechanism by which species could gradually change from one form to another. This mechanism rested upon his belief that there was a connection between an organism and its environment, a connection so intimate that a change in the environment could *drive* a change in the organism. Fossils represented dead individuals, ancestors of organisms that changed form in order to adapt perfectly to new environmental conditions.

When a young Charles Darwin set off in 1831 for a 5-year voyage of discovery aboard the HMS *Beagle*, much of the essential philosophical structure of modern science was in place. There was a very strong belief that we ought to find major theories that would explain the world, and that our data-gathering would show how particular observations were cases of these more general theories. At the same time, there was a recognition that not all theories could be true, that not all data unambiguously supported a particular theory, and that not all data were equally reliable with respect to helping us understand the natural world. Darwin had the field biologist's love of complexity and novelty. He was aware of a variety of views in geology and biology, so he had some ideas to test, including a number of evolutionary concepts. Were fossils dead individuals or dead species? Was the earth old or young? If species evolved, how did that occur—what were the roles of environmental conditions and of geographic distributions? More than anything else, he was aware of von Humboldt's (1818) assertion that:

> Identity of forms suggests an analogy of climate, but in similar climates the species may be very diversified—Alexander von Humboldt (1818)

Darwin, like von Humboldt, turned out to be too good a naturalist to be fooled by wishful thinking about a perfect fit of all organisms in all species to their surroundings, and he found ample evidence to the contrary in his explorations during the voyage of the Beagle. In 1841, just 5 years after returning to Britain, Darwin wrote "*adaptation is to make a place for yourself in nature,*" (Eldredge 2005), and 3 years later, he wrote an essay in which he asserted that, "*We cannot believe that every part of the natural economy is filled in any island, much less on a continent*" (Pearce 2010). These sentiments were clearly at odds with the prevailing idea of a place for everything and everything in its place.

Evolution was in the air, and Darwin's voluminous correspondence indicates that he knew he had seen glimpses of something innovative and of great generality, and he was determined to articulate it meticulously (see, e.g., Eldredge 2009). He first did so in 1859, when the first edition of *Origin of Species* appeared.

3.1 1859: The *Origin* Appears

We have relied on the sixth and final edition of *Origin of Species* to tell Darwin's story, rather than the first edition, which has tended to be the focus of philosophers, historians, and social scientists. This is because we are more concerned with Darwin's ideas than with his biography. If a scientist publishes more than once on a topic, the most recent version represents the best statement of that scientist's views. Furthermore, in many of the historical, philosophical, and social accounts of the *Origin*, authors have been so eager to get to Darwin's extensive discussion of the principle of natural selection that they begin halfway through whatever edition of the book they are using, then backtrack as needed. Effective scientists, however, write according to a standard formula, often summarized as, "Tell them what you are going to tell them; tell them; tell them what you told them." We present Darwin's story in that manner, literally beginning at the beginning of the last edition of *Origin of Species*.

Darwin summarized what he was going to tell his readers succinctly and explicitly in the second paragraph of the main text of the *Origin*:

> ...there are two factors: namely, the nature of the organism and the nature of the conditions. The former seems to be much more the important; for nearly similar variations sometimes arise under, as far as we can judge, dissimilar conditions; and, on the other hand, dissimilar variations arise under conditions which appear to be nearly uniform—Charles Darwin (1872)

This statement is perhaps the most underappreciated passage in the *Origin*. It is far more than Darwin's homage to von Humboldt. Suppose we picked, at random, any organism from a designated tide pool and a crab from anywhere in the world. If we then asked you readers for a list of morphological, behavioral, and ecological characteristics of the unknown organism from a given environment—say a tide pool—and of an organism—say a crab—from an undetermined habitat, more of your answers would be correct for the crab than for the unknown tide pool organism. Knowing that we are dealing with "a crab" imparts more information than the most detailed description of the tide pool, even if we collected one crab from a tide pool and the other from the top of a palm tree.

3.1.1 The Nature of the Organism

Although Darwin admitted that the laws of inheritance, of growth, and of the correlation of parts (development) were mysterious, he had a good understanding of the capacities of organisms. His conception of the *Nature of the Organism* was thus explicit. Organisms are functional wholes, capable of coping with a range of conditions, interacting with their surroundings in ways that benefited them, that enhanced their chances of survival and reproduction. Organisms, and the species they comprise, are what Darwin called "communities of descent" and most of Darwin's preoccupation with the nature of the organism concerned the evident nature of inheritance. Living systems regularly take the initiative, as each new generation imposes itself on its surroundings using inherited capacities. Darwin considered four elements of inheritance to be significant for his theoretical framework. The first is that inheritance is exuberant: it is in the nature of the organism to produce many offspring. Second, inheritance is highly conservative: it is in the nature of the organism to produce highly similar offspring due to the [then mysterious] laws of inheritance. This explained why organisms resembled each other and their parents regardless of the environmental conditions in which they were produced—the reason a squirrel looks and behaves like a squirrel is because its parents were squirrels. Third, though conservative, inheritance produces variation. All offspring are highly similar to each other and to their parents, but each one is a unique individual. And finally, inheritance is historical: not only do the members of each species resemble each other greatly, members of closely related species resemble each other, indicating that some inherited capacities remain unchanged for long periods of time.

Not only did organisms have the ability to do things on their own, they did them without regard for consequences, either to themselves or to the presumed clockwork mechanism ecosystems in which they lived. Darwin had seen for himself that the world was full of amazing biological diversity, organisms of all kinds showing substantial capacities to cope with conditions, and particularly with changing conditions. Despite such flexibility, however, Darwin realized that most organisms were not perfectly or uniquely fit to their surroundings. The Nature of the Organism thus encompasses everything that organisms are able to do for themselves, both abilities and limitations.

3.1.2 The Nature of the Organism and Darwin's Necessary Misfit

The true significance of the capacities covered by *the Nature of the Organism* is that they function *regardless of the Nature of the Conditions*. Reproduction occurs without regard for environmental conditions, and coupled with the evident exuberance of reproduction, it is in the nature of the organism to produce offspring in numbers far exceeding the resources available for their support (Malthus [1798] had shown that even humans would continue to reproduce despite being crowded together in early Industrial Revolution London with limited food availability).

This was *Darwin's Necessary Misfit* (Brooks and Hoberg 2007; Brooks 2010, 2011; Brooks and Agosta 2012; Brooks and Boeger 2019; Brooks et al. 2019). The exuberance of reproduction meant that there would always be more offspring than could survive. The conservatism and variation produced by inheritance without regard for the surroundings meant that the fit between organisms and their surroundings was imperfect. Therefore, not everything will survive, and not everything that does survive will find itself in a perfect fit with the surroundings. *Darwin's Necessary Misfit* thus broke the myth of a perfectly adapted biosphere without sacrificing the substantial ability of organisms to cope with their surroundings.

3.1.3 Natural Selection Emerges from Darwin's Necessary Misfit and the Nature of the Conditions

Now let's look at the relationship between the *Nature of the Organism* and the *Nature of the Conditions* in the context of *Darwin's Necessary Misfit*. First, reproduction is exuberant:

> There is no exception to the rule that every organic being naturally increases at so high a rate, that, if not destroyed, the earth would soon be covered by the progeny of a single pair—Charles Darwin (1872)

This does not happen, for any species anywhere. Why not? Darwin's answer was that many offspring of all species are eliminated in each generation. Second, inheritance is highly conservative: Exuberance plus conservatism in reproduction regardless of the nature of the conditions produced Darwin's *struggle for existence*. The reason many offspring are eliminated in each generation is because reproduction produces more organisms requiring the same resource than there are resources available to support them. Does this lead to a simple matter of random elimination, or are there regularities about who survives? Darwin's answer to that lay in the next element of the nature of inheritance. Third, inheritance produces variation:

> If such [variations] do occur, can we doubt (remembering that many more individuals are born than can possibly survive) that individuals having any advantage, however slight, over others, would have the best chance of surviving and procreating their kind? On the other hand, we may feel sure that any variation in the least degree injurious would be rigidly destroyed. This preservation of favourable individual differences and variations, and the destruction of those which are injurious, I have called Natural Selection—Charles Darwin (1872)

Some variants in a population will not be able to cope with the nature of the conditions in which they find themselves, either absolutely or in competition with their fellow offspring. Also, those that do survive will still be variable and will produce variable offspring.

If the conditions remain static, there will be a steady increase in the proportions of those who take the best advantage of those conditions. This does not mean everything will become the same "best" version. If reproduction produces variability as part of the nature of the organism, proportions may reflect the most favored forms, but the range of variation will reflect the overall survivability of the species. And this was a key element of Darwin's understanding of the role natural selection played in evolutionary diversification.

3.1.4 What Happens If the Conditions Change?

The inherent overproduction included variety in traits critical for coping with the conditions of life, and the varieties that were functionally superior in a particular environmental context would predominate. This explains how species tend to cope better and better with a given set of conditions through time. But that introduces a paradox: Whenever the conditions change, those organisms that already had the capacities necessary to survive would do so, whereas those lacking appropriate capacities would not. But that means that (1) organisms that were best adapted to one set of conditions were most at risk for extinction when the conditions changed, and (2) some organisms might have traits that allow them to be adequate for survival in more than one set of conditions. *This is the reason Darwin referred to natural selection as a perfecting mechanism but stopped short of saying that natural selection perfected anything.*

The production of organismal diversity thus required that organisms be at once autonomous from, and sensitive to, the environment. Darwin's perspective contrasted sharply with Lamarck's proposal that adaptation was an immediate and directed response by organisms to their surroundings. Lamarck also believed that the nature of the organism was important in the production of diversity, but only because all organisms have the same ability to change according to their needs. So, while Darwin postulated that the "nature of the organism" included autonomous, self-regulating properties, Lamarck believed that the "nature of the organism" was to be directly and completely connected to the environment. Darwinism, in contradistinction with Lamarckism, has always been a theory of coping, of doing the best one can with the capacities it inherited in the conditions in which it found itself. This is the full significance of Darwin's sentiment that *adaptation is to make a place for yourself in nature*. Natural selection is not something that exists on its own, waiting for life to evolve so it can shape it. It emerges inevitably from the interaction of the nature of the organism and the nature of the conditions, and because the nature of the organism predominates.

Darwin lacked mechanisms of inheritance and ontogeny yet understood that organisms were genealogically and developmentally cohesive. It was in the nature of the organism to produce offspring that were all highly similar to each other and their parents and other ancestors. He also postulated that reproduction produced variation without regard for environmental conditions, and therefore, it was in the nature of the organism to produce offspring in numbers far exceeding the resources available for their support. This cannot happen in a Lamarckian world, so there must be constraints on responses to the surroundings. Darwin resolved this conundrum by postulating that the nature of the organism created those constraints.

3.1.5 Natural Selection as a Blunt Instrument: Survival of the Adequate or Survival of the Fittest Collective

Darwin used Lamarckian adaptationism and domestication as *analogies* for natural selection. This was because they are aspirational forms of adaptationism that were familiar concepts to most people. After all, we had 15,000 years of human selection and adaptive evolution as a reference point. But Darwin also warned that they differed substantially from natural selection. Evolution is intimate, brutally short-sighted, and relentless, which might lead one to think that it is a mechanism of direct action. But those capacities stem from the nature of the organism, which is indifferent to the conditions. Natural selection, therefore, lacks the direct action of either Lamarckism or domestication. It leads to contingent, contextual adaptation. But Darwin realized that this was not just a good idea, it was essential. All reproducing organisms are capable of coping with the conditions of life in which they find themselves, but some cope better than others in their particular environments, where they predominate numerically over their merely adequate relatives. No matter how selection-challenged they might be, those merely adequate relatives survive and play decisive evolutionary roles. When the conditions change, the fittest in the old environment might not survive at all, whereas some of the merely adequate might flourish.

> I have hitherto sometimes spoken as if variations -so common and multiform with organic beings under domestication, and in a lesser degree with those under nature- were due to chance. This, of course, is a wholly incorrect expression, but it serves to acknowledge plainly our ignorance of the cause of each particular variation. … But the fact of variations and monstrosities occurring much more frequently under domestication than under nature, and the greater variability of species having wider ranges than of those with restricted ranges, lead to the conclusion that variability is generally related to the conditions of life to which each species has been exposed during several successive generations. … changed conditions act in two ways, directly on the whole organization or on certain parts alone, and indirectly through the reproductive system. In all cases there are two factors, the nature of the organism, which is the most important of the two, and the nature of the conditions. The direct action of changed conditions leads to definite or indefinite results. In the latter case the organization seems to become plastic, and we have much fluctuating variability. In the former case the nature of the organism is such that it yields readily, when subjected to certain conditions, and all, or nearly all the individuals become modified in the same way. … On the other hand innumerable instances are known of species keeping true although living under the most opposite climates. Such considerations incline me to lay less weight on the direct action of the surrounding conditions than on a tendency to vary, due to causes of which we are quite ignorant—Charles Darwin (1872).

Survival is both necessary and sufficient for Darwinian evolution. The issue for natural selection, therefore, is not how adaptations arise (that was to be explained by studies of the nature of the organism, specifically studies of inheritance and development), but how well they function in the conditions in which they appear. And, perhaps more importantly, how well do they need to function?

Inheritance produces variability in each generation, so it is impossible for all surviving offspring to be equally (and thus perfectly) fit to their conditions. Many variants produced in each generation are capable of basic survival. And despite selection, variability continues to be produced in every generation. Because organisms do not need to be the best to survive, what counts as "the fittest" is always a collection of all variants in each generation that were each adequate for survival and reproduction. If being "fit" in this sense means surviving long enough to participate in the species reproduction in a given generation, all variants that reproduce in each generation were "fit." This limits the ability of natural selection to promote particular variants rapidly.

> Variations neither useful nor injurious would not be affected by natural selection, and would be left either a fluctuating element, as perhaps we see in certain polymorphic species, or would ultimately become fixed, owing to the nature of the organism and the nature of the conditions—Charles Darwin (1872)

Also, organisms are cohesive wholes, with traits being subjected to different forms of selection. Some traits may not be affected by selection at a particular place and time, but their proportions may be affected by the effects of selection on traits with which they are linked by inheritance and development. Finally, there may be multiple variants that are equally fit with respect to the focus of selection, in which case there will be no net selection on any one of them.

> Hence modifications of structure, viewed by systematists as of high value, may be wholly due to the laws of variation and correlation, without being, as far as we can judge, of the slightest service to the species—Charles Darwin (1872)

In contradistinction with Lamarck, Darwin felt that extinctions of entire species had occurred, and were the result of a failure to cope with the rate, magnitude, or form of change in the conditions. This suggested that even with variation, the autonomous and conservative nature of the organism might be a hindrance to survival in changing conditions.

Compared to Lamarckism or domestication, therefore, the net results of natural selection are generally slow and incremental, indirect, and requiring long periods of time to have any permanent impact. Given his extensive discussion of the ways in which variations in the nature of the organism can make natural selection a blunt instrument, how could Darwin equate natural selection with the survival of the fittest?

One explanation is that Darwin's notion of "the fittest" meant a collective. Given that Darwin required variation as part of his explanation of the predicted outcomes of inheritance and of the way in which species cope with changing conditions, his notion of 'the fittest" may have been meant to refer to a group of variants. If you do not reproduce, you are "not fit." And if you have traits that made you unfit, those traits will diminish, or even disappear because they are not being perpetuated by inheritance. This is "non-survival of the unfit." What is left? There are many ways to cope with the conditions, and so long as your version is at least as good as others in your population, you have a chance to survive and reproduce. If your version is better than at least some others, your chances will increase. You do not need to thrive, just survive. If you thrive, that is fine, but thriving is transient in a world in which environmental conditions change. All reproducing organisms are fit, but some are fitter than others in their particular environments, where they predominate numerically over their merely adequate relatives. That is, until the conditions change, at which point the most fit might become less fit or not survive at all.

In other words, all it takes to be a member of the fittest collective in this sense is to be adequate enough to survive and reproduce. Biologists commonly teach students that you do not have to be great, just better than the competition. If you are an antelope, you do not have to be faster than the leopard, just faster than another antelope. If you are a pathogen, you do not need the best host, just a competent one.

Because natural selection is not creative, and because organisms have many demands placed on them in complex conditions, the effects of natural selection will tend to be slow and diffuse. But because organisms must interact with their surroundings in order to survive and because reproduction is exuberant, conservative, and variable, natural selection will be an ever-present property of evolution. Now we are ready for the final piece of Darwin's framework.

3.2 Darwinian Evolution: The Law of the Conditions of Existence

Without substantial autonomy from the surroundings provided by the rules of inheritance, there could be no reproductive overrun, hence no struggle for survival, thus no natural selection.

Natural selection was thus an emergent property of the inevitable conflict created by the conditions of existence and was a metaphor for the ways to resolve such conflicts, setting the stage for the resolution of conflicts yet to come. The issue is not "how do organisms survive in the current environments," which is an issue of stasis, but rather "how do organisms survive in changing environments," which is an issue of evolution. Natural selection was not the higher law it was a consequence of that law.

Darwin clearly stated that the higher law was the law governing the conditions of existence, the sum of the interactions between the nature of the organism and the nature of the conditions. Darwin may have introduced some confusion by using the term "conditions" in different ways. The "nature of the conditions" and "expression of the conditions of existence" refer to material elements in the surroundings, largely synonymous with the current use of the word "environment," any set of material circumstances with which organisms must cope. The "law of the conditions of existence" is more broadly conceptual, even metaphorical.

More than 150 pages after introducing the duality of the conditions of existence, Darwin called it the higher law of biology, underscoring the emergent nature of natural selection. The final *paragraph* of the sixth chapter of the *Origin* supplies the essential statement:

> It is generally acknowledged that all organic beings have been formed on two great laws [our emphasis] – unity of type and the conditions of existence. . .On my theory, unity of type is explained by unity of descent. The expression [our emphasis] of conditions of existence. . .is fully embraced by the principle of natural selection. . .Hence in fact the law of the Conditions of Existence [our emphasis] is the higher law; as it includes, through the inheritance of former adaptations, that of Unity of Type—Charles Darwin (1872)

Darwin recognized that the key to understanding evolution stemmed from the ways in which organisms were able to persist despite misfits between their surroundings and themselves. Finding such misfits to be common and universal, Darwin then recognized that not all viable members of all species were able to cope equally well with the nature of the conditions in which they found themselves. Nonetheless, generation after generation, all species produced offspring exhibiting varying degrees of misfit with their surroundings and in far higher numbers than could be sustained by environmental resources. This, Darwin reasoned, must lead to a struggle for survival on the part of those organisms proportional to their degree of misfit. When the inherent overproduction produced variety in traits critical for survival, organisms possessing traits that were functionally superior in that particular environmental context would survive best. Whenever an environment changed, those organisms that already had the functions necessary to survive in the new environment would do so, whereas those who lacked them would not; what is good today might not be good tomorrow.

Darwin understood that living systems were capable of acting on their own behalf and also reacting to the conditions in which they found themselves. Organisms carry so much of their history with them that species and their populations may respond to the conditions in ways that do not benefit all individuals. This is because inheritance, rather than the environment in which they find themselves, is the primary influence on how and how effectively each organism responds to selection. Natural selection may be a perfecting mechanism, but fortunately for life on a planet of changing conditions, it is such a blunt instrument that perfection will never be achieved, and the possibility of survival and evolutionary diversification lie in those less than perfect variants.

3.3 Two Powerful Visual Metaphors

The success of *Origin of Species* was due in part to the way in which Darwin wrote his book. Darwin was a master at connecting common everyday experiences and knowledge with technical observations by biologists, linking them together metaphorically to produce his panoramic synthesis of the history of life. Natural selection was easy to understand as something normative—after all, everyone knew about domestication and selective breeding of animals and plants. But Darwin's framework was much more than the simple "variation, selection, repeat until the desired endpoint is reached" that characterizes the activities of breeders. Darwin went beyond his analogies with adaptational selection to propose two rich *aspirational metaphors* to help visualize his panoramic view of evolutionary diversification—the *Tree of Life* and the *Entangled Bank.*

3.3.1 The Tree of Life

> The affinities of all beings of the same class have sometimes been represented by a great tree. I believe this simile largely speaks the truth. The green and budding twigs may represent existing species; and those produced during former years may represent the long succession of extinct species. At each period of growth all the growing twigs have tried to branch out on all sides, and to overtop and kill the surrounding twigs and branches, in the same manner as species and groups of species have at all times overmastered other species in the great battle for life. The limbs divided into great branches, and these into lesser and lesser branches, were themselves once, when the tree was young, budding twigs, and this connection of the former and present buds by ramifying branches may well represent the classification of all extinct and living species in groups subordinate to groups. Of the many twigs which flourished when the tree was a mere bush, only two or three, now grown into great branches, yet survive and bear the other branches; so with the species which lived during long-past geological periods, very few have left living and modified descendants. From the first growth of the tree, many a limb and branch has decayed and dropped off; and these fallen branches of various sizes may represent those whole orders, families, and genera which have now no living representatives, and which are known to us only in a fossil state. As we here and there see a thin straggling branch springing from a fork low down in a tree, and which by some chance has been favoured and is still alive on its summit, so we occasionally see an animal like the Ornithorhynchus or Lepidosiren, which in some small degree connects by its affinities two large branches of life, and which has apparently been saved from fatal competition by having inhabited a protected station. As buds give rise by growth to fresh buds, and these, if vigorous, branch out and overtop on all sides many a feebler branch, so by generation I believe it has been with the Great Tree of Life, which fills with its dead and broken branches the crust of the earth, and covers the surface with its ever-branching and beautiful ramifications—Charles Darwin (1872)

Prior to the advent of evolutionary thinking, natural philosophers argued that the only "real" entities were those that had immutable spatiotemporal existence. Because of their unchangeable nature, such bits of reality could be grouped into "classes" defined by the fixed properties of their components. Classic examples of such "real" entities, which at the time were called "species," are hydrogen and gold. Like these other natural kinds, biological species existed and conformed to a single hierarchical classification simply because it pleased the Creator for it to be so. Nearly 30 years before the *Origin*, Charles Lyell, the leading British geologist of the day, criticized Lamarck for postulating that biological species were mutable which, by existing professional standards, would have meant they were not real (Lyell 1832).

By the 1850s, however, accumulated evidence about fossils, geographic distributions of similar species on different continents, and his ongoing discussions with other natural philosophers led Lyell to abandon his former perspective, stating that science had reached a critical state with respect to the "species question" (Wilson 1971; Stevens 1992; Bowler 1996).

Lyell's change of heart was driven by one of his former students. Darwin (1859) suggested that our ability to group organisms naturally into species and groups of species was evidence that biological "species" were real but not immutable. Each one could change over time, and if subdivided could produce new, descendant species. Species descended from the same ancestors would naturally group together in a classification mirroring their common history of descent.

Darwin's belief in both the reality and the evolutionary nature of biological species was based on several empirical observations. First, as suggested by Lamarck, comparing fossil and living species indicated that not all of the species populating the planet today are the same as those populating the planet in the past. Second, the world today is not populated by a small number of species widely distributed in many different habitats in many different parts of the world; many different species live in specific habitats in many different parts of the world. And third, all living and extinct species fit into a single hierarchical classification that looks like a genealogy, or family tree. This led Darwin to believe that all species on this planet were related to each other through a single, common history of descent with modification—phylogeny.

The Tree of Life metaphor is thus more than an accounting scheme; it is a symbol of a major part of the evolutionary process. Metaphorically, the present is the state in which biological systems create their own futures based on their own pasts. Organisms carry so much of their history with them that their origins in space and time play integral roles in explaining the properties of organisms and the inheritance systems they form, and how they interact with their surroundings, including other inheritance systems (Brooks 1985; Brooks and McLennan 1991, 2002). In Europe, "sycamore" is a maple (*Acer pseudoplatanus*) and "plane tree" (*Platanus orientalis*) is what North Americans call "sycamore" (*Platanus occidentalis*). Darwin's metaphor of natural classification being a phylogeny enables us to understand why North American sycamores and European plane trees resemble each other so closely, why their ecologies are so similar, why they hybridize so readily.

Darwin's phylogenetic tree metaphor contrasted with a progressive view of diversity embodied in the *Scala naturae*, in which "lower forms" were replaced by "higher forms." The *Scala naturae* was a perfect visual metaphor for Lamarckian adaptationism but was not adequate for Darwin's contextual and contingent adaptationism, which produced multiple survivable outcomes linked by a common history of inheritance. The only illustrated metaphor Darwin ever provided in any edition of *Origin of Species* underscored the notion of evolution as one of selective accumulation rather than selective replacement of diversity.

Empirical and theoretical studies have combined to show that the phylogenetic tree is the most compact summary of the general and specific properties of species. Biological classification thus differs from atomic classification because of differences in the nature of the entities being classified. For living systems, the only immutable property shared is common ancestry. And they carry their history with them in abundance, which allows us to recognize many phylogenetic relationships readily, leading to the natural classification. Darwin was explicit about the significance of a phylogenetic tree in biological research.

> As it is difficult to show the blood relationship between the numerous kindred of any ancient and noble family even by the aid of genealogical trees, and almost impossible to do so without this aid, we can understand the extraordinary difficulty which naturalists have experienced in describing, without the aid of a diagram, the various affinities which they perceive between the living and extinct members of the same great natural class—Charles Darwin (1872)

Darwin expected that biologists would be able to accurately reconstruct the Tree of Life because of the conservative nature of inheritance:

> [T]he characters which naturalists consider as showing true affinity between any two or more species, are those which have been inherited from a common parent, all true classification being genealogical—Charles Darwin (1872)

The Tree of Life is thus a picture of the diversification of life through time. It is a metaphor for something dynamic, a historical snapshot of a dynamic process—evolution is a matter of selective diversification. It is not a process of winners replacing losers. Even taking extinctions into account, evolution is a matter of diversification. When extinctions occur, new diversification takes place. All extinctions are evolutionary re-sets. The nature of life on earth is that it accumulates evolved diversity.

Darwin's proposals about evolution thus created a dual role for species in biology, and he clearly recognized this duality. Species were real products of the evolutionary process, cohesive communities of descent, self-evident enough for us to document the Tree of Life.

> Why is not all nature in confusion, instead of the species being, as we see them, well defined?—Charles Darwin (1872)

Species were, however, also real units of evolutionary variation and change, so they might not always be well delimited in nature.

> Hence, in determining whether a form should be ranked as a species or a variety, the opinion of naturalists having sound judgment and wide experience seems the only guide to follow. We must, however, in many cases, decide by a majority of naturalists, for well-marked and well-known varieties can be named which have not been ranked as species by at least some competent judges—Charles Darwin (1872)

3.3.2 The Entangled Bank

Darwin realized that the phylogenetic tree was a necessary but not sufficient element of explaining biological diversity. How that diversity is arrayed on earth must also be explained, in ways that integrated the evolutionary history of life and the occurrence of diverse ecosystems in every corner of the planet. Darwin's most profound homage to von Humboldt, both in its accurate portrayal of biological diversity and in its poetic rendering, was the metaphor of the Entangled Bank:

> It is interesting to contemplate a tangled bank, clothed with many plants of many kinds, with birds singing on the bushes, with various insects flitting about, and with worms crawling through the damp earth, and to reflect that these elaborately constructed forms, so different from each other, and dependent upon each other in so complex a manner, have all been produced by laws acting around us. These laws, taken in the largest sense, being Growth with reproduction; Inheritance which is almost implied by reproduction; Variability from the indirect and direct action of the conditions of life, and from use and disuse; a Ratio of Increase so high as to lead to a Struggle for Life, and as a consequence to Natural Selection, entailing Divergence of Character and the Extinction of less improved forms. Thus, from the war of nature, from famine and death, the most exalted object which we are capable of conceiving, namely, the production of the higher animals, directly follows. There is grandeur in this view of life, with its several powers, having been originally breathed by the Creator into a few forms or into one; and that, whilst this planet has gone circling on according to the fixed law of gravity, from so simple a beginning endless forms most beautiful and most wonderful have been, and are being evolved—Charles Darwin (1872)

This clear and lyrically metaphorical statement evokes visions of selective accumulation of diversity producing complex ecosystems. It also explicitly underscores Darwin's conception of the *Law of the Conditions of Existence*, in which natural selection is a critical emergent property.

The entangled bank is partly a metaphor for species being jumbled together to form complex ecosystems, but it is equally a metaphor for Darwin's understanding of the role of geography in evolution, which is a recurring theme throughout the *Origin*. Populations of species isolated from their closest relatives in different parts of the earth would inevitably diverge into new species, but different species might arrive in a given place from different sources. Despite this potential for geographic mixing, wherever species co-occurred, they would form the complex ecosystems that fascinated von Humboldt.

So, at the end of the *Origin*, Darwin "told us what he told us," reflecting back to the second paragraph of the book, where he "told us what he was going to tell us," and in the 425 or so intervening pages, he "told us." These metaphors encapsulate the fundamental complexity of biological evolution, reinforcing Darwin's framework of great scope and generality. The *Tree of Life*, embodying the conservative nature of inheritance in the nature of the organism, explains why we find species doing similar things living in dissimilar conditions. The *Entangled Bank*, embodying the reality that biodiversity occurs in complex ecosystems comprising many species, explains why we find species doing dissimilar things living together in similar conditions.

3.4 What Was Wrong with Darwinism?

We have told a story about Darwin's theory, using his own words but arranging them in a compact manner that sounds familiar to the early twenty-first century readers. We find it remarkable how timely and comprehensive the framework appears today.

Equally remarkable is that Darwin made a theory that was so in tune with biology and yet so out of tune with what people wanted to believe at the time he published his theory.

Darwin's panoramic theory of biology rocked scientists—biologists and non-biologists alike, philosophers and social scientists, and humanists to their intellectual cores. Reactions to Darwin's new theory were generally enthusiastic within the biological community. Those who wholeheartedly accepted the new theory—notably Ernst Haeckel (1894, 1906), who befriended Darwin and became known as the German Darwin (Kutschera et al. 2019)—did so primarily because it accorded so well with practical reality, with experience. Those who equally wholeheartedly rejected the theory did so because its perceived harshness was at odds with prevailing philosophical and social expectations. Darwin's encyclopedic compilation of empirical evidence was enough to convince most intellectuals that the theory must encompass some truth, but the conflict between evidence and what was deemed to be rational and reasonable, conditioned by 150 years of European Enlightenment, produced more than a century of efforts to bring Darwinism into line with more accepted theories and social styles.

Darwin did not postulate a predictable set of outcomes for a given set of conditions and an application of external force under the control of a general and powerful natural law, as expected for acceptable scientific theories in the mid-nineteenth century. He did not suggest that evolution was a process per se but was an outcome of the interaction of two classes of phenomena (the *Nature of the Organism* and the *Nature of the Conditions*), each following their own rules but nonetheless spatially and temporally entwined. Furthermore, the two classes of phenomena were not co-equal, one was more important than the other. And yet, it is the "greater" phenomenon that inevitably creates conflict, while it is the "lesser" that leads to resolutions of those conflicts. If Darwin were writing about this today, we would easily see that he was talking about complex systems behavior, but in his day and age, his formulation seemed to many leading thinkers to be incoherent, muddled, and overly complicated; not necessarily wrong but in need of repair.

Perhaps more importantly, Darwin's message was simply not in style. The latter part of the nineteenth century was dominated by Modernism, Naturalism, and Romanticism, social styles that originated in the eighteenth century. Each was founded on long-standing elements of nineteenth-century biological and socioeconomic thought; a fascination with the apparent fit of organisms to their surroundings, and an intuitive sympathy for notions of progress, advancement, efficiency, even perfection.

3.4.1 Naturalism

By the end of the nineteenth century, the notion of a clockwork universe was beginning to wind down. But many, including many physicists, were not happy about that state of affairs. The second law of thermodynamics seemed to indicate that the fate of the universe was to use up all available energy until there was nothing left to run the clock. And that seemed to indicate that if we were not part of some kind of "Plan," then we were doomed to chaos and ruin. Darwin offered a middle ground, but with far more uncertainty than many were willing to accept. Unlike Lamarck's *absolute* adaptationism, Darwin's *relative* adaptationism included the possibility that species would sometimes be unable to cope with environmental change rapidly enough or appropriately enough to avoid extinction. Given the almost complete lack of contemporary species that were also represented in the fossil record, a traditional Darwinian would be led to believe that, sooner or later, every species met its Darwinian Waterloo. Fortunately, rarely did everything go extinct all at once. The biosphere's complement of species thus represented a system of indefinite variation in which some species were capable of surviving episodes of extinction, allowing life to persist through evolutionary diversification of the survivors.

In addition, nineteenth-century naturalists could be taken seriously as scientific thinkers only if they were working with laws. So, if you strongly believed that you were studying something "real" and important, you must tie your views to some sort of natural law or laws, even if you must simply postulate that they exist. Alfred Russell Wallace (Wallace 1855) alluded to a law linking extinctions and replacements of species to episodes of climate change, but the law was not identified. Darwin often used "law" as a synonym for "mysterious," in much the same way physicians use the word "chronic" when they exhaust treatment options. As we related above, Darwin referred to the "laws of inheritance," the "laws of the correlation of parts," and the "mysterious laws of growth," phenomena for which no causal agencies, much less any laws, were known at the time. In his entangled bank metaphor, Darwin invoked the invisible hand of "laws operating all around us," despite having written an entire book detailing the ways in which historical contingencies have shaped the biosphere. And when Darwin did refer specifically to a law, the *Law of the Conditions of Existence*, it was too complex for people to believe it was a real law. The consensus among traditionalist naturalists was that Darwin's theory was not natural, because it was not simple enough. They wanted natural selection to be a sharp-edged instrument, acting as an absolute law, leading to predictable endpoints regardless of the nature of the organism or the nature of the conditions.

3.4.2 Modernism

Darwinism suggested that evolution is not heroic or progressive, just effective and contingent. That meant it was not a modernist perspective. Modernist theories required heroic elements, something like human domination and improvement of a hostile world. Modernism meant taking charge of the imperfections of the natural world and perfecting them.

There was too much waste, not enough efficiency in the process Darwin proposed (what kind of well-oiled clockwork mechanism produces so much waste), and a negation of the sense of a drive for increasing efficiency and progress. Modernists wanted natural selection to be a sharp-edged instrument, but one that inevitably led to progress, thus mirroring and justifying domestication and other efforts to "subdue Nature."

3.4.3 Romanticism

It is ironic that Darwin was attacked most strenuously by contemporary Romanticists. The prototypical romantic naturalist was Jean-Jacques Rousseau, who lived almost a century before Darwin. Rousseau is most famous today as the naturalist who imagined ecosystems as French controlled gardens of the eighteenth century, filled with Noble Savages, all in need of benevolent protection. Another great Romantic naturalist, Alexander von Humboldt, was one of Darwin's great heroes and a darling of nineteenth-century romanticism. von Humboldt was a prototypical nineteenth-century Romantic naturalist hero. He had braved innumerable dangers in his quest to document the natural history of Venezuela and to solve the mystery of the connection between the Rio Orinoco and the Rio Negro. He was equally at home with modern technology for documenting the heights of mountains and the dimensions of the earth. He was perhaps the first member of European aristocracy to publicly denounce slavery. And he was one of the first people to recognize that some human activities might be placing species at risk; his description of the freshwater turtle egg harvest on the Orinoco includes an aside that when the Benedictine priests were in charge of the program, they always set aside a substantial portion of the nesting sites to ensure next year's harvest, but that since the Franciscans had taken over, there was no such conservation, with the predictable decline in annual output and potential for extinction of the turtles in that area. Wulf (2015) credited von Humboldt with inventing the modern sense of living nature.

Lamarckism was an excellent romanticist theory of evolution, and after Lamarck's blatant optimism, many people found Darwinism too harsh. The Darwinist perspective that not everything could survive ("Nature, red in tooth and claw"), that many individuals and indeed species would go extinct because of a failure to adapt, looked bleak indeed. Darwinian evolution is intimate, brutally short-sighted, and relentless. There is no ability to cope perfectly with changing conditions as Lamarck suggested; there is only the ability to try to cope with changing conditions. Success is contextual, diffuse, and transient; failure is absolute; and the only sense of progress is survival from one generation to the next. The apparent imperfection of the living world is not due to a lack of information, it is its true form.

Romantics wanted nature to be beautiful and fragile and in need of protection—the parts that were not meant for energetic exuberant exploitation, that is. They wanted natural selection to be like a gardener or shepherd, sharp-edged, but benevolent and the entangled bank to be a garden, not necessarily a controlled French garden but at least an English country garden. Samuel Butler wrote, in *The Deadlock in Darwinism* (1890), "To state this doctrine is to arouse instinctive loathing; it is my fortunate task to maintain that such a nightmare of waste and death is as baseless as it is repulsive." Butler was echoing a sentiment denouncing Darwin's perspective that emerged a decade before Darwin published the *Origin*.

Perhaps the best-known example of this thought is a poem written by a contemporary of Darwin's at Cambridge, Alfred Lord Tennyson, in memory of Arthur Hallam, a young poet who had just committed suicide:

> Are God and Nature then at strife, That Nature lends such evil dreams? So careful of the type she seems, So careless of the single life; That I, considering everywhere Her secret meaning in her deeds, And finding that of fifty seeds She often brings but one to bear, ... "So careful of the type?" but no. From scarped cliff and quarried stone She cries, "A thousand types are gone: I care for nothing, all shall go. ...Man, her last work, who seem'd so fair, Such splendid purpose in his eyes, Who roll'd the psalm to wintry skies, Who built him fanes of fruitless prayer, Who trusted God was love indeed And love Creation's final law-- Tho' Nature, red in tooth and claw With ravine, shriek'd against his creed—In Memoriam (1850)

3.5 Organized Resistance

In nineteenth-century society, those who could read, could afford to buy books, could afford the time to read them, and could afford the time to discuss them decided the public fate of Darwin's ideas. That wealthy and influential group of people melded naturalism, modernism, and romanticism. They wanted natural selection to be a sharp-edged instrument, like Lamarckian adaptationism and human selection; they wanted progress and achievement of perfection; they wanted heroism in the process. They wanted nature to be beautiful and fragile and in need of protection. And they wanted to know the laws of nature so they could bend them to their will. And after a century of the Industrial Revolution, all that seemed to be achievable.

Within the scientific community, these co-existing, and in some cases conflicting, social styles, coupled with a near-universal acceptance of a general notion of biological evolution, led to the emergence of several research programs aimed at "fixing" or replacing Darwin's theory. By the end of the nineteenth century, four distinct theoretical frameworks had emerged as the principal rivals in a race to replace Darwin.

3.5.1 The Geographers

> In considering the geographic distribution of organic beings over the face of the globe... [w] e see...some deep organic bond, through space and time, over the same areas of land and water, independently of physical conditions. The naturalist must be dull who is not led to enquire what this bond is. The bond is simply inheritance, that cause which alone, as far as we positively know, produces organisms quite like each other, or, as we see in the case of varieties, nearly alike—Charles Darwin (1872)

Darwin initially considered geographic isolation an essential factor in producing new species—populations of an ancestral species living in isolation from each other might well be subjected to different forms or respond in different ways to the same form of natural selection. In later writings, he explored the notion that selection alone could produce new species, ultimately adopting a broad perspective on the issue of the ways in which new species could arise.

Darwin's pluralistic perspective was too complex for some biologists, who argued that geographic isolation alone was responsible for producing new species, and all that "natural selection stuff" was irrelevant, if not simply wrong. Moritz Wagner (1868), the proponent of the latter perspective, was the catalyst for one school of thought about the evolutionary diversification of pathogen–host associations, and the earliest attempts to solve problems of general theoretical importance in biology using pathogen–host systems. Hermann von Ihering, a German-born biologist who spent his career in South America, was strongly influenced by Wagner. As an adherent of Wagner's views but ironically not of the fledgling theory of continental drift, von Ihering felt that geographic isolation rather than natural selection accounted for species formation and the disjunct distribution of related species (von Ihering 1891, 1902). He observed strong similarities in the ecological associations between some temnocephalidean (flatworm) parasites inhabiting freshwater crayfish in New Zealand and those inhabiting freshwater crayfish in the mountains of Argentina. Assuming that the temnocephalideans would never be found without the crayfish, von Ihering concluded that the contemporaneous species of both hosts and symbionts were derived from ancestors that had themselves been associated. Hence, he argued, South America and New Zealand must at one time have had freshwater connections. He later argued, in circular fashion, that such simultaneous speciation could only be explained by geographic isolation because one could not envision selective pressures operating equally on such distantly related and ecologically distinct species as hosts and parasites. He generalized his views with exuberant academic hyperbole:

> I am as much convinced of the erroneousness of this doctrine [fixity of continents] . . . as I am that the ideas of Darwin and Wallace on 'natural selection' as the cause of the origin of species will have but a historical interest in the coming centuries—Hermann von Ihering (1891)

3.5.2 *The Orthogeneticists*

A second group of biologists critical of Darwinism also developed as a critique of natural selection, in this case, the perception that Darwinism overemphasized the role of the environment (including geographic isolation) in evolution. Like the geographers' movement, the orthogenesis movement developed as a response to what many scientists saw as an overemphasis by some Darwinians on the role of natural selection in evolution. Unlike the geographers, however, orthogeneticists proposed that change resulted solely from an internal evolutionary drive, the mechanism of which was never clearly delineated. Although the term *Orthogenesis* only appeared in 1893 in a publication by the German biologist Wilhelm Haacke, Albert von Kölliker's theory of *Heterogenesis* proposed 30 years earlier had all the elements of orthogenetic thinking (see Wright 1984). It was based on three general ideas, all thought to be contrary to Darwin's views: the multiple origin of living forms, the internal causes of particular variants, and "sudden leaps" (heterogenesis) in the evolutionary process. A generation after von Kölliker, Swiss botanist Carl von Nägeli (1884) proposed that evolution was a manifestation of an internal perfecting mechanism, evidenced by trends toward increasing specialization, much like von Kölliker's heterogenesis. Perhaps because Gregor Mendel had called himself a student of von Kölliker, von Nägeli had engaged in a correspondence with Mendel from 1866 to 1873, but that did not lead to insights into the nature of the perfecting mechanism that von Nägeli was convinced existed.

3.5.3 The Neo-Lamarckians

Lamarckism after Darwin became fundamentally the opposite of orthogenesis, a theory of adaptation regardless of the nature of the organism. Adaptationists argued that the need for internal functional integrity or for an external functional fit between the organism and the environment was so strong, it was probable that an adaptive reason existed for every structure. In their worldview, functional needs shaped form; similarity of form simply reflected similarity of adaptation, which was not nearly as interesting as the divergence of form and function. The fit to the surroundings need not be perfect (they accepted a degree of misfit due to changing conditions as the driver of diversification) but so long as organisms were complex perfectly functioning entities that could both protect themselves from and take advantage of their surroundings, that was enough. Fossil assemblages were interpreted in terms of increasing perfection of adaptations to perform particular functions. Ironically, one element the adaptationists found hard to explain was the origin of new kinds of adaptations—they could explain how existing adaptations seemed to improve over time, but not how one kind of adaptation could morph into another when conditions required it.

3.5.4 The Neo-Darwinians and the Rise of "Survival of the Fittest"

Darwinism itself experienced an evolutionary transformation during the final quarter of the nineteenth century, emerging in response to a perception that the perfecting influence of natural selection was being underappreciated. Herbert Spencer's (1862, 1864, 1867, 1898) transformation of Darwinism from "survival of the adequate" to "survival of the fittest" marked the birth of *neo-Darwinism*. Few recognized what a fundamental departure from Darwinism this was. Darwin argued that many traits persist in environments in which they are viable but not particularly excellent; in other words, they are "adequate." Such traits were the source of an organisms' ability to cope with changing environments; if the environment changed in a way that favored a previously marginal form, its increasing fitness in the new environment would characterize the evolutionary change. Though Darwin spoke of natural selection as being a "perfecting" influence overall, no particular change was guaranteed to make descendants "better" in any way. For the emerging "neo-Darwinians," this was inconsistent with their perception of a high degree of functional "fit" between organisms and their environments. Spencer, perhaps drawing on his early advocacy of Lamarckism, concluded that "selection" must refer to two complementary phenomena, one of which was "traditional" Darwinian selection, while the other was a more general sense of selection that ensured that the traits available at any time would always include those that enhanced the "fit" of organisms to their environments.

In contrast to the orthogeneticists, who argued that built-in evolutionary constraints buffered organisms from the effects of natural selection, the neo-Darwinians sought to show that nothing was immune from the effects of natural selection. August Weismann showed that the production of traits and thus the production of variation, was not affected by the immediate environment, so Lamarckian inheritance could not explain Lamarckian adaptationism. If natural selection was a sharp-edged instrument and not the dull thing Darwin envisioned, we might believe that it could not only be a perfecting mechanism but that it could produce perfection. The neo-Darwinian research program became the presumption that all traits had been formed by natural selection as a perfecting mechanism and no traits were immune from its effects. This required that the exuberant and variable nature of inheritance be considered far more important than its conservative nature. It also required that there be no misfit. It restored the romantic view of perfectly fit nature and perfect adaptation while appearing to use a Darwinian mechanism. Darwin had simply underappreciated the power of selection and overestimated the degree of functional mismatch between organisms and their surroundings.

Progress and perfection were restored; optimistic adaptationism returned; natural selection was a lawlike mechanism, a sharp-edged instrument that shaved away unfit and merely adequate variants alike, leaving only the very fittest individuals. Evolution was a heroic process of achieving perfection in the functioning of organisms in their environments. It was acceptable for nature to be red in tooth and claw, so long as progress toward perfection was ensured. In other words, Spencer and his adherents produced a Darwinian-sounding theory that satisfied the social discomfort that Darwin caused. And despite the objections of traditional Darwinians, including Alfred Russell Wallace, co-originator of the principle of natural selection (see Kutschera and Hossfeld 2013 for a review), Spencer's views had "all the right stuff" for end of the nineteenth century science and society, and by the beginning of the twentieth century this was the perspective that was attributed to Darwin in student pamphlets (e.g., Hubbard 1905).

Herbert Spencer, who had begun as a Lamarckian, and clearly believed that neo-Darwinism integrated Lamarckism and Darwinism must have been surprised when his views were attacked by an American Lamarckian, the noted paleontologist Edward Drinker Cope (1886). Cope asked a seemingly simple question: If evolution is survival of the fittest, what explains the origin of the fittest? A Darwinian would have replied that changes in the nature of the conditions determined what was "the fittest," but because the nature of the organism in the form of inheritance was more important than the nature of the conditions, the fittest emerging from one transition in the nature of the conditions must have been present in low numbers—and decidedly not very fit—in the previous conditions (e.g., Wallace 1889, 1905; Haeckel 1906). The neo-Darwinians of the early twentieth century had a different explanation.

3.6 Summary

Darwin's theory of evolution was the first modern theory of complex systems dynamics, proposed more than a century before the physics community began to grapple with notions of natural complexity. His dualistic perspective of the interaction between the Nature of the Organism and the Nature of the Conditions gave rise to his Higher Law of the Conditions of Life, imbuing Nature with immense potential and exuberant free will and creativity, and combining the lawlike appearance of the Tree of Life with the mystery of the Entangled Bank. Darwin's understanding of the open-endedness of biological evolution—not the understanding of openness in the universe by twentieth century physicists and chemists—spelled an end to certainty in the natural sciences.

In retrospect, therefore, we can see that putting forth his ideas at that time was an act of immense courage on Darwin's part. Some of his colleagues—notably Wallace and Haeckel—joined him. Others, however, felt his theory tried to explain too much, or that it explained too little. Some felt it was just wrong. Some—through failure of intellect or creativity—simply did not "get it." And, of course, some felt they should have been given credit for having the idea first, once Darwin had taken most of the heat. From this nineteenth-century intellectual mélange emerged the geographic school of evolution, the orthogeneticists, and a renewal of Lamarckian views, all bickering with each other and trying to displace Darwin in the public eye.

In this venture, Darwin's biological critics were bolstered by a public mood, fostered by the narrow-mindedness and self-assuredness of late-nineteenth-century physicists and chemists, that clearly was not ready for Darwin's message. Darwin's material evidence in support of evolution was overwhelming, his scholarship was impeccable and his props as a naturalist too sound for his ideas to be dismissed out of hand. Clearly, biological theory needed to have some sense of evolution. Herbert Spencer captured this *Zeitgeist*, weaving a magical spell and convincing people that the syncretic perspectives of Lamarck and Darwin could be melded into a perspective of free will with a guaranteed future. As we will see next, this message was tailor-made not only for the end of the nineteenth century but for most of the twentieth century as well.

References

Bowler PJ (1996) Life's splendid drama: evolutionary biology and the reconstruction of life's ancestry 1860–1940. University of Chicago Press, Chicago, IL

Brooks DR (1985) Historical ecology: a new approach to studying the evolution of ecological associations. Ann Mo Bot Gard 72:660–680

Brooks DR (2010) Sagas of the children of time: the importance of phylogenetic teaching in biology. Evol Educ Outreach 3:495–498

Brooks DR (2011) The Mastodon in the room: how Darwinian is neo-Darwinism? Evol Educ Outreach 42:82–88

Brooks DR, Agosta SJ (2012) Children of time: the extended synthesis and the major metaphors of evolution. Fortschr Zool 29:497–514

Brooks DR, Boeger WA (2019) Climate change and emerging infectious diseases: evolutionary complexity in action. Curr Opin Syst Biol 13:75–81. https://doi.org/10.1016/j.coisb.2018.11.001

Brooks DR, Hoberg EP (2007) How will global climate change affect parasites? Trends Parasitol 23:571–574

Brooks DR, McLennan DA (1991) Phylogeny, ecology and behavior: a research program in comparative biology. University of Chicago Press, Chicago, IL

Brooks DR, McLennan DA (2002) The nature of diversity: an evolutionary voyage of discovery. University of Chicago Press, Chicago, IL

Brooks DR, Hoberg EP, Boeger WA (2019) The Stockholm paradigm: climate change and emerging disease. University of Chicago Press, Chicago, IL

Butler S (1890) The deadlock in Darwinism. Universal review. In: Stretfeild R (ed) Essays on life, art and science. Grant Richards, London, pp 234–340

Cope ED (1886) The origin of the fittest. D. Appleton, New York

Darwin C (1859) The origin of species by means of natural selection, 1st edn. John Murray, London

Darwin C (1872) The origin of species, 6th edn. John Murray, London

Eldredge N (2005) Darwin's other books: "Red" and "Transmutation" Notebooks, "Sketch," "Essay," and Natural Selection. PLoS Biol 3(11):e382. https://doi.org/10.1371/journal.pbio.0030382

Eldredge N (2009) Experimenting with transmutation: Darwin, the *Beagle*, and evolution. Evol Educ Outreach 2:35–54

Haacke W (1893) Gestaltung und Vererbung. Eine Entwicklungsmechanik der Organismen by the Author, Leipzig

Haeckel E (1894) Das Systematische Phylogenie. Georg Reimer, Berlin

Haeckel E (1906) Last words on evolution. (English translation of Haeckel, E. 1905. Der Kampf um den Entwickelungsgedanken). A. Owen, London

Hubbard E (1905) Darwin. In: Little journeys to homes of great scientists. The Roycrofters, Aurora, NY, pp 157–189

Kutschera U, Hossfeld U (2013) Alfred Russel Wallace (1823–1913): the forgotten co-founder of the Neo-Darwinian theory of biological evolution. Theory Biosci 25:206–214

Kutschera U, Levit GS, Hossfeld U (2019) Ernst Haeckel (1834-1919): the German Darwin and his impact on modern biology. Theory Biosci 138:1–7

Lamarck JB (1809) Philosophie zoologique, ou exposition des considérations relatives à l'histoire naturelle des animaux: à la diversité de leur organization et des facultés qu'ils en obtiennent; aux causes physiques qui maintiennnent en eux la vie et donnent lieu aux movemens qu'ils exécutent; enfin, à celles produisent, les unes le sentiment, et les autres l'intelligence de ceux qui en sont doués. Dentu, Paris

Lyell C (1832) Principles of geology, vol 2, 1st edn. John Murray, London

Nägeli C (1884) Mechanisch-Physiologische Theorie Der Abstammungslehre. Рипол Классик (Ripol Klassik)

Paley W (1802) Natural theology: or evidences of the existence and attributes of the Deity collected from the appearance of Nature. By the author

Pearce T (2010) A great complication of circumstances – Darwin and the economy of nature. J Hist Biol 43:493–528

Ruse M (1979) The Darwinian revolution. University of Chicago Press, Chicago, IL

Spencer H (1862) System of synthetic philosophy. First Principles. By the author

Spencer H (1864) System of synthetic philosophy. Principles of biology, vol I. By the author

Spencer H (1867) System of synthetic philosophy. Principles of biology, vol II. By the author

Spencer H (1898) System of synthetic philosophy. Principles of biology, revised and enlarged. vol I. By the author

Stevens P (1992) Species: historical perspectives. In: Keller EF, Lloyd EA (eds) Keywords in evolutionary biology. Harvard University Press, Cambridge, MA, pp 302–311

Tennyson, AL (1850) In memoriam

von Humboldt A (1815–1826) Personal narrative of travels to the equinoctial regions of the new continent. 7 vols. 1st ed. by the Author, Paris; English translation, London

von Ihering H (1891) On the ancient relations between New Zealand and South America. Trans Proc New Zealand Inst 24:431–445

von Ihering H (1902) Die Helminthen als Hilfsmittel der zoogeographischen Forschung. Zool Anz (Leipzig) 26:42–51

Wagner M (1868) Die Darwin'sche Theorie und das Migrationsgesetz der Organismen. Duncker und Humboldt, Leipzig

Wallace AR (1855) On the law which has regulated the introduction of new species. Ann Mag Nat Hist 16:184–196

Wallace AR (1889) Darwinism, 1st edn. Macmillan, London

Wallace AR (1905) Darwinism, 3rd edn. Macmillan, London

Wilson EO (1971) The plight of taxonomy. Ecology 52:741

Wright S (1984) Evolution and the genetics of populations: genetics and biometric foundations. University of Chicago Press, Chicago, IL

Wulf A (2015) The invention of nature: Alexander von Humboldt's new world. Vintage Books, Penguin Random House, New York

Chapter 4
Neo-Darwinism, Expansion, and Consolidation (1900–1980)

Abstract The twentieth century saw neo-Darwinism absorb its competitors and coalesce into the Hardened Modern Synthesis. Vernon Kellogg helped usher in this new era with *Darwinism Today*, proclaiming in 1907 that Darwinism was of critical historical importance but little actual scientific value and anticipating that new experimental and mathematical data would resolve the struggle among competing frameworks to replace it. Kellogg was prescient in proclaiming that the core question was how the right adaptation always arose at the right time, which became the major focus of twentieth-century evolutionary biology. Darwin's notion that preexisting variation was the fuel for coping with environmental change was abandoned in favor of something more heroic and aspirational, satisfying the ultramodernist styles of the time. In the ensuing decades, neo-Darwinian pan-adaptationism became the Modern Synthesis, permeating all corners of biology, especially ecology and behavior. By the 1980s, the Modern Synthesis hardened further. Natural selection was entrenched as a sharp-edged tool rather than Darwin's blunt instrument; a creative force from which function follows the conditions and form follows function. The Nature of the Conditions became the dominant explanation for everything, the Nature of the Organism relegated to an epiphenomenon. Evolutionary explanations became increasingly ecological and ahistorical, focused on how every trait was an adaptation to the conditions, and how the conditions were structured to accommodate organisms in the form of preexisting niches or zones. Seventy-five years on, Darwinism seemed to be as dead as Kellogg asserted.

Our story continues with someone whose career spanned the end of the nineteenth and the beginning of the twentieth centuries with considerable authority. While still in his 20s, Vernon Kellogg had begun efforts to meld geographic isolation and natural selection into a more pluralistic view of pathogen–host evolution, based on studies of bird lice. By the final year of the nineteenth century, Kellogg joined fellow Stanford University professor David Starr Jordan in promoting these ideas in a more general manner (Jordan and Kellogg 1900). Though only 40 years old, Kellogg was thus already a distinguished Stanford University biologist when he published *Darwinism Today* in 1907.

S. J. Agosta, D. R. Brooks, *The Major Metaphors of Evolution*, Evolutionary Biology – New Perspectives on Its Development 2,
https://doi.org/10.1007/978-3-030-52086-1_4

Darwinism Today helped set the tone for twentieth-century evolutionary biology. Kellogg described Darwinism as a reigning monarch, sitting back complacently as contenders for the throne squabbled, not realizing that one of those contenders would surely replace it. Kellogg wrote that Darwinism was of critical historical importance but little actual scientific value in light of the ultramodernism of the early twentieth century. According to Kellogg, Darwin had only two ideas, the Tree of Life and Natural Selection, and the former was not original with Darwin while the ability of the latter to produce species remained controversial and might be overrated.

Having quickly dismissed Darwinism (*Darwin is dead, let's get modern*), Kellogg discussed a number of aspirants for the soon-to-be-vacated crown. He did a masterful job of summarizing the different perspectives, rarely taking sides, always careful to let the reader know when he was injecting his personal perspective. He suggested that each of the perspectives he discussed had some merit, though he felt some had less merit than others. Kellogg did not offer a grand new vision of a new evolutionary biology, choosing instead to focus on the common unanswered questions that served as flashpoints among the various aspirants to replace Darwinism. Kellogg's perspective mirrored the zeitgeist of the day. At the beginning of the twentieth century, modernism and naturalism merged into an optimistic and powerful outlook on life. Science was uncovering the laws of nature and technology was permitting us to take advantage of those laws to improve the quality of human life.

Two particular advances that took hold at the end of the nineteenth century influenced twentieth-century biology. Like other progressive thinkers of his day, Kellogg anticipated that mathematical/statistical studies of variation and experimental studies of inheritance would eventually produce enough information to resolve ambiguities in inferences drawn from the relatively old-fashioned pursuits of developmental biology, taxonomy (including what he considered affiliated areas of field biology, such as biogeography and ecology), and paleontology. After all, Mendel's work had recently been reinvigorated by Hugo DeVries and Thomas Hunt Morgan was beginning his experiments that would lead to a Nobel Prize for his pioneering work on the nature of inheritance. Kellogg believed that accumulating experimental and mathematical data would resolve the struggles among advocates of competing theoretical frameworks to determine which one would eliminate the others.

We ended the last chapter with a conundrum. Cope (1886) focused on one key issue—if evolution is survival of the fittest, how do you explain the origin of the fittest? By the dawn of the twentieth century, the Darwinian notion that what was necessary for survival must already be present—in low numbers, in the background or on the margins but present—or extinction would result was not a sexy enough answer. Biologists craved something more direct, more powerful-sounding, more heroic, mythic, even magical. And it was in that vein that Kellogg wrote that the various aspirants to replace Darwinism all agreed that the core question in evolution was understanding how the right adaptation always arose at the right time. Kellogg was prescient: this question became the major focus of twentieth-century evolutionary biology.

Despite this, Kellogg is rarely mentioned in discussions of the history of evolutionary biology. He left Stanford for a year in 1915–1916 and then permanently in

1921 to participate in programs of social justice, having become a strong opponent of what he considered to be the misapplication of evolutionary ideas to human societies espoused by the German High Military Command just before WWI. He became the head of the US National Research Council and a trustee of what became the Society for Science and the Public, before dying in 1937.

4.1 Low Hanging Fruit: The Geographers

Kellogg was also progressive in his efforts to integrate the geographers' school of evolution with natural selection. Therefore, even though he was skeptical about the power of selection alone to produce new species, he felt it must play a role in the process. He proposed that while geographic isolation initiated speciation by partitioning variation, natural selection completed the process. A year after *Darwinism Today* appeared, Kellogg teamed again with David Starr Jordan to propose the *law of geminate species* (Jordan and Kellogg 1908). Geminate species were considered to be pairs of what we now call sister-species, the descendants of a unique common ancestor. This underscored an implication of Darwin's Tree of Life; evolution is not a process of elimination and replacement, it is a process of diversification. Speciation produces sister-species, not just replacements. Jordan and Kellogg found that geminate species most commonly occurred adjacent to each other geographically, separated by some kind of physical barrier, but often appearing to cope with their conditions of life in the same way (presumably inherited from their common ancestor). This supported the geographers' notion that their common ancestor once inhabited at least part of both areas where the descendants live today. In instances when members of these species come together, however, they tended not to interact with each other in a competitive manner, and they did not reproduce with each other. Jordan and Kellogg interpreted this as an indication that the sister-species had been subjected to divergent natural selection in their respective areas; thus geographic separation had set the stage for speciation, but natural selection completed the process.

The geographers noted that both members of Jordan and Kellogg's geminate species lived in areas with similar conditions and appeared mostly to do the same thing in each area, presumably the same thing their common ancestors did. They interpreted this as evidence of species formation—denoted by a lack of reproduction between the two geminate species—in the absence of divergent selection leading to divergent adaptations in the geographically isolated populations of a common ancestor. Adaptive divergence in sister-species, therefore, did not tell us anything about the origin of the species, according to the geographers. *Why* the geographically separated populations diverged into separate species without the assistance of natural selection, the geographers could never muster a definitive answer.

Antipathy to the notion of natural selection by traditionalist geographers kept their research program alive into the first quarter of the twentieth century, buoyed by Alfred Wegener's theory of continental drift (Wegener 1912, 1924). The notion that the major surface features of the earth were in motion provided the geographers with a means of allowing ancestral species to expand geographically into connected habitats, then become subdivided into different species by additional alterations of the earth. A serious problem for the geographers was that the world's geologists directed substantial hostility toward Wegener's ideas, and we do not overstate the issue when we use the word *hostility* (see Leviton and Aldrich 1985). The widespread condemnation of Wegener's ideas by the world's geologists in the early part of the twentieth century undermined the cause of geographic speciation advocates. Dispersal across land bridges or fixed geographical barriers mediated by climate (Matthew 1915, 1939; Mayr 1942; Darlington 1938, 1943, 1957; Udvardy 1969) were the only allowed mechanisms accounting for the geographic disjunction of ancestral species, and many felt that they alone could not account adequately for the ultimate diversification of the geographically isolated portions of an ancestral species into divergent sister-species. Evolutionary biologists interested in the geographic context of speciation were quickly absorbed into the maturing neo-Darwinian framework in the late 1930s and early 1940s.

4.1.1 Speciation by Reinforcement

Theodosius Dobzhansky (1937) and Ernst Mayr (1942) sided with Kellogg by asserting that the critical question to be answered about species formation lay in understanding how speciation was completed, and they produced a plausible answer to the question of how sister-species isolated under quite similar conditions nonetheless became different species that did not reproduce with each other even when they came into contact. This became known as *speciation by reinforcement.*

Classical speciation via reinforcement begins with geographic isolation (allopatry) and is completed when the isolated populations come into secondary contact with one another (sympatry) (Dobzhansky 1937, 1951; Mayr 1942, 1954). Following Darwin's notions that competition is likely to be most intense between close relatives, we would expect competition in areas of overlap that would lead to selection magnifying even slight variations into divergent forms that would no longer compete. Reinforcement happens when the two groups have diverged so much in allopatry that there is a penalty for making mating mistakes in sympatry; otherwise, we would have no diversification. The penalty is the occurrence of postzygotic incompatibilities between the two populations due to changes in the fertilization system that have evolved in allopatry (e.g., as a correlated response to natural selection). These incompatibilities manifest themselves along a line of decreasing cost, from the absolute failure of fertilization to reduced hybrid fitness under certain environmental conditions. Selection for prezygotic isolating mechanisms then occurs as a response to these costs. Combined, these factors would inevitably complete the divergence of differentiated species.

4.1.2 Peripatric Speciation Plus Reinforcement

By the 1940s, research in population genetics had shown that natural selection would be most effective in small isolated populations. Mayr (1942, 1954) incorporated this concept into a refinement of speciation by reinforcement, which he called *peripatric speciation*. Peripatric speciation postulates that a new species may arise from a small, isolated population that is usually, but not always, on the periphery of a larger ancestral population. The small, isolated population results from active dispersal by members of the ancestral species into a geographical locality not previously occupied by the ancestor. The small isolated population is separated from the rest of the ancestral species by a preexisting geographical barrier sufficiently strong to permit the establishment of novel phenotypes in the isolated populations but not so strong as to preclude secondary contact leading to reinforcement. When confronted with data suggesting that geminate species had similar sized geographic ranges, Mayr (1942) relegated those observations to secondary range expansion occurring after the business of speciation had been concluded. While this resolved one problem in the geographer's framework, it required some magical thinking. The non-competition in secondary contact required a *ghost of competition past*—close relatives do not compete today because they competed before, even if there is no evidence for it.

4.1.3 Changing the Nature of Species

Following the publication of *Origin of Species* taxonomists, by and large, continued to concentrate on species as units of classification and tended to ignore evolutionary principles, generally invoking them only to justify taxonomic decisions made following traditional pre-evolutionary practices (Stevens 1992). For example, at the end of the nineteenth century, the primary taxonomic guidelines for dealing with species were (1) they should be easily recognizable to laymen, (2) taxonomic nomenclature should be stable, and (3) there should not be too many names (Greene 1910; Brooks and McLennan 2002 called this *the Amadeus criterion*). This led some biologists to assert that species were whatever good taxonomists said they were (Regan 1926; Dobzhansky 1937; Gilmour 1940), a paraphrase of Darwin's (1872) assertion that for cases in which the separation of closely related species was difficult to ascertain, the consensus of competent taxonomists was the best course of action.

Darwin (1872) had asserted four properties of species. First, they are real. This was not a trivial assertion in the mid-nineteenth century. At that time, the term "species" was used to denote collections of identical entities with spatiotemporal boundaries, thus entities that did not change over time; elements like Hydrogen were considered species, so were certain minerals. Second, species are self-evident but with fuzzy boundaries. Third, they form classification groups and geographical assemblages due to inheritance. And fourth, they exist as communities of descent. Species were important because they showed us the trail of common inheritance that connected them all into the Tree of Life and dispersed them all over the planet in complex ecosystems (entangled banks).

Neo-Darwinism changed all that. The widespread acceptance, by the 1930s, of the notion that natural selection must be involved in speciation, either promoting divergence between allopatric populations or making the process irreversible through reinforcement, meant that the way in which species are formed plays a significant role in thinking about the nature of species. It also meant changing the definition of what species are, from what kept them together (Darwin's communities of descent) to what kept them apart (reproductive isolation). Dobzhansky (1937) and Mayr (1942) popularized what became known as the *Biological Species Concept*: species are populations of interbreeding, or potentially interbreeding organisms that are reproductively isolated from all other such groups of populations. This species concept quickly gained center stage in discussions of what became known as "the species problem" despite being restricted to sexually reproducing (eliminating most microbial life), non-hybridizing (eliminating most vascular plants) species with some degree of mate recognition (eliminating all broadcast spawners, such as many marine species that release gametes into the water where fertilization takes place in clouds of eggs and sperm).

4.1.4 Yes, but

Among the North American luminaries of neo-Darwinism, only George Gaylord Simpson (1944, 1951, 1953, 1961, 1963) insisted on a sense of evolutionary history in his *evolutionary species concept: species are reproductive lineages (communities of descent) each having their own evolutionary tendencies and fate separate from other such lineages.* As a paleontologist, Simpson was particularly interested in making certain that evolutionary biology integrated the "deep history" evidence provided by fossils. To Simpson, the evolution of a single species was analogous to the development of a single organism; just as an organism changes its appearance without losing its identity during development, so a species can change its appearance without losing its identity during evolution. The formation of new species was analogous to asexual reproduction, in which new individuals are distinct from the old individual because they form independent evolutionary lineages. Over time, distinct historical trajectories emerge from the speciation process, each differing to some degree from its ancestor and closest relatives but retaining some of its ancestry.

4.2 The Big Enchilada: Pan-adaptationism

Darwin had cast into doubt the myth of perfect adaptation as the cornerstone of evolutionary theory. Neo-Darwinians were intent on erasing that doubt. They strove for an explanatory framework that embodied the direct, heroic, progressive, optimistic, and powerful-sounding zeitgeist of that period. Nothing could be more heroic than the right adaptation showing up at the right time. Just a few years after

Darwinism Today appeared, North America had joined Europe in the race to provide the heroic account of evolution that Kellogg saw as inevitable, stemming from a combination of experiments on heredity and mathematical analysis. And indeed, this is where neo-Darwinism had its clearest vision, strongest focus, and greatest success.

The challenge was to provide a robust explanatory framework integrating Spencer's combination of Darwinian natural selection and Lamarckian pan-adaptationism. How could natural selection be a rapid, sharp-edged tool, progressive and possibly even creative? Weismann's studies at the end of the nineteenth century had already allowed neo-Darwinians to reject Lamarck's notion that there was a mechanism whereby efforts to adapt to novel conditions by an organism could become incorporated into its inheritance system and passed on to its descendants. But there was more work to be done.

On a conceptual level, Darwin's sense of adaptation as *coping with* the nature of the conditions had to be modified into Lamarck's sense of adaptation as *adapting to* them. The notion that the nature of the organism determined the ways in which and the extent to which organisms could adapt to the conditions of life in which they found themselves began to be replaced by the idea that the nature of the conditions determined which elements of the nature of the organism would predominate. If the nature of the conditions changed to something new, we would expect organisms to cope as best they could, but we would also expect that a novel trait that functioned better than the previous traits in the new conditions—the right adaptation—would replace the old. Such a novelty had to be immediately and directly connected to the novel conditions. In that way, the advantage it gave to organisms possessing it would be translated into differential survival, and offspring with that adaptation would "be favored." This provided a direct connection between how well organisms "fit" their surroundings and the relative success of their descendants, which was dubbed "fitness."

This perspective dovetailed nicely with the emergence of modernism in architecture and social planning at the beginning of the twentieth century. In the USA, Louis Henry Sullivan (1856–1924) became known as the father of skyscrapers, and the father of modernist architecture. His architecture and industrial design was based on the concept that *form follows function*; the form of any building or any object must enhance the function for which it was built. The neo-Darwinian complement of this principle would come to be encapsulated in the notion that *function responds to the conditions* and that *form follows function*.

This conceptualization of neo-Darwinian pan-adaptationism was still incomplete. Still to be addressed was the paradox of believing in a material basis for inheritance (refuting Lamarckism) but rejecting Darwin's notion that adaptations to new conditions were already in existence, being part of the history of inheritance. The neo-Darwinian answer to that paradox was that accumulated history was part of the problem the new "right adaptation at the right time" had to overcome. The past never contained the answer, only impediments to getting to the answer quickly. In modernist terms, if it is not modern, it is something to replace. The final piece of the puzzle was determining the mechanism by which the right adaptation appears.

Answering that question became the focus of three parallel and broadly overlapping lines of research.

4.2.1 Mathematics

The origins of mathematical approaches to studying lawlike behavior in selection and adaptation lie in late-nineteenth-century Great Britain, with Francis Galton. One of Darwin's nephews, Galton was an early eugenicist and a friend of Spencer's. He is credited with formulating correlation analysis and regression to the mean, as well as the concept of *nature versus nurture*. His most influential publication was *Natural Inheritance* (Galton 1889). A decade after *Natural Inheritance* appeared, Mendel's work was rediscovered and given close scrutiny by the scientific community. The appropriate mathematics for a fusion of Mendelism and neo-Darwinism would include the notion that in the absence of selection evolution would not occur, with the population or species existing at equilibrium with respect to itself and its surroundings. Within 3 years of the rediscovery of Mendel's work, Castle (1903) in the USA and Pearson (1903) in Great Britain suggested that without selection, genotype frequencies would remain stable. And just a year after Kellogg's book was published, Hardy (1908) in Great Britain and Weinberg (1908) in Germany independently formulated the general mathematical solution for that notion, now known as the Hardy–Weinberg Equilibrium equation.

For neo-Darwinians, this perspective fit well with the romanticist sense of a balance of nature. Natural selection could be described mathematically as an external force moving a species or population from one equilibrium state to another, each one determined by the nature of the conditions. In principle, selection could move a population from one equilibrium point to another at one time, then change in such a way that the population could move back to the original state. If you were shown a videotape of this process, you could tell that natural selection was occurring, but you could not tell whether you were seeing the videotape in its actual sequence or in reverse.

Galton died in 1911, but his legacy was carried on by a number of researchers working on mathematical aspects of neo-Darwinian pan-adaptationism, often with strong eugenics connotations. Most notable were R.A. Fisher and J.B.S. Haldane.

Fisher began publishing in the *Eugenics Review* in 1914 and produced a prodigious number of advances in the following decade, including the refinement of regression analysis.

Less than a decade after Kellogg's book appeared, Fisher referred to Darwinism as a theory of adaptation, not a theory of evolution (Fisher and Stock 1915). In that same article, Fisher expressed clear disdain for Lamarckian principles, including Cuenot's (1911) recently proposed *preadaptation* (see McLennan 2008 for an extended discussion of Cuenot's ideas).

In 1924, Haldane produced the first book-length treatment of the mathematics of neo-Darwinian selection and adaptation, making it clear that he did not support the distinction between natural and artificial selection articulated by Darwin. These first efforts at statistical analysis of selection were integral for Haldane's effort to demonstrate that mathematical models combining natural selection with Mendelian genetics could explain adaptation.

Haldane (1924) presented the first effort to quantify the influence of natural selection in a natural setting. In 1896, J.W. Tutt linked the increased frequency of melanic peppered moths (*Biston betularia*) to natural selection. He proposed the "differential bird predation hypothesis," as a mechanism of natural selection. The melanic morphs were better camouflaged against the bark of trees without foliose lichen, whereas the *typica* morphs were better camouflaged against trees with lichens. As a result, birds would find and eat those morphs that were not camouflaged with increased frequency (Tutt 1896). In 1924, Haldane, using a simple selection model, calculated the selective advantage necessary for the recorded natural evolution of peppered moths. If the frequency of dark-colored moths in 1848 was 2%, and by 1895 it was 95%, the dark-colored, or melanic, form would have been 50% more fit than the typical, light-colored form.

Fisher's magnum opus, *The Genetical Theory of Natural Selection*, appeared 6 years later (Fisher 1930). The book gathered together Fisher's contributions, refined regression analysis and some of Haldane's earlier ideas, and laid out the fundamental principles for what would become known as population genetics. Chief among these principles were the assertion that "Mendelism therefore validates Darwinism" and "The vast majority of large mutations are deleterious; small mutations are both far more frequent and more likely to be useful." The first statement was meant to be the final nail in the coffin of Lamarckism while the second was meant to do away with a short-lived perspective that Kellogg (1907) had discussed briefly, called *mutationism*, whose advocates believed that new species were formed when genetic mutations produced changes in members of species of a quality and magnitude sufficient to lead directly to species formation (see Theunissen 1994; Olby 2000 for reviews).

Fisher's explanation of how the right adaptation appears at the right time focused on the statistical properties of Mendelian inheritance in small populations. Many small populations increase the chances that multiple novelties will arise, while natural selection operating on small populations increases the chances of rapid amplification and fixation. If most novelties are deleterious, small, and incremental, it is likely that they would be eliminated by selection unless they enhanced survival in the particular conditions in which they emerged. What was important was not how the trait was created, but how the adaptation occurred. And so long as that displayed some kind of lawlike behavior, the origin of the novel traits was irrelevant. This boldly swept Cope's (1886) complaints about the "origin of the fittest" under the carpet.

Fisher's perspective set the stage for what has been called "populational thinking," that adaptations are an outcome of phenomena occurring within populations. The expected outcome of natural selection operating on novelties within populations—quickly eliminating deleterious traits and equally rapidly amplifying the rare beneficial ones—would lead to high fitness/low variance population structure. Darwin's necessary mismatch became a thing of the past. Naturalist neo-Darwinians quickly internalized this message:

> ... the organic world is segregated into more than a million separate species each of which possesses its own limited supply of variation which it does not share with the others...
> —Theodosius Dobzhansky (1937)

4.2.2 Yes, but

4.2.2.1 Hardy–Weinberg Equilibrium Assumptions

Seven assumptions underlie the Hardy–Weinberg equilibrium formulation: organisms are diploid; only sexual reproduction occurs; generations are nonoverlapping; mating is random; population size is infinitely large; allele frequencies are equal in the sexes; there is no migration, gene flow, admixture, mutation, or selection. Most of those assumptions are violated regularly in natural systems and attempting to compensate for those effects created many research opportunities for biostatisticians. While those refinements were being produced, most researchers were content to believe that the extent to which a population was not in Hardy–Weinberg equilibrium was proportional to the impact of selection being exerted on it. No one doubted that populations were being subjected to selection, it was simply a matter of how much. And if everyone was ignoring the same assumptions, perhaps their results were all flawed in the same way and therefore were comparable.

4.2.2.2 Genetic Drift

In 1929, Sewall Wright noted a problem with theoretically expected gene frequencies related to the issue of sampling (Wright 1929). Small isolated populations are non-random samples of all genetic variation within a species, and in some cases, a trait that otherwise would not have survived or become widespread would have a chance, just by chance. This was a cautionary tale about the significance of small populations. The ability of selection to promote adaptation could be enhanced or diminished depending on which genes were represented, and in what frequencies in each small population, as well as the particular conditions affecting each population. The extent to which genetic drift could play a role in species formation was the same as for selection and adaptation—small populations favored non-random sampling, which could affect the chances of a positive innovation arising, as well as the rate of response to selection. Once again, biostatisticians had many new research opportunities, producing protocols for distinguishing the effects of selection from the effects of drift. In practice, most considered drift to be a default explanation to be invoked if no particular form of environmental selection could be identified (or imagined).

4.2.2.3 Heritability

Heritability is a concept developed to provide a quantitative assessment of historical conservatism both in terms of genetic correlations among organisms within a species and also correlations between organisms and their previous environments. Not surprisingly, heritability has been of particular importance in agriculture and livestock breeding. The term itself seems to have been in common use among geneticists but perhaps not appearing in print until Lush (1939, 1940) used the term in discussing issues involved in selective breeding of dairy cattle. He himself originally intended the concept to be broadly synonymous with Darwin's notion of "hereditary tendencies" (Bell 1977), based on the notion that the nature of the organisms for each species differed according to which traits varied greatly—and were thus susceptible to natural selection—and which traits did not. For Lush, this explained why some breeds of cattle could be enhanced for dairy production but not meat production, while for others it was the reverse. Neo-Darwinians accepted the material basis of inheritance and therefore acknowledged the existence of historically persistent characteristics in all organisms. They believed, however, that the past never contained the right adaptation for new conditions. Retained history was part of the problem a new adaptation had to overcome, the load of impediments to getting to the answer quickly. Heritability was thus a blanket term for all the historical baggage that selection needed to overcome, an indicator of how strong selection needed to be for the "right adaptation at the right time" to succeed "on time."

4.2.2.4 Experimental Genetics

> ...biology is no longer simply a branch of history. It is now a science—Thomas Hunt Morgan (1932)

With the emergence of sophisticated statistical tools, it was now possible to conduct laboratory experiments guided by first principles. At the forefront of this research was the American Thomas Hunt Morgan, who won a Nobel Prize in 1933 for his work. Modernists like Morgan believed that if you cannot control and reproduce particular phenomena in the laboratory, you are not studying anything real. But even more, if you get repeating results, you are not just telling a story of how something happened in the past, you are one step closer to the actual mechanism of adaptation. Now you could begin to say, "it is this way because it had to be this way." Morgan felt that studying evolution in a Darwinian sense, with its emphasis on historical contingencies, was not science, but simply another branch of history. Neo-Darwinism, with its emphasis on selection and adaptation, was far more appealing. Morgan did not deny that there had been an evolutionary history. He just considered that irrelevant for a modern, early-twentieth-century scientist.

The first target was a reevaluation of the rules of inheritance put forward by Gregor Mendel. Morgan was aided by his choice of experimental organisms, the now-famous fruit fly *Drosophila melanogaster*. Morgan's work, and that of many others following in rapid succession, demonstrated that selection in the laboratory can be fast and time-symmetric, confirming what the mathematicians were saying in their models. He confirmed Mendel's basic rules of inheritance but went far beyond, showing that the situation was far more complex than Mendel imagined. In doing so, he drove the final nails into the coffin of Lamarckism. He also reinforced Haldane's (1924) idea that the only difference between natural and artificial selection was the speed of the process.

4.2.2.5 Natural History

Neo-Darwinism gave rise to a research program known as "the comparative method." Taking advantage of biostatistics, the comparative method sought to provide biology with rigorous evidence of lawlike behavior that could be ascribed to the effects of natural selection. In order to show statistically that selection acted like a natural law, each trait of each species was assumed to represent an evolutionarily independent variable, and thus an independent outcome of selection. The same or similar traits in each species could then be correlated with the same or similar environmental variables and a common lawlike cause inferred. Boas (1896) sounded a note of concern with this new approach when he discussed the limitations of the comparative method. He suggested that evolutionary biology was being studied in two different ways, which he denoted as the "historian" and the "evolutionist" perspectives. The historians assumed that similarities between species might be the result of common ancestry, the criterion upon which Darwin suggested biological classifications should be based. This assumption, however, lowered the number of independent origins of traits, weakening statistical arguments for the lawlike behavior of selection. The evolutionists' assumption that every trait was evolutionarily independent conformed to the prevailing view that the best scientific explanations were statistical in nature, and also allowed adaptations to be explained in terms of "something new." This anticipated Kellogg's assertion 11 years later that Darwin's notion that new adaptations must have been in existence, in low numbers, prior to the change in conditions that made them significant was absurd. Boas was not an advocate of the historian view, but he also noted that the evolutionists comparative method produced explanations that sounded remarkably Lamarckian. This was an early indication that the neo-Darwinian desire to co-opt Lamarckian pan-adaptationism was working.

By accepting the mathematics and experiments attesting to lawlike behavior on the part of selection and adaptation, naturalists no longer needed to demonstrate selection and adaptation in particular cases. Instead, they could make a number of simplifying assumptions.

First, by accepting the notion that function responds to the conditions and form follows function, behavior and ecology could be assumed to owe nothing to persistent ancestral attributes. One could use the increasingly sophisticated mathematical tools to assess the degree of functional fit between organisms and their surroundings and explain that if a given trait functioned in a particular way in a given environment, it must have evolved as an adaptation to that environment, regardless of historical background. Second, if the right adaptation had to be new—repetition is okay, but persistence is not—we should expect a lot of convergent and parallel evolution, all of which would represent independent episodes of adaptation. At the same time, homologous traits shared between species could not be adaptations because they were the historically persistent traits that Kellogg had disavowed, forming the historical baggage of heritability that selection had to overcome.

Armed with permission to find neo-Darwinian adaptations everywhere, field biologists filled the pages of their journals with accounts of such phenomena. At one extreme, the explanations were based solely on qualitative observations of what organisms did in their environments. At the other extreme were cases in which sophisticated mathematical tools were used to assess the degree of functional fit between organisms and their surroundings, thus adding a quantified sense of "how well adapted" organisms were to those evolutionary success stories.

The naturalists were most successful in modifying the geographers' program so that natural selection was required to complete the process, portraying selection as creative in a certain sense. And they were industrious in filling the pages of scientific journals with tales of putative adaptation based on natural history observations. But their greatest service to the emergent neo-Darwinism was in documenting a real case of natural selection in the wild.

By the 1950s, neo-Darwinism needed an unambiguous example of selection and adaptation in nature. We now return to the peppered moth. Haldane (1924) had assessed Tutt's proposal mathematically, but more than 30 years passed before Bernard Kettlewell (1958) investigated the biology of peppered moth coloration in a series of experiments under natural conditions. In experiments in the field from 1953 to 1956, Kettlewell found that a light-colored body was an effective camouflage in environments unaffected by industrial pollution, while the dark color was beneficial in polluted settings. This selective survival was due to birds that easily caught dark moths on clean trees, and white moths on trees darkened with soot. The story, supported by Kettlewell's experiments, had all the right stuff for neo-Darwinian adaptationism. It showed selection to be time-independent, fast, sharp-edged, responsive directly to changes in the conditions. For that reason, it became the poster child for neo-Darwinism, occupying a place in virtually every introductory biology textbook in the world (see Majerus 2009 for a review).

4.3 Co-opting Orthogenetic Adaptationism

What could be more heroic than a single species coming up with the right adaptation at the right time? What about entire groups of species resulting from such heroic efforts? A single species, having survived an environmental onslaught by coming up with the right stuff, then becoming the progenitor of a long-lasting lineage living happily ever after by producing lots of descendant species? This vision was the core of the orthogeneticists framework for evolutionary diversification—any adaptation that showed up and allowed progressive specialization was by definition a right adaptation at the right time, and selection played no role.

Henry Fairfield Osborn was one of the key leaders of the orthogenetic movement that swept through North America and Europe in the early to mid-1900s. Although adaptation was generally not a key component of orthogenetic explanations, a form of it was an important component of Osborn's research. For example, he described five different "lines of specialization" in mammalian teeth, diet, and limbs and feet (fitting their possessors for different habitats and types of locomotion), then combined characteristics from each line to reconstruct various mammalian orders. From this, Osborn proposed the following:

> [I]t is a well-known zoölogical principle that an isolated region, if large and sufficiently varied in its topography, soil, climate, and vegetation, will give rise to a diversified fauna according to the law of adaptive radiation from primitive and central types. Branches will spring off in all directions to take advantage of every possible opportunity of securing food. The modifications animals undergo in this adaptive radiation are largely of mechanical nature, they are limited in number and kind by hereditary, stirp, or germinal influences, and thus result in the independent evolution of similar types in widely separated regions under the law of parallelism or homoplasy. This law causes the independent origin not only of similar genera but of similar families and even of similar orders. Nature thus repeats herself on a vast scale, but the similarity is never complete or exact—Henry Fairfield Osborn (1910)

According to Osborn (1910, 1934), each unique set of characteristics was an *adaptive mode*, and all exhibiting the same adaptive mode occupied an *adaptive zone*. Osborn believed that every lineage would travel through various adaptive zones, each marked by increasing ecological specialization, eventually becoming so specialized that it went extinct. He believed that adaptive radiation was driven by internal principles of diversification rather than responses to external demands or opportunities, so he rejected a role for either Lamarckian inheritance or natural selection. Osborn's sense of adaptation was very similar to Darwin's use of adaptation simply to mean "functional ability." It was not, however, connected to Darwin's notion of adaptation as "selective fit to the environment." In *Nine Principles of Evolution Revealed by Paleontology*, Osborn (1932) noted: "All that we can say at present is that Nature does not waste time or effort with chance or fortuity or experiment, but that she proceeds directly and creatively to her marvelous adaptive ends of biomechanism." In accordance with other orthogeneticists, Osborn attached special significance to associations between different species, including pathogen–host systems. His "seventh law of adaptive radiation in the external body form" was *Symbiotic Adaptation*: "where vertebrate forms exhibit reciprocal or interlocking adaptations with the form evolution of other vertebrates or invertebrates. It is these two principles of too close adjustment to a single environment and of the non-revival of characters once lost by the chromatin which underlie the law that the highly specialized, and most perfectly adapted types become extinct, while primitive, conservative, and relatively unspecialized types invariably become the centres of new adaptive radiations."

For the first half of the twentieth century, neo-Darwinians were content to deal with orthogenesis mostly by denying the existence of orthogenetic trends, be they formal or functional. Neo-Darwinians accepted the orthogenetic notion of a progressive drive toward specialization, but they explained adaptations on a case-by-case basis, how they emerged and how they were amplified by selection at the right time in every case.

George Gaylord Simpson did his graduate work with Osborn. There is some evidence that Osborn and Simpson disagreed substantially about the social implications of some interpretations of evolution. Like Kellogg, Simpson disagreed with evolutionary interpretations imputing differing intrinsic values to different groups of humans. Unlike Kellogg, Simpson did not leave evolutionary biology for the struggle for human rights. Rather, he exerted his prodigious intellect in an all-out effort to co-opt Osborn's concepts, portraying adaptive radiations as a neo-Darwinian, rather than orthogenetic, research program.

> In adaptive radiation and in every part of the whole, wonderful history of life, all the modes and all the factors of evolution are inextricably interwoven. The total process cannot be made simple, but it can be analyzed in part. It is not understood in all its appalling intricacy, but some understanding is in our grasp, and we may trust our own powers to obtain more—George Gaylord Simpson (1944).

Simpson (1944) argued that the environment could be divided into "a finite and more or less clearly delimited set of zones or areas," similar to Wright's (1932) concept of *adaptive landscapes*. He envisioned adaptive radiation as a three-step phenomenon, beginning slowly with small, isolated populations slowly accumulating evolutionary novelties that would gradually make them increasingly more poorly adapted to their environments (Fig. 4.1). This accumulation would either drive a population to extinction or allow it to make a "quantum leap" into a new adaptive zone.

Once in the new zone, the ancestor and its descendants would diversify along one or both of two different pathways: (1) geographic expansion + speciation, producing many species with the same *evolutionarily conservative ecologies* in different places, and (2) ecological diversification + speciation, producing many species with *divergent ecologies* in the same place (Fig. 4.2).

Simpson's ideas about adaptive radiations emerged at the same time physicists were popularizing quantum mechanics. Quantum mechanics emerged from observations on the double life of photons (energy/wave versus matter/particle), which seems to introduce a fundamental duality into the universe. Simpson built a phylogeny/ecology duality into his adaptive zones, which could be specified simultaneously by the group that occupied them (e.g., the "felid zone") and by the adaptive trait that matched the environment (e.g., the "carnivore zone") (Fig. 4.3). Different zones could thus have different degrees of environmental and taxonomic complexity, which made it difficult to determine exactly what constituted a zone.

Although Simpson's ideas were later modified by a variety of authors, including himself, his basic components remained: a *key innovation* (Miller 1949) moved a group into a new zone, where the group diversified (radiation of species and

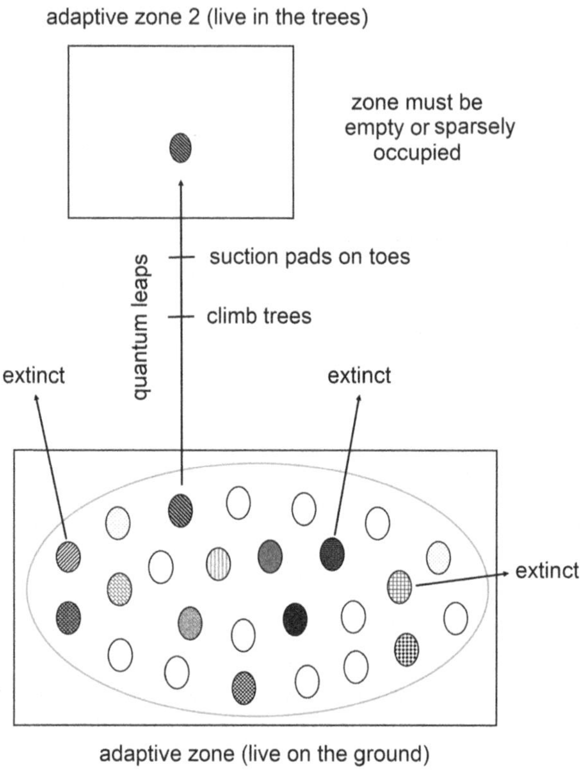

Fig. 4.1 Stage 1 in an adaptive radiation à la Simpson. *Circles* = populations (different patterns = different mutations); *boxes* = adaptive zones. Over time, small isolated populations accumulate evolutionary novelties that gradually make them more poorly adapted to their current environment (adaptive zone) which may lead to extinction or a transition to a new adaptive zone. Redrawn and modified from Brooks and McLennan (2002)

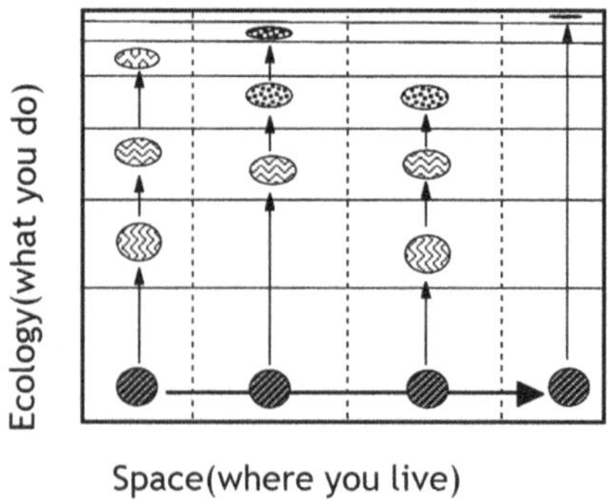

Fig. 4.2 Stages 2 and 3 in an adaptive radiation à la Simpson. Once in a new adaptive zone, the population will diversify along one or both of two different dimensions. Diversifying in space (geographic expansion) leads to many species with shared ecologies in different places. Diversifying ecologically leads to many species with different ecologies in the same places. Redrawn and modified from Brooks and McLennan (2002)

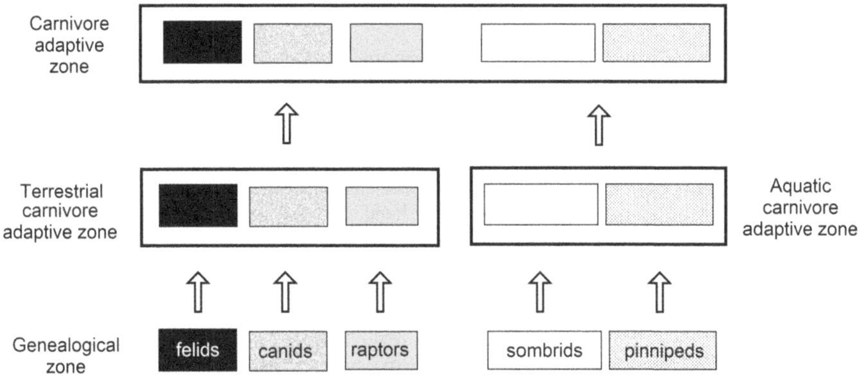

Fig. 4.3 Ambiguities associated with defining the term "zone." For example, felids (cats) are simultaneously part of the felid genealogical zone, the terrestrial carnivore zone, and the carnivore zone. Redrawn and modified from Brooks and McLennan (2002)

adaptations) to an extent limited only by the amount of available space (unoccupied niches) in the zone (Huxley 1942, 1953; Lack 1944; Wright 1949; Mayr 1963).

These various formulations of adaptive radiation had three significant problems. First, the idea that speciation produces species that always and progressively partition the environment into smaller and more specialized subunits is orthogenetic, predating even Osborn, and Simpson was not able to explain why this should always be the case in a consistently neo-Darwinian manner (Brooks and McLennan 2002). Second, there was no real definition of what qualified as a key innovation or how that innovation caused a population to jump from one zone into another. Traditionally, a key innovation was considered to be a novel feature characterizing a group so ecologically or morphologically distinct that it could be placed in its own "higher" taxon (e.g., order or class: Simpson 1944, 1953, 1960; Miller 1949). There was no necessary link between the key innovation and speciation; only between the key innovation and its putative adaptive function, which in turn was related to the distinctive nature of the new "adaptive zone." Third, the term "zone" was not easily defined. Was it an area, an ecology, or a taxon? Moreover, if the zone was a predetermined component of the environment, how could it be subdivided?

The neo-Darwinian solution to this problem marked an important point in melding neo-Darwinism with ecology, something that ecologists had been suggesting for almost 25 years:

> There have been three stages in the development of the biological sciences: first, a period of general work, when Darwin, Agassiz and others amassed and gave their knowledge of such natural phenomena as could be studied with the limited methods at hand; next, men specialized in different branches, and gradually built up the biological sciences which we know today; and now has begun the third or synthetic stage. Since the biological field has been reconnoitered and divided into its logical parts, it becomes possible to see the interrelations and to bring these related parts more closely together. Many sciences have developed to the point where contact and cooperation with related sciences are essential to full development.

> Ecology represents the third phase—Barrington Moore (1920)

During the next quarter-century, calls for integration and cooperation emerged gradually, but the timing was good in the early 1940s:

> Since variations in ecological conditions (physical or biotic) markedly effect the lives of individual organisms, and through this, of species, it follows that there is a broader line between the usual ecological emphasis upon succession of communities to the climatic or edaphic climax of a given region, on the one hand, and the taxonomic and geographic distributional emphasis of taxonomists and biogeographers on the other. The study of habits of animals, interpreted in the light of both ecology and taxonomy is, thus, an aid--indeed an absolute essential--to a complete understanding by either group of workers of the peculiar problems of either—Arthur N Bragg and Charles Clinton Smith (1943)

Ecologists had long used a descriptive term—niche—to encompass all the ways in which organisms interacted with their surroundings. Neo-Darwinians added an explanatory dimension to this concept. Henceforth, the nature of the conditions would be called an organism's or a species' niche. Simpson's perspective on adaptive radiations informed this expansion of neo-Darwinism. Simpson began with the assertion that adaptation was the result of interactions between organisms-as-random-variation and a structured environment. In order to study adaptation, we must have some way of quantifying the environment. But the environment is complex:

> ...including not only the physical conditions, average and variant, of the organism's geographic surroundings but also all existing foods, competitors and enemies, all forms of life affecting the given organism in any way whatever, other members of the same group, and even the organism itself—George Gaylord Simpson (1944)

Therefore, in order to study adaptation:

> ...we need to divide the environment into a finite and more or less clearly delimited set of zones or areas—George Gaylord Simpson (1944)

From this perspective, niches are slots in the surroundings that species adapt to. For neo-Darwinism, they need no explanation for how they came into existence, how they disappear, or how they are replaced. All that is needed is (1) the environment must be structured as a set of interconnected niches, (2) niches must pre-date any species that fill them, and (3) traits explained as adaptations to particular niches must not predate the niches (for reviews, see Korb and Dorin 2011; Winemiller et al. 2016).

Prominent ecologists like Grinnell, Elton, and Gause agreed with the emerging neo-Darwinian perspective. The niche was defined as a property *of the environment* that "might be filled at one time or place by a certain species, and at a [sic] another time or in another place by some different species, or might even be empty in one place and filled in another" (Colwell and Rangel 2009). In a seminal article, Grinnell (1917) wrote that the California thrasher occupied "one of the minor niches which their occupants all together make up the chaparral association." Similarly, Elton (1927) concluded that since the arctic fox and spotted hyena both eat eggs and carrion; they, therefore, occupy "the same two niches." For both Grinnell and Elton, the niche was something external to organisms that they somehow came to occupy (Colwell and Rangel 2009). For Elton at least, this meant two or more species could occupy, or at least attempt to occupy, the same niche. This perspective fit the neo-Darwinian narrative. The external conditions drive biological systems and organisms simply need to come up with the "right adaptation" at the "right time" to fill "niches" or "adaptive zones" in the surroundings.

Interestingly, the flaw in this thinking was articulated by another prominent ecologist and limnologist, G. Evelyn Hutchinson. Hutchinson (1957, 1959, 1978) attempted to (re)define the niche as something inherent to the organism, not to the conditions (see Colwell and Rangel 2009 for an excellent historical synopsis). Hutchinson's famous metaphor of a niche being a *multidimensional hypervolume* invoked an organism-centric visualization of a space in which the axes were "conditions" (e.g., temperature, rainfall, salinity, prey abundance, etc.) that define a "niche space" (i.e., combination of conditions) in which individuals/populations/species can "make a living" or have *fitness*. For Hutchinson, the capacity to survive and reproduce (to have fitness) under certain sets of conditions arises from the nature of the organism—without an organism there is no "niche." By reproducing, diversifying, and evolving, life elaborates and grows its own "niche space." This perspective negates the idea of preexisting niches or "zones" in the surroundings that evolution somehow "pulls" organisms into or "fills."

Simpson's efforts to shunt orthogenesis aside by co-opting the concept of adaptive radiation for neo-Darwinism were largely successful. Most biologists today assume that the concept of adaptive radiations originated with Darwin. But Simpson's efforts to offer a grand unified vision for adaptive radiations within neo-Darwinism fragmented evolutionary studies of large-scale evolutionary diversification. All the various formulations produced by that fragmentation foundered, and efforts like Hutchinson's were casualties. It was never possible to define objectively what qualifies as a key innovation or, more importantly, how that innovation caused a population to "leap" from one zone to another. Nor could the term "zone" ever be defined objectively. Basically, if you shut your "ecology eye," adaptive zones appear as phylogenetically coherent groupings that may not be ecologically cohesive; and if you shut your "phylogeny eye," adaptive zones appear as ecological groupings that may not be phylogenetically cohesive. Simpson may have recognized this conundrum when he coined the term quantum evolution, raising visions of the wave/particle duality. Ultimately, there was no link between the key innovation and speciation; only between the key innovation and its adaptive function, which in turn was related to the distinctive nature of the new "adaptive zone" as part of the nature of the conditions.

4.4 Act 2: The Hardened Synthesis (1959–1980)

> Our most recent, still continuing, period has been dominated by reaction against an earlier perspective considered too sweeping, too ambitious in scope, too weak in data and method.

In outline caricature, the devolution from generalizations of bold scope has been first to drop the generalizations, and then the scope—Dell Hymes (quoted in Stocking 1982)

Herbert Spencer provided the *inspirational* basis for neo-Darwinism, catching the *Zeitgeist* of the late nineteenth century with a powerful modern, naturalistic, and romantic reformulation of Darwin's theory. Kellogg (1907) articulated the *aspirational* framework for neo-Darwinism with his assertion that Darwinism was dead, and several competitors would henceforth battle it out for the privilege of being the next general theoretical framework for evolution. By the 1930s and early 1940s, influential general texts by Fisher (1930), Dobzhansky (1937), Mayr (1942), Huxley (1942), and Simpson (1944) made it clear that neo-Darwinism had achieved *operational* success.

Gould (1983) proposed the term "hardening of the modern synthesis" for what he perceived as a progressive commitment to pan-adaptationism by Dobzhansky beginning in 1937, but Kellogg's articulation of the commitment to pan-adaptationism predated Dobzhansky by more than a generation. Eldredge (1985, 1995) suggested that the hardening of the synthesis began with Julian Huxley, editor of *The New Systematics* (Huxley 1940) and author of *Evolution; The Modern Synthesis* (Huxley 1942), who argued strongly for taxonomy to be replaced by ecology. And yet, Huxley was by no means the first neo-Darwinian to suggest this. Almost half a century earlier, Boas (1898) suggested that biologists abandon efforts to provide a link between taxonomy and evolution by advocating purely empirical classification procedures identifying species as entities having no particular connection with evolutionary principles. Huxley's similar suggestion came at a time when neo-Darwinism's transition from aspirational to operational provided fertile ground for those ideas to take root, in the form of numerical taxonomy, or phenetics (Sokal and Sneath 1963; Sokal and Crovello 1970).

If making the transition from aspirational to operational represents the hardening of an explanatory framework, then we agree that the hardening of neo-Darwinism began in the late 1930s and early 1940s. We believe, however, that the fully hardened synthesis emerged roughly 20 years later, around the centenary of the publication of the *Origin*. That was the final triumph of neo-Darwinism, and Kellogg's prophecy seemed to have been fulfilled. We had an explanation for how the right adaptation shows up at the right time. The robust portions of the frameworks espoused by the geographers and orthogeneticists were being subsumed by neo-Darwinism. Most of the biological community had converged on a simple and compact view of evolution: function follows the nature of the conditions and form follows function, blurring the distinction between Darwinian and Lamarckian explanations, thus representing the complete assimilation of the geographers and neo-Lamarckians and nearly complete assimilation of the orthogeneticists into neo-Darwinism. With a final act of co-option, progressive evolutionary biologists would announce that evolution was a simple matter of "variation—selection—adaptation" or, even more concisely, "changes in gene frequencies in changing environments."

4.4.1 *Absorbing the Final Holdout: Co-opting Coevolution*

The version of orthogenesis that characterized twentieth-century research on inter-species relationships emerged near the end of the nineteenth century. The most notable proponent of this aspect of orthogenesis was Theodor Eimer (1897, 1898). Like other orthogeneticists, Eimer assumed that variation and evolutionary options were constrained at the origin of life to such an extent that the pathway and outcome were predetermined. Eimer, however, rejected orthogenetic ideas of inner growth laws and a *perfecting* internal drive. Rather, he insisted that there was an internal *adaptive* drive built into living systems. Eimer documented various trends in which no one stage seemed particularly better-adapted than another, trends that showed something like increasing perfection, and trends in which the transformations seemed to become increasingly maladaptive. As a result of his studies, Eimer believed that a full picture of evolution could only be achieved by linking all these different kinds of trends together into a unified narrative, and it was in ecology that he found the key.

Eimer believed that orthogenetic adaptation always led to progressive ecological specialization. Ongoing orthogenetic evolution in different lineages at different stages in their orthogenetic progression produced various adaptations characterizing the complex interplay of ecological associations characteristic of biodiversity. At some point in the process, however, *progressive ecological specialization turned into overdependence on other species, a loss of evolutionary independence, secondary simplification indicated by presumed reduction and loss of structures, and finally to self-imposed extinction.*

It is with this latter part of Eimer's narrative that pathogens became seen as key examples of orthogenetic evolution. Pathogens' apparent extreme dependence on their hosts was taken as an indication that they had reached the penultimate stage in evolution—overdependence and overspecialization prior to self-imposed extinction. The propensity to create disease was, therefore, a by-product of evolution, indicative that pathogens had become maladaptive in their environments and would soon eliminate themselves.

The neo-Darwinian alternative to the orthogenetic perspective on coevolution is linked to a familiar name: Vernon Kellogg. Kellogg's primary empirical research focus, beginning in the mid-1890s, was melding geographic isolation and natural selection into a pluralistic view of pathogen–host evolution, based on studies of bird lice. Kellogg wanted selection to play a role in evolution but he also wanted host-specific parasites to be infallible indicators of geographic speciation. His solution to this paradox was twofold: (1) parasites are highly adapted to their hosts, and (2) the host environment is so uniform it buffers parasites from the effects of natural selection. Kellogg attributed correlated parasite–host associations to phylogeny in a very Darwinian way, writing:

> I do believe that it is a commonness of the genealogy rather than a commonness of adaptation that is the chief explanation of this restriction of certain parasite groups to certain host groups—Vernon Kellogg (1913)

Kellogg found instances in which both the hosts and parasites had been geographically isolated but only the hosts had speciated. He hypothesized that if the environment was different enough in the geographically isolated regions to establish selection pressures leading to differentiation of the hosts, while the parasites' environment, consisting of parts of the host, did not change to the same extent, then the parasites would not speciate because they were not subjected to the same magnitude of selection. This was the basis for Kellogg's views that the *presumed uniformity of the host environment buffered parasites from selection.* All types of associations were the result of a single phenomenon—the pathogens were strongly adapted to their hosts which, in turn, buffered them from the effects of natural selection. This *neo-Darwinian* approach to understanding the evolution of inter-species associations found its most energetic researchers among those studying the interactions between plants and phytophagous insects. Those associations often showed pronounced specificity with respect to hosts on a species-by-species basis yet showed no clear phylogenetic component with respect to relatedness among host species or geographic distributions. This meant that the host *species* was not the focus of evolution; rather, *traits that the hosts possessed* to which the insects were adapted was the focus. These researchers began the process of changing research into inter-species associations from the host species *is* the evolutionary resource to the host species *has* the evolutionary resource (see Brooks and McLennan 1993, 2002).

Just 3 years after Kellogg published his book, Dutch biologist Émile Verschaffelt (1910) implicated the chemical basis of plant odor as a trait to which insects are adapted for eating and egg-laying. Within a decade, Charles Thomas Brues (1920, 1924), an American entomologist with a background in both tropical disease and agricultural research, suggested that a growing body of evidence supported Verschaffelt's findings. Less than 20 years later, another American researcher Vincent Dethier began amplifying this line of research (Dethier 1941, 1953, 1954), finding that these chemical features of plants repelled some insects and attracted others. Furthermore, he asserted that while all attractants are chemical in nature, some repellants can be physical as well. And finally, he noted that the chemicals acting as attractants may be very specific in nature and yet widespread among plant species. The arena within which insects adapted to particular plant traits was becoming extensive and complex.

The findings by those working with insect–plant systems did not alter the orthogenetic pathway that researchers studying pathogens—especially parasites of vertebrates—and their hosts followed (Brooks and McLennan 1993, 2002; Brooks et al. 2019). A singular article published by D.H. Wenrich in 1935 attempted to reintroduce some basic Darwinian principles to temper the overwhelming support for orthogenesis among pathogen specialists. Wenrich argued that capacities associated with changing hosts or changing from a free-living to a pathogenic lifestyle must pre-exist the transition:

> The associated habit, once established, the spread from host to host reflects the principle of opportunism, just as does the geographic dispersal of most animals. Any species of parasite

> will, presumably, establish itself upon or within any kind of host in association with which it finds favorable conditions for survival—DH Wenrich (1935)

These traits allow a pathogen to colonize a new host (or a free-living species to adopt a host), which *then* lays the foundations for the evolution of new adaptations through selection associated with the new host. His account was remarkable not just because it represented an argument for Darwinian principles supposedly outmoded and discarded by the turn of the century, but more impressively because he argued evolutionary principles based on the evolution of the parasite, not of the host.

Few listened to Wenrich's message because parasitologists liked orthogenesis and neo-Darwinians accepted the orthogenetic notion of a progressive drive toward specialization but with a different spin about its causes and consequences. They co-opted the orthogeneticists' argument that changes in behavior set the stage for new selection that led to the emergence of new traits. Both schools thus still believed in form following function and in the centrality of explaining how the right adaptation shows up at the right time. Kellogg's perspective that a pathogen species must adapt to its host while its host need not adapt to it and, once that adaptation has happened the host provides protection from natural selection became known as *Resource Tracking*, or *Sequential Colonization* models (see references and discussion in Brooks and McLennan 2002).

The final blow to the orthogeneticists program came a year before the centenary of the publication of the *Origin*, when the distinguished American mathematician Charles Mode was a young scholar studying the genetic interaction between wheat and rust fungus pathogens. Mode (1958) published the first mathematical model of mutual evolutionary modification resulting from mutual antagonism between parasites and hosts, introducing a concept called *coevolution*. Mode received little recognition for this accomplishment within evolutionary biology until recently (e.g., Brooks and McLennan 2002; Brooks et al. 2019), although that publication marked a critical juncture in our understanding of the evolutionary dynamics of inter-species associations. For the first time, scientists had to take into account not only the impact of the pathogen on the host but also of the host on the pathogen. Just 6 years later, Paul Ehrlich and Peter Raven (1964) proposed that the evolutionary diversification of plants and insects had been fueled by complex interactions involving mutual modification—notably of plant chemicals and insect olfactory and gustatory preferences. They suggested that such coevolutionary dynamics might have a general phylogenetic context, but the fine details need not parallel the evolutionary history of the specific taxa involved. The distribution of insects among plants followed the evolution of host resources and the evolution of insects' abilities to utilize those resources, rather than the evolution of host species themselves. Ehrlich and Raven's presentation of the ideas quickly caught the imagination of evolutionists and ecologists, producing what became known as *Coevolutionary Arms Race* models (see Janz 2011 for an excellent review). The primary assumption in these models is that coevolving associations are maintained by mutual adaptive responses. Variations on this theme have similar beginning assumptions: (1) pathogens reduce the fitness of their hosts; (2) hosts which, by chance, evolve traits that make them resistant to the pathogens (defense mechanisms) will increase their fitness relative to their undefended brethren, and the new defense mechanism will spread throughout the species; and (3) some pathogens will, in turn, evolve a counter-defense allowing them to survive in association with the previously protected hosts. If this trait confers a fitness advantage to those individuals, the counter-defense mechanism will spread throughout the pathogen species. The pathogen species would then enjoy unhampered dining on the previously protected members of the host species until the cycle began anew.

Constant conflict between many pathogens and hosts creates an environment in which the random appearance of the right adaptation at the right time would have high immediate adaptive value. The traits that enable pathogens and hosts to maintain specialized associations must in some way be connected with how they change host allegiances. If that is true, however, all of the traits associated with the specialized relationship must be unique to each coevolutionary episode. That is, each host switch must be accompanied by the right adaptation(s) showing up at the right time. Mode's, and Ehrlich and Raven's concept of species and their hosts adapting to each other killed orthogenetic views of coevolution, and also killed Kellogg's notion that pathogens are buffered from selection by their hosts. Coevolution became *neo-Darwinian adaptationism with constant conflict*, in which each participant is the "nature of the conditions" for the other.

4.5 Reinforcing the Cornerstones

4.5.1 Speciation

By the time the hardened synthesis emerged, peripatric speciation with reinforcement had become accepted as the primary mode of species formation (Mayr 1963). There was now no need to ask how speciation was initiated; researchers assumed that dispersal across a barrier established geographic isolation from the central population, stopping gene flow. Research focused more on the ways in which and the speed with which this mode of speciation could complete the speciation process. Peripatric speciation was generally thought to occur rapidly if strong selection was involved, and reinforcement fulfilled the bill as a form of strong selection. Research during the period from 1960 to 1980 demonstrated that the rate at which speciation in the peripheral isolate is completed depends upon a complex interaction among the rate at which gene flow decreases (Hennig 1950, 1966; Mayr 1963; Brundin 1966); the size of the founding population, including its genetic composition and the effects of mutation plus drift (Carson 1975, 1982; Wright 1978a, b; Templeton 1980, 1981; Lande 1981; Carson and Clague 1995); and the strength of divergent selection, including sexual selection, in the new area (Hennig 1950, 1966; Mayr 1954, 1963, 1982).

The hardened synthesis also encouraged renewed interest in modes of speciation requiring that speciation be initiated and completed while the diverging sister-species maintain reproductive contact with one another (called divergence with gene flow mechanisms) (Futuyma and Mayer 1980). *Parapatric speciation* (Smith 1965, 1969; Endler 1977) occurs when two populations of an ancestral species differentiate into descendant species despite the maintenance of some gene flow and geographical overlap during the process. Stochastic events (e.g., drift), adaptive responses to different local selection pressures along an ecological cline, or both, initiate the differentiation; low vagility among members of the populations (decreasing gene flow even when overlapping), a decrease in heterozygote/hybrid fitness leading to positive assortative mating, or both, promote it. *Sympatric speciation* (Darwin 1872; Fisher 1930, 1958; Bush 1966, 1969, 1974, 1975a, b, 1982, 1994; Maynard Smith 1966; Dickinson and Antonovics 1973; Tauber and Tauber 1977a, b; Rosenzweig 1978, 1997; White 1978) occurs when one or more new species arise without geographical segregation of populations. Although this was the mode originally preferred by Darwin, support for sympatric speciation wavered when population geneticists demonstrated that the homogenizing effects of recombination during meiosis coupled with gene flow among populations would tend to destroy the nonrandom associations of traits (the double-variation model) arising via divergent selection (Felsenstein 1981; see also Mayr 1963; Futuyma and Mayer 1980; Paterson 1981). The lack of theoretical support notwithstanding, one of the most passionate advocates of nonallopatric speciation modes, Guy Bush, argued:

> the future holds many surprises…I suspect that macro-mutations and rapid nonallopatric mechanisms of speciation will prove to be far more important in many groups of organisms than previously imagined—Guy Bush (1982)

4.5.2 Species

Despite the insights of the founders, many biologists interested in seeing "evolution in action" concentrated more and more on local inbreeding populations, or *demes*, claiming that these groups of "replicators" were the most inclusive level of evolutionary processes (e.g., Ehrlich and Raven 1964; Brown and Gibson 1983, 1998; Brown 1995). Populational thinking was critical in developing this view (Mayr 1942, 1963, 1982, 1988). Advocates of populational thinking treat species as assemblages of organisms held together by reproductive bonds exclusive to them, that can develop like individual organisms (but do not have to die of old age) and can "reproduce" by something analogous to binary fission. This approach allows biologists to slip comfortably into a transformational or evolutionary mode, moving away from a static or typological view, because it treated species as collections of organisms characterized by both common and variable traits. Emphasis on populational thinking, however, led many biologists (notably Huxley 1942) to conclude that only demes and populations were real, species being artificial constructs.

> Diversity is idiosyncratic. It is impossible to reconcile idiosyncrasy with preconceived ideas of diversity. The search for hidden likenesses is unlikely to yield a unifying species concept—Donald A Levin (1979).

[S]pecies lack reality, cohesion, independence, and simple evolutionary or ecological roles ... The concept that is most operational and utilitarian ... is a mental abstraction—Donald A Levin (1979)

Many had already abandoned the idea that species were useful, focusing on smaller units, populations, that were presumably more susceptible to the effects of selection. Ironically, one of the earliest and most influential advocates of this perspective was Julian Huxley, a descendant of Darwin's most influential and vocal advocate, Thomas Huxley. Lots of polymorphic widespread species were needed, so taxonomists were encouraged to combine previously named species into single entities; they were called "lumpers" by taxonomists who resisted the trend, who in term were disparaged with the term "splitters" by their more progressive colleagues. Following logically from the lumping exercise is the recognition that large polymorphic species cannot be high fitness/low variance systems, so populations within the species (many of which had previously been named as species) must be the real evolutionary entities because they can be high fitness/low variance systems.

Dobzhansky (1937), Mayr (1942, 1954, 1963), and Simpson (1944, 1953, 1961), however, argued that the species (and speciation) was the "keystone of evolution." They argued that species must be real entities for two reasons. First, the collection of demes (local populations where most reproductive activity occurs) construed as representing a species often exhibits more geographical and ecological cohesion and temporal persistence than the demes themselves (demes can disappear and reform without destroying the species). Species are cohesive entities having properties beyond any one population; a species is more than the sum of its parts, a sentiment with which Levin (1979) disagreed. If one population goes extinct, the species does not go extinct (unless that was the last population alive). Second, reducing evolution solely to reproductive exchange within individual demes (changes in gene frequencies in populations) does not explain the origins of the shared ecological and behavioral similarities among the collections of demes that make up what we call species; that calls for outbreeding among demes.

The hardened synthesis also included those who believed that species were real but, true to the neo-Darwinian framework, tried to define species without resorting to evolutionary history (phylogeny, the tree of life). Such *nondimensional or relational species concepts have the general form of*, "A is a species relative to B and C if it maintains its identity as a distinct entity relative to B and C." The prevailing nondimensional species concept for more than a quarter-century was the *biological species concept*, proposed by Dobzhansky (1937, 1940, 1970, 1976) and championed most strongly by Mayr (1942, 1963, 1976, 1982) that we discussed earlier. The hardened synthesis opened the door to additional perspectives. Most notably for the hardened synthesis, Van Valen (1976) changed the focus of attention from reproduction to ecology, proposing that each species has its own unique ecological niche and it is the niche that defines the species; this is the *ecological species concept*.

4.5.3 Adaptationism and the Hardened Synthesis

Neo-Darwinian adaptationism works best if there is a clearly defined environmental target to which the species (or population) can adapt. The hardened synthesis, therefore, made the niche concept a crucial element of adaptation. Niches were embodied by a "problem posed by the environment" to which species had to adapt or go extinct. And in the case of more than one species adapting to the same niche, the speed with which adaptation occurred determined which species won the competition for the niche. This helped emphasize the idea of a direct connection between "fit to the environment" and "fitness." By the Darwinian centennial, however, ecologists had a bit of a nomenclatural problem. There were three major categories of niches, Grinnellian, Eltonian, and Hutchinsonian, named after the ecologists who proposed them. All had many things in common, each emphasizing a different element of the ways in which living systems interacted with their surroundings. All of them could be used in adaptational stories, so their individual distinctions tended to be lost.

The American Leigh Van Valen proposed a truly innovative solution, the *Red Queen Hypothesis* (Van Valen 1973). Environmental changes create conflict between organisms and their surroundings, to which successful organisms respond by adapting. This makes *changes in niches* the source of conflict driving adaptive evolution. The faster the environment changes, the faster evolution will occur, but evolution will always lag behind. Speciation and extinction are locked together in a dynamic of birth and death mediated by never-ending environmental conflict, so adaptation is an eternal ongoing process.

This ultimate neo-Darwinian vision of niches and of organism/environment interactions harkened back to the Romanticism of the eighteenth and nineteenth century, especially the French Enlightenment view that nature (society) is a garden that needs to be controlled and ordered by a benevolent gardener (monarch). In such a controlled environment, all species are assumed to coexist in a harmonious "balance of nature" where every species has its own specific niche to which it is perfectly adapted (a place for everything and everything in its place).

4.6 The Hardened Synthesis and Ecology: The Rise of Evolutionary Ecology

Robert MacArthur (1958) and Evelyn G. Hutchinson (1959) set the tone for ecological studies of species coexistence and the search for correlations between changes in a species' ecology and changes in the environment. This research program was primarily concerned with attempting to answer the general question, "Why are there so many species?" and its corollary, "How do these species manage to coexist?" Darwinian answers to these questions had been sought within a phylogenetic framework, emphasizing the nature of the organism, but the expansion of neo-Darwinism led to an increasing emphasis on searching for answers in the nature of the conditions, such as Cain's (1938) species–area proposal, suggesting that the size of a geographic area determined how many species lived there, and Huxley's (1940, 1942) assertion that taxonomists needed to focus on ecology. MacArthur's perspective formalized these notions, "Ecological investigations of closely-related species then are looked upon as enumerations of the diverse ways in which the resources of a community can be partitioned" (MacArthur 1958). King emphasized the importance of searching for competitive exclusion within a closely related group of organisms in his critique of MacArthur's broken stick model of species abundance:

> As realized by Darwin, the principle of competitive exclusion is most applicable to closely related sympatric species (that is, to species of high taxonomic affinity) having similar but not identical niches. This may be related to the MacArthur model since when competitive exclusion has taken place, the species of high taxonomic affinity that remain may be expected to have niches which are nonoverlapping but contiguous. Hairston suggests that tests of these species should display better fits to the MacArthur model than do tests of all species occurring in the habitat. That these predictions are valid was first indicated by the striking fits obtained by Kohn when only members of the genus Conus were examined. Subsequent investigations of fresh-water fishes. .. reveal that in one collection from a single locality members of the class do not fit well, but when members of the same family are considered the fit is much better—Charles E King (1964)

In this statement, we again see the transition from aspirational to operational neo-Darwinism. Darwin said that competition was by far the most intense among members of the same species. He assumed that competition among closely related species was rare, mostly due to the presumption of allopatry, which formed the basis of Jordan and Kellogg's concept of geminate species. It was only when peripatric speciation by reinforcement became entrenched as the mode of species formation for neo-Darwinism that the need to postulate a strong case for inter-specific competition—past or present—became necessary.

The evolutionary ecology perspective was that "closely related" species are either competing now or they competed in the past; and we know they competed in the past because they do not compete now. Was this a fundamental truth or an unfortunate adventure in circular reasoning? The Darwinians, now disparaged as "historians," would have suggested comparative phylogenetic analyses. By the 1960s, however, some prominent systematists had concluded that explicit and robust protocols for inferring evolutionary history were not possible (e.g., Sokal and Sneath 1963). As a result, the number of historically based studies began to decrease within the rapidly burgeoning field of ecology paralleling an increase in the number of studies concerned with examining ecology within an adaptational context. Along with this came the search for surrogate measures of history, particularly geographic distributions (MacArthur 1960, 1965, 1969, 1972; MacArthur and Wilson 1963, 1967). This represented yet another ecological element of the hardened synthesis (see e.g., Brown 1981 for an update of Hutchinson's 1959 article).

4.6.1 Geography as a Proxy for History

Philip J. Darlington spent his career doing intensive natural history observations—mainly of carabid beetles—focused around an interest in how species came to live in the places where he found them. His foundational publication (Darlington 1943) was a monograph about the insect faunas of mountaintops and islands. His accumulated observations led him to a general framework in which new species and new ecological characteristics arose in "centers of diversification" and their subsequent distributional ranges periodically fluctuated around a more stable, continuously occupied center as a result of environmental perturbations. An advocate of classical geographic speciation, Darlington proposed that these biotic fluctuations might be interrupted by the formation of barriers to dispersal, producing episodes of isolation in which new species arose and ecological diversification might occur. Breakdown of those barriers produced new episodes of biotic expansion, setting the stage for yet more episodes of isolation. Populating the planet with species was thus a matter of sequential colonization emanating from the center of origin for any given taxon. Darlington's enormous database of natural history observations supported Cain's (1938) view that larger islands have greater species richness than smaller islands. He found similar patterns held for different-sized mountaintops, and thus suggested that there might be a relationship between the size of any area and the number of species living there.

Darlington's observations about the size of an area and the number of species living in it convinced another Harvard researcher, E.O. Wilson, to focus on the number of species living in a particular area more than on the manner in which they were formed. Wilson (1959, 1961) used the term *taxon cycles* to refer to this extension of Darlington's framework. Groups of species living together and affected by short-term, small-scale, environmental perturbations, disperse actively and colonize new areas when environmental changes allow expansion, and then contract their ranges when environmental changes reduce the amount of suitable habitat. These taxon cycles occur without producing new species and without new species arriving in the stable, continuously occupied center of the geographic distribution of the species involved. Taxon cycles occurring on intermediate time scales and affecting whole communities are often associated with *succession*. Successional changes occur when species residing in an area alter their surroundings to such an extent that they can no longer live there; think about the transitions leading to old-growth forests. The way in which they alter their surroundings, however, leads to their being superseded by species taking advantage of the changes brought about by the previous residents. In its longest time-scale version, a series of successional events eventually brings the area back to the initial state. Thus, even on long time scales, there is still no net diversification inherent in the framework. Nor is there any thought given to what the excluded species do once they exclude themselves, and yet they must survive somewhere if they are to make a reappearance at the end of a successional cycle.

4.6.2 Geography as a Means of Eliminating the Confounding Effects of History

A collaboration with Robert MacArthur formalized Wilson's elaboration of Darlington's ideas. Their seminal book *Equilibrium Theory of Island Biogeography* (MacArthur and Wilson 1967) is based on the view that dispersal from "source" areas to "islands" (actual or metaphorical), mediated by island size and distance, produces a linear log-normal relationship (called the *species–area relationship*) between species richness and the size of an island. Changes in the species–area relationship result from a dynamic balance between immigration, that is, colonization from a source area, and extinction. In other words, any given "island" can support only a certain number of species, called the equilibrium number of species. When the island has fewer than the equilibrium number of species, it is open to colonization. When the equilibrium number of species is reached, no new species may colonize the area without displacing one already in residence. MacArthur and Wilson acknowledged that historical effects, though real, would confound the species–area relationship and adaptational predictions, but suggested that for most cases it was probably safe to omit the production of new species within an area from the model as its effect on the species–area relation is "probably significant only in the oldest, largest, and most isolated islands."

4.7 The Hardened Synthesis and Ethology: Behavioral Ecology Emerges

Ethology, as a science, was founded upon a tradition of investigating behavior within an explicitly phylogenetic framework. Darwin started the ball rolling when he compared, among other things, the behavior of two species of ants within the genus *Formica* in an attempt to trace the evolution of slave-making in ants. Following this example, the "founding fathers" of ethology, Oskar Heinroth and Charles O. Whitman, proposed that there were discrete behavioral patterns which, like morphological features, could be used as indicators of common ancestry. Whitman's (1899) views mirrored Darwin's: "Instinct and structure are to be studied from the common viewpoint of phyletic descent" (Darwin 1872: 262). This perspective served as the focal point for a plethora of studies in the early twentieth century. Behavioral data were examined with an eye to their phylogenetic significance for birds, including anatids (ducks and their relatives) (Heinroth 1911; Herrick 1911a, b, c), weaver birds (Chapin 1917), cowbirds (Friedmann 1929), and birds of paradise (Stonor 1936); and for insects and spiders, including wasps of the family Vespidae (Ducke 1913), bumblebees (Plath 1934), caddisfly larvae (Milne and Milne 1939), termites (Emerson 1938), social insects in general (Wheeler 1919), and spiders (Petrunkevitch 1926). Wheeler reaffirmed the basis of ethological studies at the time:

> Of late there has been considerable discussion ... as to the precise relation of biology to history ... and what most of us older investigators have long known seems now to be acceded, namely that biology in the broad sense and including anthropology and psychology is peculiar in being both a natural science and a department of history (phylogeny)—William Morton Wheeler (1928)

This view would bring ethologists increasingly into conflict with the neo-Darwinian mantra begun by Morgan (1932).

Comparative behavioral studies flourished under the direction of Konrad Lorenz and Niko Tinbergen during the 1940s and 1950s. Both of these ethologists repeatedly emphasized two distinct but related points: behavioral patterns are as useful as morphology in assessing phylogenetic relationships, and behavior does not evolve independently of phylogeny. Lorenz stated that "all forms of life are, in a way, phylogenetic attainments whose special objects would have to remain completely obscure without the knowledge of their phylogenetic development" (1941), and "every time a biologist seeks to know *why* an organism looks and acts as it does, he must resort to the comparative method" (1958; see also Lorenz 1950). Tinbergen outlined his conception of the comparative method:

> The naturalist... must resort to other methods. His main source of inspiration is comparison. Through comparison he notices both similarities between species and differences between them. Either of these can be due to one of two sources. Similarity can be due to affinity, to common descent; or it can be due to convergent evolution. It is the convergences which call his attention to functional problems... The differences between species can be due to lack of affinity, or they can be found in closely related species. The student of survival value concentrates on the latter differences, because they must be due to recent adaptive radiation—Nikko Tinbergen (1964)

In other words, the phylogenetic relationships among species provide the platform from which explanations of processes responsible for behavioral evolution within species must be derived. Although this comparative approach to studying behavioral evolution flourished during the 1950s and 1960s, skepticism mounted about Lorenz's assertion that species-specific behavioral characters were valuable systematic characters. By the centenary of the publication of Darwin's book, two widely divergent viewpoints had emerged:

> To assume evolutionary relationships on the basis of behavior patterns is not justifiable when such findings clearly contradict morphological considerations. The methods of morphology will therefore remain the basis for the natural system [of classification]—Dietrich Starck (1959) (cited in Eibl-Eibesfeldt 1975)

> If there is a conflict between the evidence provided by morphological characters and that of behavior, the taxonomist is increasingly inclined to give greater weight to the ethological evidence—Ernst Mayr (1960)

Lorenz had marked the hardening of the synthesis in ethology when he wrote, "I am quite aware that biologists today (especially young ones) tend to think of the comparative method as stuffy and old-fashioned – at best a branch of research that has already yielded its treasures, and like a spent gold mine no longer pays the working. I believe that this is untrue" (1958:69). However, in a scathing review of the ethologists' research program, Atz made only a cursory reference to these successes when he concluded:

> The number of instances in which behavior has provided valuable clues to systematic relationships has continued to grow but it should be made clear that the establishment of detailed homologies was seldom, if ever, necessary to accomplish this... Functional, and especially behavioral, characters usually do not involve demonstrable homologies, but depend instead on resemblances that may be detailed and specific but nevertheless cannot be traced, except in a general way, to a common ancestor... Until the time that behavior, like more and more physiological functions, can be critically associated with structure, the application of the idea of homology to behavior is operationally unsound and fraught with danger, since the history of the study of animal behavior shows that to think of behavior as structure has led to the most pernicious kind of oversimplification—James W Atz (1970)

Following Atz's review, few intrepid souls maintained the belief that phylogenetic components of behavior could be studied scientifically and were important to explanations of the evolution of behavior (Dunford and Davis 1975; Radinsky 1975; Drummond 1981; Greene 1983).

Neo-Darwinian thinking took advantage of the vacuum left by the collapse of the ethologists' tradition, and a new methodology emerged based on arranging behavioral characters to tell stories of "plausible series of adaptational changes that could easily follow one after the other" (Alcock 1984). Joining forces with E.O. Wilson's elaboration of Sociobiology as an adaptational pursuit (Wilson 1975), these researchers created the field of Behavioral Ecology as a replacement for Ethology. For them, the details of specific behaviors mattered little; what was critical was understanding the environment in which the largest number of offspring were produced as a result of the behaviors they performed.

4.8 Yeah, but

4.8.1 Genetic Drift and Shifting Balance

Genetic drift became a potential irritant within the hardened synthesis, raising the possibility that species could originate without the need for reinforcement. As a result, Dobzhansky and Mayr denounced it as Lamarckian. In response, Wright (1955, 1956) suggested that his shifting balance framework (Wright 1931, 1932, 1940) showed how alternations between drift and selection were completely compatible with speciation by reinforcement. Both sides claimed victory, but Wright was not invited to a symposium honoring the founders of the New Synthesis organized by Mayr (Mayr and Provine 1980), prompting him to reiterate the potential role of genetic drift in speciation (Wright 1982), and Provine (1986) to edit a book reaffirming Wright's place in the pantheon of neo-Darwinism. More recently, Yoshida (2017) reviewed the links between Wright's shifting balance theory and the hardened synthesis.

4.8.2 Epigenetic Landscapes

More difficult for neo-Darwinism to cope with was developmental biology. The British developmental biologist Conrad Waddington was the lightning rod for these difficulties. In 1935, Waddington worked in Morgan's laboratory. In 1939, he proposed envisioning development as a kind of landscape, analogous to Wright's (1932) *fitness landscape* and *adaptive landscape* concepts (for a review, see Pigliucci 2008), which he called the *epigenetic landscape*. The cohesion of development (anticipated by Darwin in his term "correlation of parts" and supported by emerging studies documenting complex interactions among genes during development) produced *developmental stability*. Waddington's *epigenetic landscape* was populated by the developmental stability of all phenotypes, each represented by a trough, indicating that during development, organisms may have different interactions with their surroundings, while retaining their essential cohesion; Waddington called this *canalization*. The dimensions of each trough indicated how much flexibility a developing organism had in coping with changing conditions during development. Waddington called that *genetic assimilation*, perhaps the earliest formalization of what we now call *phenotypic plasticity*. For Waddington, the epigenetic landscape was similar to heritability; the degree of canalization in one trough was proportional to how much selection would be necessary to make a transition to a new one.

Dobzhansky and Mayr once again assumed the role of thought police and leveled the charge of Lamarckian thinking. They dismissed genetic assimilation as simply selection on previously existing rare genotypes. Waddington responded that neo-Darwinians badly neglected the phenomenon of gene interactions in development, which constrained the production of new variants necessary to ensure that the right adaptation showed up at the right time. Waddington became one of the first "insiders" of the hardened synthesis to call for an extended evolutionary synthesis (Waddington 1957, 1966, 1968, 1977). To the extent that we can see Waddington's perspective to be an attempt to inject some sense of the nature of the organism to neo-Darwinism, it is understandable that, while he became critical of the neo-Darwinian synthetic theory of evolution, he described himself as a Darwinian (see Bard 2008 for a review).

4.9 Summary

By the 1980s, neo-Darwinism was firmly entrenched as the only game in town. Natural selection had been developed into a sharp-edged, creative tool. The nature of the conditions, in the form of *ecology explains everything*, had taken over and the nature of the organism had been relegated to an afterthought. The evolutionary ecology revolution (MacArthur 1958, 1960, 1965, 1969, 1972; MacArthur and Wilson 1963, 1967), in which evolution became a synthesis of genetics and ecology, completed the hardened synthesis. The modernist view that function follows the environment and form follows function became firmly entrenched. As would be expected, evolutionary explanations became increasingly ecological, explaining every trait of every species as an adaptation to the current nature of the conditions. At its most fundamental level, neo-Darwinism became a process in which environmental conditions determined the births and deaths of species (Van Valen 1973). In addition to sounding very modern, this framework fed easily into our legacy as a story-telling species.

There is a dark side to this success story. The notion of conflict resolution has been replaced by never-ending conflict as the raison d'etre for evolution and for evolutionary biology.

In keeping with the focus on never-ending conflict, the scientific literature began to fill with aggressive anthropomorphisms—selfishness, cheating, attacking, defending, invading, sneaking, arms races, aliens, defense-counter defense, enemies, deceit, sexual conflict. In parallel, there were many personal attacks on almost any new idea that deviated from the simple core. This was nothing new:

> Scientists have odious manners, except when you prop up their theory; then you can borrow money of them—Mark Twain (1906)

Neo-Darwinism did not win the Kellogg Sweepstakes by eliminating its rivals, it absorbed them. As a result, it still retains vestiges of Lamarckism (not surprising given that Herbert Spencer was originally a Lamarckian) and orthogenesis. For example, statements like "the first land creatures developed lungs *so* they could breathe air" (Acot 1997) are Lamarckian, while the widely held belief that parasites are "secondarily simplified, highly specialized and degenerate" is, besides being incorrect, taken directly from orthogenesis (Brooks and McLennan 1993, 2002; Brooks et al. 2019). Even terms like "optimality" embody the upbeat perspective of Lamarck. The complicated new vistas now being revealed by molecular geneticists have led some researchers to propose that a complete theory of evolution would eventually require both Darwinian and Lamarckian processes (discussed in Barker 1993) but neo-Darwinism had already done that.

What neo-Darwinism *replaced* was *Darwinism*. To underscore this conclusion, we present the following statements about evolution expressed by self-described neo-Darwinians, preceded by statements consistent with Darwin's views in the sixth edition of the *Origin*. We believe that each reader's conceptual framework will comprise a combination of Darwinian and neo-Darwinian assertions.

1. Evolution is the interplay of the nature of the organism and the nature of the conditions, the nature of the organism being far more important *became* evolution is adaptation by random variation to changing environments.
2. Phylogeny is a critical part of causal explanations *became* phylogeny is a passive record of past selection events (more recently, analogous to the error term in an ANOVA model).
3. Ecology is played out on an evolutionary stage *became* evolution is played out on an ecological stage.

4. Evolutionary outcomes are generally (but not necessarily always) gradual because evolution is the net outcome (the resultant vector to use today's terms) of organisms responding to many different aspects of the nature of the conditions (so many selection vectors) *became* all evolution is inherently gradual.
5. Most evolutionary dynamics are the result of simple inheritance, organisms with non-zero fitness wandering through fitness space (fitness space is "sloppy": Agosta 2006; Agosta and Klemens 2008) *became* fitness space is highly optimized with fuzzy boundaries, and organisms do not change fitness space without eliminating a less fit occupant (the Spencerian influence is strongly shown by the use of optimization and optimality theory, from the same etymological root as optimism).
6. High levels of variation are expected; all variants with non-zero fitness in the environments in which they are born survive *became* high levels of variation are not expected; the variant with the highest fitness replaces all others (if there is variation, all variants have the same fitness).
7. The conservative nature of inheritance and insensitivity of reproduction to the nature of the conditions produce more organisms needing the same resources than there are resources available *became* limited environmental resources create conflicts.
8. Survival of the fit *became* survival of the fittest (optimality theory).
9. Survival (he who lives longest wins) is paramount *became* optimality (he who dies with the best toys wins) is paramount.
10. *Necessity* is the nature of the organism (material inheritance) and *chance* is the nature of the conditions became *chance* is the nature of the organism ("random variation") and *necessity* is the nature of the conditions.
11. Conflict is resolved by mutual accommodation (division of labor sensu Maynard Smith and Szathmàry 1995) *became* old conflict is replaced by new conflict.
12. Persistent traits are the result of inheritance regardless of past or present conditions *became* persistent traits are the result of selection in the past, so the actual history is not relevant.
13. Heritability is inherited tendencies *became* heritability is a measure of the necessary strength of selection needed to achieve the right adaptation at the right time.
14. Historical effects illuminate understanding about evolution *became* historical effects confound understanding about evolution.
15. Coping with the conditions *became* adapting to the conditions.
16. Adaptation is finding a place for yourself in nature *became* adaptation is filling a niche.
17. Species are real and speciation is caused *became* species are not real and speciation is a demographic accident.
18. Geography is a proxy for evolutionary history *became* geography is a means of eliminating the confounding effects of evolutionary history.
19. Every species that can make a place for itself survives *became* the environment determines the maximum carrying capacity.

Kellogg (1907) was prescient when he predicted that Darwinism was going to be replaced soon. He did not anticipate, however, that by the end of the twentieth century the replacement would be the subject of similar scrutiny and skepticism or that such scrutiny would eventually take a decided "back to the future" appearance.

References

Acot P (1997) The Lamarckian cradle of scientific ecology. Acta Biotheor 45:185–193
Agosta SJ (2006) On ecological fitting, plant-insect associations, herbivore host shifts, and host plant selection. Oikos 114:556–565
Agosta SJ, Klemens JA (2008) Ecological fitting by phenotypically flexible genotypes: implications for species associations, community assembly and evolution. Ecol Lett 11:1123–1134
Alcock J (1984) Animal behavior, 3rd edn. Sinauer Assoc, Sunderland, MA
Atz JW (1970) The application of the idea of homology to behavior. In: Aronson LR, Tobach E, Lehrman DS, Rosenblatt JS (eds) Development and evolution of behavior. W. H. Freeman & Co, San Francisco, CA, pp 53–74
Bard J (2008) Waddington's legacy to developmental and theoretical biology. Biol Theory 3:188–197
Barker G (1993) Models of biological change: implications of three studies of "Lamarckian" change. Perspec Ethol 10:229–248
Bell AE (1977) Heritability in retrospect. J Hered 68:297–300
Boas F (1896) The limitations of the comparative method in anthropology. Science 4:901–908
Boas F (1898) A precise criterion of species. Science 7:860–861
Bragg AN, Smith CC (1943) Observations on the ecology and natural history of anura, IV. The ecological distribution of toads in Oklahoma. Ecology 24:285–309
Brooks DR, McLennan DA (1993) Parascript: parasites and the language of evolution. Smithsonian Institution University Press, Washington DC
Brooks DR, McLennan DA (2002) The nature of diversity: an evolutionary voyage of discovery. University of Chicago Press, Chicago, IL
Brooks DR, Hoberg EP, Boeger WA (2019) The Stockholm paradigm: climate change and emerging disease. University of Chicago Press, Chicago, IL
Brown JH (1981) Two decades of homage to Santa Rosalia: toward a general theory of diversity. Am Zool 21:877–888
Brown JH (1995) Macroecology. University of Chicago Press, Chicago, Chicago, IL
Brown JH, Gibson AC (1983) Biogeography, 1st edn. Mosby, St. Louis, MO
Brown JH, Gibson AC (1998) Biogeography, 2nd edn. Mosby, St. Louis, MO
Brues CT (1920) The selection of food-plants by insects, with special reference to lepidopterous larvae. Am Nat 54:313–332
Brues CT (1924) The specificity of food-plants in the evolution of phytophagous insects. Am Nat 58:127–144
Brundin L (1966) Transantarctic relationships and their significance, as evidenced by chironomid midges. Kungl Svenska Vetenskap Handl 11:1–472
Bush GL (1966) The taxonomy, cytology, and evolution of the genus *Rhagoletis* in North America (Diptera, Tephrytidae). Bull Mus Comp Zool 134:431–562
Bush GL (1969) Sympatric host race formation and speciation in frugivorous flies of the genus *Rhagoletis* (Diptera, Tephrytidae). Evolution 23:237–251
Bush GL (1974) The mechanism of sympatric host race formation in the true fruit flies (Tephrytidae). In: White MJD (ed) Genetic mechanisms of speciation in insects. Australia and New Zealand Book Co, Sydney, pp 3–23
Bush GL (1975a) Modes of animal speciation. Annu Rev Ecol Syst 6:339–364

Bush GL (1975b) Sympatric speciation in phytophagous parasitic insects. In: Price PW (ed) Evolutionary strategies of parasitic insects and mites. Plenum Press, New York, pp 187–206
Bush GL (1982) What do we really know about speciation? In: Milkman R (ed) Perspectives on evolution. Sinauer Associates, Sunderland, MA, pp 119–128
Bush GL (1994) Sympatric speciation in animals: new wine in old bottles. Trends Ecol Evol 9:285–288
Cain SA (1938) The species-area curve. Am Midl Nat 19:573–581
Carson HL (1975) The genetics of speciation at the diploid level. Am Nat 109:83–92
Carson HL (1982) Speciation as a major reorganization of polygenic balances. In: Barigozzi C (ed) Mechanisms of speciation. Alan R. Liss, New York, pp 411–433
Carson HL, Clague DA (1995) Geology and biogeography of the Hawaiian islands. In: Wagner WL, Funk VA (eds) Hawaiian biogeography: evolution on a hot spot archipelago. Smithsonian Institution Press, Washington, DC, pp 14–29
Castle WE (1903) Mendel's law of heredity. Science 18:396–406
Chapin JP (1917) The classification of the weaver-birds. Bull Am Mus Nat Hist 37:243–280
Colwell RK, Rangel TF (2009) Hutchinson's duality: the once and future niche. Proc Natl Acad Sci 106:19651–19658
Cope ED (1886) The origin of the fittest. D. Appleton, New York
Cuènot L (1911) La genèse des espèces animales. Ancienne Librairie Germer Baillière et Cie, Paris
Darlington PJ Jr (1938) The origin of the fauna of the greater Antilles, with a discussion of dispersal of animals over water and through air. Q Rev Biol 13:274–300
Darlington PJ Jr (1943) Carabidae of mountains and islands: data on the evolution of isolated faunas, and on atrophy of wings. Ecol Monogr 13:37–61
Darlington PJ Jr (1957) Zoogeography: the geographical distribution of animals. Wiley, New York
Darwin C (1872) The origin of species, 6th edn. John Murray, London
Dethier VG (1941) Chemical factors determining the choice of food plants by *Papilio larvae*. Am Nat 75:61–73
Dethier VG (1953) Host plant perception in phytophagous insects. Trans IXth Int Congress Ent Amsterdam 2:81–88
Dethier VG (1954) Evolution of feeding preferences in phytophagous insects. Evolution 8:33–54
Dickinson H, Antonovics J (1973) Theoretical consideration of sympatric divergence. Am Nat 107:256–274
Dobzhansky T (1937) Genetics and the origin of species, 1st edn. Columbia University Press, New York
Dobzhansky T (1940) Speciation as a stage in evolutionary divergence. Am Nat 74:312–321
Dobzhansky T (1951) Genetics and the origin of species, 3rd edn. Columbia University Press, New York
Dobzhansky T (1970) Genetics of the evolutionary process. Columbia University Press, New York
Dobzhansky T (1976) Organismic and molecular aspects of species formation. In: Ayala FJ (ed) Molecular evolution. Sinauer Associates, Sunderland, MA, pp 95–105
Drummond H (1981) The nature and description of behaviour patterns. In: Bateson PPG, Klopfer P (eds) Perspectives in ethology, vol 4. Plenum Press, New York, pp 1–33
Ducke A (1913) Über Phylogenie und Klassification der sozialen Vespiden. Zool Jahrb Abt Syst Oekol Geogr Tiere 36:303–330
Dunford C, Davis R (1975) Cliff chipmunk vocalizations and their relevance to the taxonomy of coastal sonoran chipmunks. J Mammal 56:207–212
Ehrlich PR, Raven PH (1964) Butterflies and plants: a study in coevolution. Evolution 18:586–608
Eibl-Eibesfeldt I (1975) Ethology: the biology of behavior, 2nd edn. Holt, Rhinehart & Winston, New York
Eimer T (1897) Orthogenesis Der Schmetterlinge: Ein Beweis Bestimmt Gerichteter Entwicklung Und Ohnmacht Der Natürlichen Zuchtwahl Bei Der Artbildung. Zugleich Eine Erwiderung an August Weismann, vol 2. Engelmann, Leipzig

Eimer T (1898) On orthogenesis: and the impotence of natural selection in species formation. Open Court Publishing Company, Chicago, IL
Eldredge N (1985) The ontology of species. In: Vrba E (ed) Species and speciation, Transvaal Mus. Monogr, No, vol 4. Transvaal Museum, Pretoria, pp 17–20
Eldredge N (1995) Reinventing Darwin: the great debate at the high table of evolutionary theory. Wiley, New York
Elton CS (1927) Animal ecology. Sidgwick and Jackson, London
Emerson AE (1938) Termite nests- a study of the phylogeny of behavior. Ecol Monogr 8:247–284
Endler JA (1977) Geographic variation, speciation, and clines.. Monographs in population biology no. 10. Princeton University Press, Princeton, NJ
Felsenstein J (1981) Skepticism towards Santa Rosalia, or why are there so few kinds of animals? Evolution 35:124–138
Fisher RA (1930) The Genetical theory of natural selection. The Clarendon Press, Oxford
Fisher RA (1958) The Genetical theory of natural selection, 2nd edn. Dover, New York
Fisher RA, Stock CS (1915) Cuènot on preadaptation: a criticism. Eugen Rev 7:46–61
Friedmann H (1929) The cowbirds. Charles C. Thomas, Springfield, IL
Futuyma DJ, Mayer GC (1980) Non-allopatric speciation in animals. Syst Zool 29:254–271
Galton F (1889) Natural inheritance. Macmillan, London
Gilmour JSL (1940) Taxonomy and philosophy. In: Huxley J (ed) The new systematics. Oxford University Press, Oxford, pp 461–477
Gould SJ (1983) The hardening of the modern synthesis. In: Grene M (ed) Dimensions of Darwinism. Cambridge University Press, Cambridge, pp 71–93
Greene EL (1910) Certain aspects of the species question. Am Midl Nat 1:245–263
Greene HW (1983) Dietary correlates of the origin and radiation of snakes. Am Zool 23:431–441
Grinnell J (1917) The niche relationships of the California thrasher. Auk 34:427–433
Haldane JBS (1924) A mathematical theory of natural and artificial selection. Part II the influence of partial self-fertilisation, inbreeding, assortative mating, and selective fertilization on the composition of Mendelians populations, and on natural selection. Biol Rev 1:158–163
Hardy GH (1908) Mendelian proportions in a mixed population. Science 28:49–50
Heinroth O (1911) Beiträge zur Biologie, namentlich Ethologie und Psychologie der Anatiden. Verh Ver Int Ornithol Kongr (Berlin) 1910:589–702
Hennig W (1950) Grundzüge einer Theory der phylogenetischen Systematik. Deutscher Zentralverlag, Berlin
Hennig W (1966) Phylogenetic systematics. University of Illinois Press, Urbana, IL
Herrick FH (1911a) Nest and nest-building in birds. Part I. J Anim Behav 1:159–192
Herrick FH (1911b) Nest and nest-building in birds. Part II. J Anim Behav 1:244–277
Herrick FH (1911c) Nest and nest-building in birds. Part III. J Anim Behav 1:336–373
Hutchinson GE (1957) Concluding remarks. Cold Spring Harb Symp Quant Biol 22:415–427
Hutchinson GE (1959) Homage to Santa Rosalia or why are there so many kinds of animals? Am Nat 93:145–159
Hutchinson GE (1978) An introduction to population biology. Yale University Press, New Haven, CT
Huxley JS (ed) (1940) The new systematics. Oxford Univ. Press, Oxford
Huxley JS (ed) (1942) Evolution, the modern synthesis. Allen and Unwin, London
Huxley JS (ed) (1953) Evolution in action. Harper, New York
Janz N (2011) Ehrlich and Raven revisited: mechanisms underlying codiversification of plants and enemies. Annu Rev Ecol Evol Syst 42:71–89
Jordan DS, Kellogg VL (1900) Animal life: a Firstbook of zoology. Appleton and Co., New York
Jordan DS, Kellogg V (1908) The law of geminate species. Am Nat 42:73–80
Kellogg VL (1907) Darwinism today. Holt, New York
Kellogg VL (1913) Distribution and species-forming of ectoparasites. Am Nat 47:129–158
Kettlewell HBD (1958) A survey of the frequencies of *Biston betularia* (L.) (Lepidoptera) and its melanic forms in Great Britain. Heredity 12:51–72

King CE (1964) Relative abundance of species and MacArthur's model. Ecology 45:716–727
Korb KB, Dorin A (2011) Evolution unbound: releasing the arrow of complexity. Biol Philos 26:317–338
Lack D (1944) Ecological aspects of species-formation in passerine birds. Ibis 86:260–286
Lande R (1981) Models of speciation by sexual selection on polygenic traits. Proc Natl Acad Sci U S A 78:3721–3725
Levin DA (1979) The nature of plant species. Science 204:381–384
Leviton AE, Aldrich ML (eds) (1985) Plate tectonics and biogeography: earth sciences history. J Hist Earth Sci 4:91–196
Lorenz K (1941) Vergleichende Bewegungstudien an Anatien. J Ornithol 89:194–294
Lorenz K (1950) The comparative method in studying innate behaviour patterns. Symp Soc Exp Biol 4:221–268
Lorenz K (1958) The evolution of behaviour. Sci Am 199:67–78
Lush JL (1939) Methods of measuring the heritability of individual differences among farm animals. Proc. 7th inter. Genetics congress
Lush JL (1940) Intra-sire correlations or regressions of offspring on dam as a method of estimating heritability of characteristics. Proc Am Soc Anim Prod 1940:293–301
MacArthur RH (1958) Population ecology of some warblers of northeastern coniferous forests. Ecology 39:599–619
MacArthur RH (1960) On the relative abundance of species. Am Nat 94:25–36
MacArthur RH (1965) Patterns of species diversity. Biol Rev 40:510–533
MacArthur RH (1969) Patterns of communities in the tropics. Biol J Linn Soc 1:9–30
MacArthur RH (1972) Geographical ecology. Harper & Row, New York
MacArthur RH, Wilson EO (1963) An equilibrium theory of insular zoogeography. Evolution 17:373–387
MacArthur RH, Wilson EO (1967) The theory of island biogeography. Princeton University Press, Princeton, NJ
Majerus MEN (2009) Industrial melanism in the peppered moth, *Biston betularia*: an excellentteaching example of Darwinian evolution in action. Evo Edu Outreach 2:63–74
Matthew WD (1915) Climate and evolution. Ann N Y Acad Sci 24:171–318
Matthew WD (1939) Climate and evolution. Climate and evolution. Special publication 1 of the New York Academy of Sciences
Maynard Smith J (1966) Sympatric speciation. Am Nat 100:637–650
Maynard Smith J, Szathmàry E (1995) The major transitions in evolution. W.H. Freeman, Oxford
Mayr E (1942) Systematics and the origin of species. Columbia University Press, New York
Mayr E (1954) Change of genetic environment and evolution. In: Huxley J, Hardy AC, Ford EB (eds) Evolution as a process. Allen and Unwin, London, pp 157–180
Mayr E (1960) The emergence of evolutionary novelties. In: Tax S (ed) Evolution of life. University of Chicago Press, Chicago, IL, pp 349–380
Mayr E (1963) Animal species and evolution. Harvard University Press, Cambridge, MA
Mayr E (1976) Is the species a class or an individual? Syst Zool 25:192
Mayr E (1982) Processes of speciation in animals. In: Barigozzi C (ed) Mechanisms of speciation. Alan R. Liss, New York, pp 1–19
Mayr E (1988) Toward a new philosophy of biology: observations of an evolutionist. Belknap Press, Cambridge, MA
Mayr E, Provine WB (eds) (1980) The evolutionary synthesis: perspectives on the unification of biology. Harvard University Press, Cambridge, MA
McLennan DA (2008) The concept of co-option: why evolution often looks miraculous. Evo Edu Outreach 1:247–258
Miller AH (1949) Some ecologic and morphologic considerations in the evolution of higher taxonomic categories. In: Mayr E, Schüz E (eds) Ornithologie als Biologische Wissenschaft. Carl Winter, Heidelberg, pp 84–88

Milne MJ, Milne LJ (1939) Evolutionary trends in caddis-worm case construction. Ann Entomol Soc Am 32:533–542
Mode CJ (1958) A mathematical model for the co-evolution of obligate parasites and their hosts. Evolution 12:158–165
Moore B (1920) The scope of ecology. Ecology 1:3–5
Morgan TH (1932) The scientific basis of evolution. W.W. Norton and Co, New York
Olby RC (2000) Horticulture: the font for the baptism of genetics. Nat Rev Genet 1:65–70
Osborn HF (1910) The age of mammals in Europe, Asia and North America. Macmillan, New York
Osborn HF (1932) Nine principles of evolution revealed by paleontology. Am Nat 66:52–60
Osborn HF (1934) Aristogenesis, the creative principle in the origin of species. Am Nat 68:193–235
Paterson HEH (1981) The continuing search for the unknown and the unknowable: a critique of contemporary ideas on speciation. S Afr J Sci 77:119–133
Pearson K (1903) Mathematical contributions to the theory of evolution. XI On the influence of natural selection on the variability and correlation of organs. Phil Trans R Soc A 200:1–66
Petrunkevitch A (1926) The value of instinct as a taxonomic character in spiders. Biol Bull 50:427–432
Pigliucci M (2008) Sewall Wright's adaptive landscapes: 1932 vs 1988. Biol Philos 23:591–603
Plath OE (1934) Bumblebees and their ways. MacMillan, New York
Provine WB (ed) (1986) Evolution: selected papers by Sewall Wright. University of Chicago Press, Chicago, IL
Radinsky L (1975) Primate brain evolution. Am Sci 63:656–663
Regan CT (1926) Organic evolution. Rep British Assoc Advance Sci 1925:75–86
Rosenzweig ML (1978) Competitive speciation. Biol J Linn Soc 10:275–289
Rosenzweig ML (1997) Tempo and mode of speciation. Science 277:1622–1623
Simpson GG (1944) Tempo and mode in evolution. Columbia University Press, New York
Simpson GG (1951) The species concept. Evolution 5:285–298
Simpson GG (1953) The major features of evolution. Columbia University Press, New York
Simpson GG (1960) The meaning of evolution. Yale University Press, New Haven, CT
Simpson GG (1961) Principles of animal taxonomy. Columbia University Press, New York
Simpson GG (1963) Biology and the nature of science. Science 139:81–88
Smith HM (1965) More evolutionary terms. Syst Zool 14:57–58
Smith HM (1969) Parapatry: sympatry or allopatry? Syst Zool 18:254–255
Sokal RR, Crovello TJ (1970) The biological species concept: a critical evaluation. Am Nat 104:127–153
Sokal RR, Sneath PHA (1963) The principles of numerical taxonomy. W. H. Freeman, San Francisco, CA
Stevens P (1992) Species: historical perspectives. In: Keller EF, Lloyd EA (eds) Keywords in evolutionary biology. Harvard University Press, Cambridge, MA, pp 302–331
Stocking GW Jr (1982) Race, culture and evolution: essays in the history of anthropology. University of Chicago Press, Chicago
Stonor CR (1936) The evolution and mutual relationships of some members of the Paradiseidae. Proc Zool Soc London 106:1177–1185
Tauber CA, Tauber MJ (1977a) Sympatric speciation based on allelic changes at three loci: evidence from natural populations in two habitats. Science 197:1298–1299
Tauber CA, Tauber MJ (1977b) A genetic model for sympatric speciation through habitat diversification and seasonal isolation. Nature 268:702–705
Templeton AR (1980) The theory of speciation by the founder principle. Genetics 92:1011–1038
Templeton AR (1981) Mechanisms of speciation – a population genetic approach. Ann Rev Ecol Syst 12:23–48
Tinbergen N (1964) On aims and methods of ethology. Z Tierpsychol 20:410–433
Theunissen B (1994) Closing the door on Hugo de Vries' Mendelism. Ann Sci 51:225–248
Tutt JW (1896) British moths. George Routledge, London
Twain M (1906) What is man. Published anonymously by the author

Udvardy MDF (1969) Dynamic zoogeography with special reference to land animals. Van Nostrand Reinhold, New York
Van Valen L (1973) A new evolutionary law. Evol Theory 1:1–30
Van Valen L (1976) Ecological species, multispecies, and oaks. Taxon 25:233–239
Verschaffelt E (1910) The cause determining the selection of food in some herbivorous insects. Proc Acad Sci Amsterdam 13:536–542
Waddington CH (1957) The strategy of the genes. Allen and Unwin, London
Waddington CH (1966) New patterns in development and genetics. Columbia University Press, New York
Waddington CH (1968) Towards a theoretical biology. Nature 218:525–527
Waddington CH (1977) Tools for thought. Paladin, St. Albans
Wegener A (1912) Die Entstehung der Kontinente. Peterm Mitt 58:185–195
Wegener A (1924) The origin of continents and ocean basins. Methuen, London
Weinberg W (1908) Über den Nachweis der Vererbung beim Menschen. Jahreshefte des Vereins für vaterländische Naturkunde in Württemberg 64:368–382
Wenrich DH (1935) Host-parasite relations between parasitic protozoa and their hosts. Proc Am Philos Soc 75:605–650
Wheeler WM (1919) The parasitic Aculeata, a study in evolution. Proc Am Philos Soc 58:1–40
Wheeler WM (1928) The social insects: their origin and evolution. Harcourt, Brace & Co, New York
White MJD (1978) Modes of speciation. W. H. Freeman, San Francisco
Whitman CO (1899) Animal behavior. In: Whitman CO (ed) Biological lectures. Wood's Hole, Ginn and Co, Boston, pp 285–338
Wilson EO (1959) Adaptive shift and dispersal in a tropical ant fauna. Evolution 13:122–144
Wilson EO (1961) The nature of the taxon cycle in the Melanesian ant fauna. Am Nat 95:169–193
Wilson EO (1975) Sociobiology: the new synthesis. Harvard University Press, Cambridge
Winemiller KO, McIntyre PB, Castello L, Fluet-Chouinard E, Giarrizzo T, Nam S, Baird IG et al (2016) Balancing hydropower and biodiversity in the Amazon, Congo, and Mekong. Science 351:128–129
Wright S (1929) The evolution of dominance. Am Nat 63:556–561
Wright S (1931) Evolution in Mendelian populations. Genetics 16:97–159
Wright S (1932) The roles of mutation, inbreeding, crossbreeding and selection in evolution. In: Proceedings of the 6th International Congress of Genetics, pp 356–366
Wright S (1940) Breeding structure of populations in relation to speciation. Am Nat 74:232–248
Wright S (1949) Adaptation and selection. In: Jepson GL, Simpson GG, Mayr E (eds) Genetics, paleontology, and evolution. Princeton University Press, Princeton, pp 365–389
Wright S (1955) Classification of the factors of evolution. Cold Spring Harb Symp Quant Biol 20:16–24
Wright S (1956) Modes of selection. Am Nat 90:5–24
Wright S (1978a) Modes of speciation. Paleobiology 4:373–379
Wright S (1978b) Evolution and the genetics of populations. IV. Variability within and among populations. University of Chicago Press, Chicago
Wright S (1982) The shifting balance theory and macroevolution. Annu Rev Genet 16:1–19
Yoshida I (2017) Sewall Wright, shifting balance theory, and the hardening of the modern synthesis. Stud Hist Phil Biol Biomed Sci 61:1–10. https://doi.org/10.1016/j.shpsc.2016.11.001

Chapter 5
Criticism, Resistance, a Glimmer of Hope

Abstract By the end of the twentieth century, *neo-Darwinism* had morphed from the *New Synthesis* to the *Hardened Synthesis*. At the same time, major contributions from molecular genetics and developmental biology made shortcomings and contradictions in the hardened synthesis evident. More fundamentally, insights from two seemingly disparate sources, phylogenetic systematics and complex systems theory, threatened to erode the foundation of the hardened synthesis by re-elevating the *Nature of the Organism* to its Darwinian status and returning history to its essential role in biological explanations. Yet, as the twenty-first century dawned, none of those insights were able to penetrate the core of the hardened synthesis. Staunch proponents maintained the status quo, arguing that any aspect of the nature of the organism, including their historicity and cohesive inheritance and developmental natures, could be subsumed by the consensus framework. As a result, none of the core shortcomings recognized in the last 20 years of the twentieth century have been resolved; they have simply been shunted aside by what has become the *Extended Hardened Synthesis*. Evolutionary theory cannot move forward without the true integration of novel insights. Three independent proposals for conceptual frameworks from the period 1980–1995 make virtually identical core assertions, have complementary foci of attention, and most importantly are radical in the sense that they returned to the roots of Darwinism. They sowed the seeds then for where we are today—on the brink of going "back to the future" to reset the evolutionary narrative, rediscovering and extending the panoramic and inclusive framework that Darwin proposed.

> The wind blew. And the shit flew. And for weeks the vision was blurred.—Evan Hunter (1953)

Seventy-five years after Kellogg published *Darwinism Today*, biology commemorated the 100th anniversary of Darwin's death. As Kellogg had anticipated, Darwinism was nearly extinct, and the celebratory events had the air of Catholic priests celebrating Mass in a church built on top of the rubble of an Aztec temple. But all

S. J. Agosta, D. R. Brooks, *The Major Metaphors of Evolution*, Evolutionary Biology – New Perspectives on Its Development 2,
https://doi.org/10.1007/978-3-030-52086-1_5

was not calm inside the cathedral. The old ways had not died completely, and there were rumblings of discontent.

In addition to the ideas of Waddington and Wright we mentioned at the end of the last chapter, Lewontin (1966), Eldredge and Gould (1972), and Gould (1980) worried about the lack of attention paid to the role of history in evolutionary explanations, in particular the historical nature of species. Eldredge and Gould (1972) suggested that perhaps evolution was not uniformly a process of slow and gradual diversification, and Trivers (1972) was concerned about the exclusion of group selection from neo-Darwinian explanations, and thus the rejection of cooperation and altruism as possible evolutionary phenomena.

Ross (1972a, b) reported that only 1 out of every 30 speciation events in Caribbean insect groups was correlated with some form of ecological diversification, suggesting that adaptive change was much less frequent than phylogenetic diversification. He could not uncover any predictable patterns to explain the shifts that did occur, so he proposed that adaptive change manifests itself as a biological "uncertainty principle" in evolution. Brundin (1972) and Riedl (1978) extended Waddington's ideas by suggesting that many adaptations have been inherited relatively unchanged for long periods of time due to developmental dynamics that are buffered from selection. They suggested that such slowly evolving traits might influence the scope of the adaptively possible at every point in evolution. Finally, Boucot (1975a, b, 1981, 1982, 1983, 1990) reported that the majority of adaptive changes leaving some trace in the fossil record lagged behind phylogenetic diversification, or "evolution takes place in an ecological vacuum" (Boucot 1983).

In North America, Lewontin (1978), Gould and Lewontin (1979), and Gould (1983) objected to adaptive "just so stories" that seemed to invoke magic. In a complementary way, in Great Britain, John Maynard Smith (1968, 1972, 1974, 1978, 1981, 1986, 1988, 1989, 1993; Maynard Smith and Price 1973) was busy introducing game theory, optimality theory, evolutionary lagload, and the Gambler's Ruin to evolutionary biology in an (ultimately unsuccessful) effort to convince neo-Darwinians that evolution did not necessarily produce the best of all possible worlds.

Even the iconic peppered moth example was viciously attacked and vigorously defended. And although the basic story of the peppered moth color morphs was eventually upheld, the controversy left a sense that the number of documented cases of natural selection in the wild was pretty small. John Endler (1977) undertook to report the number of rigorously documented examples of natural selection in the wild, which for some was astonishingly few, but which for advocates of the hardened synthesis was persuasive. Perhaps not surprisingly, then, the greatest challenges to neo-Darwinism came from sources outside the mainstream of the hardened synthesis.

5.1 The Return of History

> Do we study history to have a more complete record of the past or to understand the present better?—George W Stocking (1968)

5.1.1 The Phylogenetics Revolution

Systematics, with its focus on phylogeny and evolutionary history, was not a core element of neo-Darwinism, much less the hardened synthesis. Boas (1898) suggested that biologists abandon efforts to provide a link between taxonomy and evolution by advocating purely empirical classification procedures identifying species as entities having no particular connection with evolutionary principles. Huxley (1942) argued strongly and effectively for taxonomy to be replaced by ecology. A generation later, numerical taxonomy (Sokal and Sneath 1963) promised to make systematics a purely mechanical operation. Stenseth (1984) would soon write—with great hyperbole—that ecology had been the hand-maiden of taxonomy until the 1960s, when it became an independent science of its own, an homage to evolutionary ecology. Systematists, who felt *they* were the handmaidens, did not pay a lot of attention to these rumblings because they were free to pursue what they wished, having been pushed to the margins of evolutionary biology, and they wanted to document phylogeny even more explicitly. This explains the rapidity with which numerical taxonomy was shunted aside for the codified method of phylogenetic analysis published by the German entomologist Will Hennig (1950, 1966) (see analysis by Hull 1988).

Wiley (1981, 1986) listed the minimal set of evolutionary assumptions necessary to proceed with attempts to determine phylogenetic patterns as: (1) We must assume that evolution has occurred, that it (2) produces phylogeny in the form of internested sets of descendant species, and that (3) the history of phylogenetic diversification leaves its trace in the characters of species, living and fossil (this will be true so long as replication rates are higher than mutation rates in inheritance) (see also Felsenstein 1982, 1988; Sober 1988; Donoghue 1990; Swofford and Olsen 1990; Wiley et al. 1991; Beatty 1994; Brooks and McLennan 1991, 2002). Wiley's classic (1981) textbook established Hennig's insights as follows: (1) presume a priori that similarity equals homology, based on the Darwinian assumption that (evolutionarily) all homologous traits covary with phylogeny

> Homologous parts tend to vary in the same manner, and homologous parts tend to cohere—Charles Darwin (1872)

(the *evolutionary homology criterion*: Wiley 1981; Patterson 1982; Roth 1984, 1988, 1991, 1994; Gould 1986; Rieppel 1992; McKitrick 1994; Brooks and McLennan 2002); (2) use comparisons with species outside the study group to determine which traits are older than the group under study (*outgroup comparisons*); (3) group species according to shared traits that are not older than the study group

(*group by synapomorphy*); (4) in the event of conflicting evidence, choose the phylogenetic relationships supported by the largest number of synapomorphies (*epistemological parsimony*); and (5) interpret inconsistent results, post hoc, as cases of independent origins (*homoplasies*). So, homologies, which indicate phylogenetic relationships, are determined *without* reference to a phylogeny, while homoplasies, which are inconsistent with phylogeny, are determined as such *by* reference to the phylogeny. Phylogenetic systematics marked a return to the position advocated by Darwin.

> Community of descent is the hidden bond which naturalists have been unconsciously seeking, and not some unknown plan of creation, or the enunciation of general propositions and the mere putting together and separating of objects more or less alike—Charles Darwin (1872)

Unrecognized at the time, Wiley's criteria for phylogenetic analysis resurrected the Darwinian metaphor of the nature of the organism. Also, phylogenetics' technical criterion for choosing the best working hypothesis for phylogeny—parsimony—also suggested that historical persistence of traits was more important than repeated origins. *This represented an existential threat to Neo-Darwinism.* It is therefore mildly ironic that the earliest proposal of a parsimony criterion for phylogeny reconstruction was published by E.O. Wilson (1965).

Producing more robust estimates of evolutionary history was only the beginning of the troubles phylogenetics would make for the hardened synthesis.

5.1.2 Speciation

The belated acceptance of the theory of *plate tectonics* and continental drift by geologists (e.g., Dietz and Holden 1966; Windley 1986) led to a revival of Wegener's original ideas and reemergence of interest in the geographers' school of speciation, especially the nineteenth century versions that denied natural selection any role in the process. A group of phylogeneticists suggested that it was not the dispersal of existing species, but the formation of new species from them that created generalities in biogeography. They postulated that geographic fragmentation of entire biotas caused by the formation of barriers catalyzed geographic speciation in many species isolated at the same time in the same way. This kind of geographic speciation was permitted by geologists' newly rediscovered support for continental drift. Darlington did not support continental drift, based on the opinions of geologists of his day, and thus had to assume that all geographic speciation occurred as a result of populations dispersing across preexisting barriers. The form of geographic speciation advocated by the phylogeneticists that came to be called "vicariance" (and the research program vicariance biogeography: Croizat et al. 1974; Rosen 1975, 1978, 1979, 1985; Platnick and Nelson 1978; Nelson and Rosen 1980; Nelson and Platnick 1981) was a resurrection of David Starr Jordan's *law of geminate species* proposed half a century earlier.

The phylogenetics revolution restored the geographers' notion that the way speciation is initiated is at least as important as how it is completed. This represented another threat to the hardened synthesis, which had settled comfortably into the notion that speciation—if species were even real—was synonymous with adaptation.

In addition, the phylogenetic revolution contributed to a growing recognition that there are multiple plausible modes of species formation, with geographic isolation at one end of the spectrum and classic peripatric speciation with reinforcement at the other. These modes may be initiated with or without geographic isolation, and with or without ongoing gene flow during the process. Speciation may be completed by facultative or obligate reproductive isolation from closest relatives. Theoretical input from quantitative genetic models and empirical data from both field and laboratory studies have demonstrated that *there is no one general mechanism for completing speciation. Nor is there any particular combination of these factors unique to a particular speciation mode* (summarized in Brooks and McLennan 2002).

5.1.3 Species

Eldredge and Gould (1972) referred to species as homeostatic systems held together by the cohesion of common evolutionary history. Wiley (1978, 1980, 1981) suggested that any general species concept must (1) encompass species persistence through time as well as their divergence, (2) recognize that species are cohesive wholes, and (3) recognize their inherent historicity (see also Wiley and Brooks 1982; Eldredge 1979, 1985; Brooks and Wiley 1986, 1988).

Wiley identified Simpson's (1944) Evolutionary Species Concept (ESC), which was a refinement of Darwin's "species are communities of descent," as the most consistent ontological view of species for phylogeneticists. Löther (1990) considered species to be material supraorganismal systems, spatiotemporally organized, forming an integrated whole. Frost and Kluge (1994; see also Kluge 1990) considered species to be genealogical systems existing as lineages of reproductive connections extended through time preserving a unique mixture of genetic information. Kornet (1993a; see also Kornet 1993b; Kornet and McAllister 1993; Kornet et al. 1995) extended these variations on the Darwin–Simpson theme, suggesting that species function in evolutionary theory as mutually exclusive, nonarbitrary historical entities (O'Hara 1993; Frost and Kluge 1994). As a result, she concluded that the historically irreversible splitting of cohesive lineages provided the only universal criterion for recognizing species.

Finally, by making evolutionary history the core element of biological species, phylogeneticists allowed evolutionary biologists to stop worrying about species boundaries and return to another Darwinian proposition

> Because evolutionary history is something we are still in the midst of, it will not always be possible for us to determine which varieties—which distinctive populations in nature—are temporary and which are permanent, and so our counts of species across space and through time will always have some measure of ambiguity in them that we cannot escape. If there is

any consolation in this, it must be that the very existence of this ambiguity—the very fact that some organisms in nature cannot easily be grouped into species—is itself, as Darwin recognized, one of the most important pieces of evidence for the historical process we call evolution.—Robert J O'Hara (1994)

5.1.4 The Orthogeneticists Return: Co-speciation

Following the geographers, some phylogeneticists began to ask how phylogenetic information could be used in the study of pathogen–host systems. The orthogenetic view that, with few exceptions, pathogens occurring in specific hosts would be related to each other in the same way that the hosts were related was still strong, especially in parasitology. Brooks (1979) discussed two elements of parasite evolution that could be examined phylogenetically, co-speciation and co-accommodation (associated with notions of host specificity and host–pathogen co-adaptation). Co-speciation, instances in which pathogens and hosts speciate together, provided common ground between geographers and orthogeneticists. Brooks suggested that phylogenetic analysis could be especially useful in determining cases in which the orthogeneticists' expectations were not met but, like Maynard Smith's cautionary use of optimality theory, this was not the way the message was interpreted.

The perspective that pathogens are so highly specialized on specific hosts that they are effectively inherited with their hosts and thus parasite and host phylogenies can be superimposed on each other without a mismatch, remained strongly attractive to some research groups, especially parasitologists. The belief is so strong that an array of models and analytical methods have been developed to "fix" data that do not correspond to this notion of *maximum cospeciation*. This perspective seemed an ideal way to unite the geographers and the orthogeneticists, at least superficially. The belief in co-speciation would be reinforced if pathogens and hosts experienced concomitant episodes of geographic speciation. Secretly, the geographers would think co-speciation was due to geographic isolation, and the orthogeneticists would think it was due to the inherent orthogenetic drive. So long as comparisons of hosts and geography and parasites were all congruent, however, everyone could be happy. Neo-Darwinians and phylogeneticists rightly pointed out that the degree of congruence or incongruence between host and pathogen phylogenies neither corroborated nor falsified any particular theory of coevolution (see Brooks and McLennan 2002 for an extended discussion), but the research program persists (see Brooks et al. 2019 for a discussion).

5.1.5 Adaptationism Questioned

Coddington (1988, 1990, 1992, 1994) proposed the first protocol for studying adaptations in a phylogenetic context, asking two questions. First, is the same trait in different species the result of persistence (Darwin's preferred means) or independent repeated origins (the neo-Darwinian approach)? And second, how well does the diversification of adaptations follow phylogeny rather than local environments (common inheritance rather than repeated independent episodes of parallel selection leading to the same adaptive outcome)? These questions amounted to a reemergence of Darwin's Necessary Mismatch that threatened the pan-adaptationist core of the *hardened synthesis*. It was now possible to document cases of the same trait used in the same way in the same or in different environments by two closely related species due to common inheritance (persistence rather than independent origin), and the same trait used in different ways (co-option) in the same or in different environments by two closely related species due to common inheritance. Phylogenetic analysis consistently corroborated Darwin's assertion that persistent inheritance is more important than recurring novel innovation in common environments in explaining the occurrence of all traits and that the phylogenetically conservative diversification of adaptations follows phylogeny closely (e.g., McLennan et al. 1988; Maddison 1990, 1994; McLennan 1991, 1993, 1994, 1996, 2000; Sillén-Tullberg 1993; Armbruster 1994; Pagel 1994; Fitter 1995; Read and Nee 1995; Werdelin and Sillen-Tullberg 1995; Crespi 1996; Hart et al. 1997; Maddison and Maddison 2000; Brooks and McLennan 2002).

> The results of phylogenetic studies of adaptation suggested firmly that Darwin's view about how "the right adaptation shows up at the right time" was correct and Kellogg's assertion was simply wrong.

5.1.6 The Return of History to Comparative Biology

Swedish botanist Hans-Erik Wanntorp (1983) published the first phylogenetic study highlighting Darwin's suggestion that some traits that emerged as adaptations in one set of conditions might later be co-opted for another, also adaptive, function later. He showed that deciduousness in *Nothofagus* originated in the southern hemisphere in semitropical, seasonally xeric habitats, but functioned also in the northern hemisphere in temperate, seasonally cold conditions. Shortly thereafter, Brooks (1985) proposed that the research of Ross and Boucot, as well as Wanntorp's article and his own work with phylogenetic studies of parasite–host systems, all could be encompassed in a research program he called "historical ecology." Within 5 years, Wanntorp et al. (1990) published a short review suggesting multiple applications of phylogenetic information in ecology and behavior. A year later, Brooks and McLennan (1991) provided the first book-length synthesis attempting to expand the boundaries of historical ecology to include two general evolutionary processes, speciation and adaptation, and to explore the macroevolutionary effects of these processes in the production of both clades and multispecies ecological associations. These efforts struck a resonant chord with many biologists, resulting in an explosion of publications incorporating phylogenetic information in a broad range of ecological and behavioral studies by the end of the twentieth century, which Brooks and

McLennan (2002) summarized a decade later, linking that research program explicitly to Darwin's notions of the primacy of the nature of the organism and the importance of history in evolutionary explanations.

The single most important insight from this expansion of phylogenetic studies was that phylogenetic information was often part of the explanation for evolutionary phenomena, not simply part of the baggage. Beginning with Coddington's (1988) article, evolutionary biologists realized that phylogenetically informed studies could test, and potentially falsify the evolutionary just-so stories criticized so strongly by Gould and Lewontin (1979). The resurgence of interest in, and appreciation of, the significance of phylogenetic history for evolutionary explanations turned Morgan's (1932) assertion on its head:

> Evolutionary Biology is no longer just a discussion of how things function in current environments, it now includes explanations based on the history of origins and persistence through changing environments.

5.2 Evolution Meets Complex Systems Analysis

Having spent 75 years modifying evolutionary theory into a simple and compact naturalistic form as advocated by physicists, advocates of the hardened synthesis must have been bemused, if not befuddled, when in the 1980s physicists announced that they—and they alone—had discovered that the universe was complex and not simple. Having stated that the world was complex, the physicists then fell to bickering about nomenclature, institutes were formed and many books appeared. For some, complexity was simply enormous amounts of simplicity. For others, complexity was something qualitatively different. Some of the attributes of complex systems are: they tend to be self-stabilizing with respect to their surroundings; they are sensitive to initial conditions, so how they begin is important to explaining what they become; they are characterized by some kind of boundary between themselves and their surroundings such that flows into and out of the system are out of phase with fluxes of matter and energy within them, leading to nonlinear behavior; nonlinear behavior is manifested in emergent as well as systemic properties; they exhibit complicated and persistent structure, often at least partially hierarchical; they are able to impose distinctions on their surroundings and not just the reverse (so context is causally important: Juarrero 1999, 2002); and they have a high degree of historicity.

Living systems immediately caught the attention of complex systems researchers as well as some biologists. For the last 20 years of the twentieth century, a loose collaboration among these researchers produced a lot of discussion and perhaps more heat than light, but substantial insights also emerged. We will briefly discuss three examples.

5.2.1 A Complex Systems View of the Nature of the Organism

Organisms, and the various systems they produce, are functional wholes with respect to the way they engage their surroundings as well as with respect to their internal organization (e.g., Newman 1970; Layzer 1978, 1980; Zotin and Zotina 1978; Kjellstrom and Taxen 1981; Wiley and Brooks 1982; Lewontin 1983; Salthe 1985, 1993, 1998; Brooks and Wiley 1986, 1988; Collier 1986, 1988, 1990, 1998, 2000; Kauffman 1986, 1993; Ulanowicz 1986, 1997; Wicken 1987; Odling-Schmee 1988; Schneider 1988; Smith 1988, 1998; Weber et al. 1988; Wimsatt and Schanck 1988; Brooks et al. 1989; Csanyi 1989; Matsuno 1989, 1995, 1998; Brooks 1990, 1992, 1994, 1997, 1998, 2000; Brooks and McLennan 1990, 1991, 1997, 2002; Kampis 1991, 1998; Maurer and Brooks 1991; Cowan et al. 1994; Corning 1995; Depew and Weber 1995; Gladyshev and Kitaeva 1995; Holland 1995; Maynard Smith and Szathmary 1995; Bak 1996; Gladyshev 1996; Kjellstrom 1996; Odling-Schmee et al. 1996; Raff 1996; Landweber et al. 1998; Rocha 1998; Van de Vijver et al. 1998; Vogel 1998; Collier and Hooker 1999; Hewzulla et al. 1999; Juarrero 1999, 2002; Niklas 1999; Callebaut et al. 2007; Brooks and Agosta 2012; for a recent review, see Soto et al. 2016). Orderliness and organization in biological systems results from the interaction of selection processes with three aspects of the nature of the organism: (1) historical uniqueness, (2) cohesive properties, and (3) functional integration and hierarchical organization.

Historical uniqueness manifests itself in two ways. One is historical contingency; events that take place in the past may have an effect on the subsequent behavior and fate of the system without being deterministically connected to each other (i.e., the earlier event does not cause the later event). The other is temporal irreversibility; biological processes such as reproduction, development, aging and death, speciation, and extinction, appear to be temporally irreversible on any time scale. Spontaneous irreversible behavior in biological systems always involves two things: (1) growth and increasing complexity, and (2) physical manifestations of at least some of the systems' history (Brooks and Wiley 1986, 1988; Wimsatt and Schanck 1988; Brooks et al. 1989). Or, to paraphrase Heraclitus:

> When you step in the river, understanding what is happening downstream requires that you know what happened upstream.

Cohesive properties range from cell–cell adhesion and recognition, to sexual reproduction and specific mate recognition systems, to common phylogenetic history. Cohesion is especially important to evolutionary explanations because the cohesive properties of living systems limit the ways in which and the extent to which populations can respond to environmental selection, and are the "glue" of functional integration and hierarchical organization that are so characteristic of biological systems (Wake and Roth 1989). Many biological processes that demonstrate irreversible behavior manifest such changes as a result of interactions among cohesive factors, which tend to keep species and populations together, and diversifying factors, which tend to split them into separate systems. Speciation in sexually reproducing species, for example, results when the developmental and reproductive constraints acting as cohesive forces maintaining an ancestral species as a single lineage are overridden by environmental forces acting to split it apart into descendant species (Wiley 1981; Wiley and Brooks 1982; Brooks and Wiley 1986, 1988; Brooks and McLennan 1991, 2002). This is the reason Maynard Smith and Szathmary (1995) associated the origin of sex with the origin of true biological species. In a complementary fashion, the environmental boundaries within which each species lives might, in some cases, be pronounced enough to be considered extrinsic cohesion in the form of stabilizing selection (Collier 1998, 2000). The major evolutionary transitions of Maynard Smith and Szathmary (1995) are all associated with the emergence of a novel form of cohesion.

Biological systems are not just complex, they are *complexly organized* (Collier and Hooker 1999). *Functional integration* and *hierarchical structure* are hallmarks of biological organization (e.g. Salthe 1985, 1993; Brooks and Wiley 1988; Wake and Roth 1989; Kauffman 1993; Niklas 1999; McShea and Changizi 2003). A major component of *functional integration* is the interdependence of parts. Evolutionary changes in biological systems do not occur all at the same time; thus, when such changes occur, only part of the system changes. All changes, from point mutations on a chromosome to alteration of part of a complex mating ritual, must integrate with the rest of the system in such a way that viable organisms result. The most prominent research program in this regard is called evolutionary developmental biology or developmental systems theory, more commonly known as evo-devo. Tracing their immediate professional ancestry to early advocates of complex systems theory (e.g., Brundin 1972; Riedl 1978, Goodwin 1982; Goodwin and Trainor 1983), evo-devo also benefited from early input by phylogeneticists (Wiley 1981; Mabee 1993, 2000; Mabee and Humphries 1993; Reilly et al. 1997), those who sought to integrate experimental developmental biology with phylogeny (Wray and Lowe 2000), and paleontologists (Raup and Gould 1974; Gould 1977; Alberch et al. 1979). An early concerted effort to forge a broad coalition representing multiple points of view, including the hardened synthesis (Maynard Smith et al. 1985) underscored the potential importance of evo-devo but produced a summary of different perspectives rather than an integrated framework.

Hierarchical structure, from the genealogical relationships of individuals and species to the behavioral relationships of complex social systems and the trophic interactions among species in ecosystems, plays an important role in biological evolution. Salthe (1985, 1993) concluded that hierarchies provide stability, reinforce boundaries between system and surroundings, allow increasing amounts of complexity without losing organizational coherence, and provide a way in which causation and control can be tied together. He suggested that hierarchical structure can be decomposed into sets of "triads," comprising (1) upper level (causal or initiating), (2) lower level (control or boundary) elements impinging on (3) a focal level, from which emerges a particular form of structure and/or organization. Complex hierarchical systems are combinations of linked triadic units. Within a given hierarchical system, relatively random lower-level effects are screened off by the cohesive properties of the higher-level effects. In addition, the various levels in the hierarchy have diminishing effects on any given level in proportion to the remoteness of their interactions.

5.2.2 A Complex Systems View of Microevolution and Macroevolution

When neo-Darwinians began to question seriously whether evolutionary dynamics operated at the level of species or populations (Huxley 1942), the debate was quickly cast in terms of *microevolution* and *macroevolution*, the former associated with populations and the latter with species. In keeping with the integration of ecology and evolution moving forward, many considered the issue either irrelevant (macro-evolution is just historical record) or differentiated the phenomena in terms of spatial scale (microevolution occurs on small spatial scales, macroevolution on larger spatial scales). Just "how small" or "how large" fueled many debates around drinks at conferences as well as published articles, and the absence of any consensus, much less any integration, allowed the debates to continue for a very long time.

The integration of macroevolution and microevolution became a possibility with the emergence of the field of phylogeography (Avise 1989, 2000). Evolving species are commonly viewed as collections of populations distributed horizontally across geography (Avise 1989, 2000; Futuyma 1989). Kornet (1993a, b; Kornet and McAllister 1993; Kornet et al. 1995) added an important dimension to this perspective by suggesting that species are collections of historical lineages subdivided into smaller historical lineages extending through time. When one of these smaller lineages experiences a permanent, or irreversible, split from the others, the relationship between that lineage and the others changes from a microevolutionary to a macroevolutionary one. The divergent lineage is now called a macro-species to indicate that it has experienced a permanent split from its ancestor and may now produce its own complement of smaller lineages extending through time, which constitute its own micro-species. We expect a host of demographic phenomena, such as local extinctions and fusions with other micro-species as a result of dispersal and gene flow to limit the number of micro-species that become macro-species in their own right. This is exactly the pattern that is starting to emerge from the explosion of studies in intraspecific phylogeography. Phylogeographers are interested in describing the "deployment of genetic variation within species" (Zink 1996: 308). This deployment is uncovered by reconstructing phylogenetic relationships among populations, then examining the effects of relatedness and geography on differences in the genetic structure of those populations (Chernoff 1982; Avise 1989, 2000). One of the major generalizations arising from these studies is that the relationships among populations within a species are complex and reticulated, often showing only moderate to very little resolution (e.g., Ellsworth et al. 1994; Klein and Brown 1994; Phillips 1994; Seutin et al. 1993; Zink 1994, 1996; Riddle 1996; Shaffer and

McKnight 1996; Hedin 1997; Barrowclough et al. 1999; Lessios et al. 1999; Avise 2000).

Kornet's concept of *micro-species* encompasses all those assemblages of conspecific organisms that we have called subspecies, differentiated populations, geographical races, or incipient species. These terms are all linked by two implications. First, the group of organisms you are studying can be distinguished objectively in some manner. Second, there is a probability that this group will become a macro-species in its own right, but it has not done so yet, even though you can identify it objectively. Micro-species thus represent the interface between the realm of what is happening right now and the realm of possibilities for what might happen in the future. The micro-species of any given macro-species should be relatively numerous and locally differentiated yet highly similar due to their close common history, so naturally replicated exemplars will abound. Kornet's distinction between nonpermanent (reversible) and permanent (irreversible) splits provided a clear view of a major component differentiating micro- and macroevolution. That component is not magnitude, but quality. It involves assessing the quality of the cohesion holding the micro-species together as a single complexly organized entity. More fundamentally, it embodies time and history, because it is the difference between reversible and irreversible phenomena (Eldredge 1979; Wiley 1981; Wiley and Brooks 1982; Brooks and Wiley 1986, 1988; Maynard Smith and Szathmary 1995).

> Macroevolution is survival in the face of large-scale inevitable conflict; irreversibility is the marker of such survival events.—Koichiro Matsuno, Presentation at International Society for Systems Science Congress (2000)

5.2.3 *Niches and Niche Construction*

Lewontin (1983) wrote that the "error is to suppose that because organisms construct their environments they can construct them arbitrarily in the manner of a science fiction writer constructing an imaginary world. . .Where there is strong convergence is in certain marsupial-placental pairs, and this should be taken as evidence about the nature of constraints on development and physical relations, rather than as evidence for pre-existing niches." Niches in the hardened synthesis are magical because we have no mechanism for the transitions from one to another. They cannot be produced prior to speciation but must predate the species that fill them, and somehow there cannot be more of them than there are species. A neo-Darwinian can say that dinosaurs evolved into birds because dinosaur niches disappeared and bird niches appeared, then natural selection turned some dinosaurs into birds. This is magical thinking, and it is an easy trap to fall into.

Taking the lead from Lewontin and their understanding of complex systems, Brooks and Wiley (1986, 1988) proposed that organisms made their own niches by imposing themselves on their surroundings using their inherited capabilities. They, like Lewontin, knew that this perspective went counter to the prevailing view of the hardened synthesis that niches existed independently of organisms filling them, posed problems in the nature of the conditions which organisms had to adapt to, and preexisted the adaptations themselves. They did not realize, however, that their perspective resurrected Darwin's primacy of the nature of the organism, as well as his notion that adaptation was finding a place for yourself in nature. Had they made that connection, Hutchinson's (1957) original conception of the niche and Maynard Smith's (1976) trenchant observation that most of the nature of the conditions for any given living system are other living systems would have given them conceptual continuity with Darwinian thought. The appearance that the surroundings are constructed as niches is analogous to a kind of optical illusion that the nineteenth-century philosopher David Hume warned about. The more historically conservative organisms are, the more likely it is that each generation will be characterized by many organisms all preferring the same resources, those resources being the same as the ones preferred by the previous generation. The repeated exploitation of the same resources could fool some into believing the environment was doing the structuring when in reality evolutionary history was responsible. Organisms impose themselves on their surroundings, existing where they occur and doing what they do to survive because of their particular natures and history. They are not pulled into unoccupied niches—there is no external force drawing species to their destinies.

At nearly the same time, Odling-Schmee (1988; Odling-Schmee et al. 1996) proposed the *niche construction hypothesis*. This perspective borrowed from the same complex systems thinking that inspired Brooks and Wiley, but in a way that extends and modifies the hardened synthesis. Organisms do impose themselves on their surroundings but do so in such a way that they construct modifications in their surroundings that increase their fitness. Fitness did not increase solely as a result of the conditions eliminating the unfit, organisms had the ability to increase their fitness directly and actively (Day et al. 2003; Korb and Dorin 2011; Laland et al. 2015).

To be sure, members of the hardened synthesis allowed organisms, species, and their interactions to be complex

> Now my own suspicion is that the Universe is not only queerer than we suppose, but queerer than we can suppose.—J.B.S. Haldane (1927)

> ...in every part of the whole, wonderful history of life, all the modes and all the factors of evolution are inextricably interwoven. The total process cannot be made simple, but it can be analyzed in part. It is not understood in all its appalling intricacy, but some understanding is in our grasp, and we may trust our own powers to obtain more.—George Gaylord Simpson (1953)

But acknowledging that organisms exhibit a range of complex properties is not the same as incorporating those properties into your explanatory framework. Also, the notion of distinguishing microevolution and macroevolution based on shared history and irreversibility is difficult to reconcile with the hardened synthesis, which excluded history a century ago and whose mathematical foundation (Fisher's Fundamental Theorem) embodies no arrow of time (Maynard Smith 1970), the signature of irreversible processes.

By the end of the twentieth century, neither the phylogenetics revolution nor the emergence of complex systems dynamics led to new explanatory frameworks in evolutionary biology. As well, the complex systems people and phylogenetictists who saw broadly overlapping interests were unable to muster sufficient interest from the rest of their respective research groups to create a critical mass of cooperating specialists to forge a grand unification of biology. The first 20 years of the twenty-first century, therefore, has witnessed the emergence of many specialized perspectives, connected only by the belief that each is privy to research insights unique to their research area. We will not list them, but they all have evidence to support their claims, and many of them invoke a mixture of phylogenetic information, complex systems dynamics principles, and hardened synthesis nomenclature. Failing to cooperate and learn from each other on the part of almost every research group, we begin the third decade of the twenty-first century with more knowledge and less integration than ever before. That is not to say there is no consensus framework.

5.3 Extending the Hardened Synthesis

During the past 40 years, there has been no shortage of researchers who maintained the belief that all aspects of the nature of the organism, including their historicity and their cohesive inheritance and developmental natures, could be subsumed by the hardened synthesis. This prompted Gould to refer to the hardened synthesis as "the Blob," a cinematic giant amoeba-like creature that absorbed everything in its path. From the perspective of the hardened synthesis, this was reasonable. All such phenomena were simply lumped together in a blanket notion of heritability in which the constraints provided by the nature of the organism were simply the properties possessed by organisms in the conflict arena not yet overcome by natural selection (Charlesworth et al. 1982; Maynard Smith et al. 1985; Tuomi et al. 1988; Wagner 1984, 1985; Wake et al. 1983; Schwenk 1995; Wagner and Schwenk 2000; Schwenk and Wagner 2001). And yet, the phylogeneticists seemed to be producing a consistent (and growing) body of results indicating that this might not be true.

5.3.1 Renewed Interest in Galtonian Comparative Biology

Not all biologists were ecstatic about the reintroduction of phylogenetic information into evolutionary explanations. A group of dedicated disciples of Galton wished to use phylogenies to identify the historically correlated elements of evolution and partition them out of statistical analyses. The first approach originated when Clutton-Brock and Harvey (1977) who, echoing Boas (1896), noted that treating each species as an independent variable in statistical analyses was tantamount to assuming that all interspecific similarities were due to convergent or parallel evolution. This could overestimate the effects of selection and possibly obscure real patterns in the data. A number of studies dedicated to disentangling variation due to common history from variation due to ongoing processes emerged (Clutton-Brock and Harvey 1984;

Gittleman 1981; Harvey and Mace 1982; Ridley 1983; Felsenstein 1984, 1985; Cheverud et al. 1985; Dunham and Miles 1985; Harvey and Clutton-Brock 1985). Harvey et al. (1995a, b) summarized the comparative method as "a constructive attempt to extract information from non-experimental data which is riddled with non-independence" (1995a: 535); *phylogenetic information* thus became *phylogenetic error* (e.g., Fitter 1995; Westoby et al. 1995a, b).

This program's development culminated with the publication of a book by Harvey and Pagel (1991), which provided the launch point for an explosion of comparative studies using statistical methods during the last decade of the twentieth century and continuing today. Because this research program relies on statistical analysis, it has been referred to as "the comparative method," in reference to the historical roots of the statistical approach in the late nineteenth century. There is a constellation of methods available to researchers interested in this type of comparative approach, but recent studies suggest that many of these "methods" are variants of each other (Garland et al. 1999; Garland and Ives 2000). Huey et al. (2019) recently designated Felsenstein's (1985) article a "key innovation" in evolutionary biology. To the extent that such methods helped the hardened synthesis cope with phylogenetic information without modifying its central core, this is indeed an apt description.

5.3.2 Evolutionary Ecology

Ricklefs (1987) suggested that an "eclipse of history" had a profound and adverse effect on the field of community ecology. He argued that community ecology had relied mostly on local-process theories for explanations of patterns that are strongly influenced by regional processes. Local explanations rely on the action of competition, predation, and disease to explain patterns of species diversity in small areas, from hectares to square kilometers. According to this perspective, the community is maintained at a saturated equilibrium by biotic interactions. However, independent lines of evidence from different communities suggest that regional diversity plays a strong role in structuring local communities. Brown and Maurer (1989; Maurer et al. 1992; Brown 1995; Maurer 1999) reinforced this conclusion with their suggestion that general statistical regularities in ecological associations occur on much larger spatial scales than previously considered. They proposed a research field, called *Macroecology*, in which the emphasis is on large-scale rather than small-scale studies. Like Ricklefs, they recognized that enlarging the spatial scale of evolutionary ecological studies would increase the amount of phylogenetic influence in the systems under investigation. Given the existence of these effects, then, Brown and Maurer called for ways to incorporate them into the explanatory framework of macroecology. *History* for these researchers was still a surrogate for *space*, so macroecology became a call—in part—to examine the elements of evolutionary ecology that MacArthur and Wilson wished to avoid rather than actually incorporating phylogenetic information directly into explanations about the assembly and functioning of communities, ecosystems, and biotas (Brooks and McLennan 1991, 1993a, b, 1994, 2002).

5.4 Why Does the Hardened Synthesis Still Exist, and Is Even Being Extended?

Why, in the face of all the phylogenetics information, complex systems theory, evo-devo and molecular genetics, and input from multiple individual research programs documenting shortcoming and contradictions in the framework, has the hardened synthesis been able to persist self-assured (Losos et al. 2013; Wray et al. 2014) and even talk confidently about extending itself (Niklas 2004; Pigliucci 2007, 2009; Rose and Oakley 2007; Pigliucci and Muller 2010; Huang 2011; Weber 2011; Laland et al. 2013, 2014, 2015; Mesoudi et al. 2013; Myers and Saupe 2013; Muller 2017; Febregas-Tejeda and Vergara-Silva 2018)?

The reason is that humans are storytellers and will not abandon a comforting story—even if flawed—unless something better comes along. The hardened synthesis is a flawed story, but it is familiar and comforting. Darwinism foundered because people did not like the implied message that Darwin brought and abandoned it as soon as a more upbeat story consistent with their own aspirations appeared. Gould and Lewontin's argument against just-so stories foundered for the same reason Darwinism foundered. People do not want a world without stories, they wanted better—meaning simple and elegant—stories. The relatively lukewarm reception of numerical taxonomy by the systematics community, despite the fact that it had a lot of modern technological buzz going for it, and the rapid adoption of the codified method of phylogenetic analysis documented so extensively by Hull (1988) showed that systematists did not want high-tech, non-phylogenetic systematics; they wanted systematics that was more explicitly phylogenetic, that told the story of evolution better.

Anything anthropomorphic will be attractive to a general audience, so people who are unabashedly anthropomorphic create a positive feedback for just-so storytelling, even if the anthropomorphisms evoke magic. In this case, the magic is called teleology—*trait x evolved in order to give the species magical powers to adapt to threatening conditions*. From Spencer to the hardened synthesis, neo-Darwinism has always been rife with such magical explanations and that is the source of its enduring attraction. Speaking of an extended synthesis seeks to reassure people that no matter how the framework changes, the old, familiar, comforting stories will be intact. Thus, at the end of the day, the *Extended Hardened Synthesis* is not powerful because it is true; it is true because it is powerful. And anyone who thinks the hardened synthesis needs to be replaced rather than simply extended must understand the emotional and psychological dimensions of the audience as well as the facts of the matter. Remember:

If you want to change the world, you have to change the metaphor—Joseph Campbell (interview with Bill Moyers 1987)

5.5 Back to the Future

Next, we summarize briefly three cases from the period of 1980–2000, each of which, judging by the amount of twenty-first century rebranding they have experienced, contributed in a nontrivial way to evolutionary thinking, all of which fell short of being recognized as a replacement framework for the hardened synthesis at the time. This is not meant to belittle any other proposed frameworks during that same time period (e.g., Gould 1980). We chose these because they seem to be wildly divergent perspectives that nonetheless have a fundamental connection that has been overlooked. Those connections could not be found if we were not looking for them specifically in an effort to recover the fundamental Darwinian metaphors.

5.5.1 Eldredge and Salthe (1984)

Eldredge and Salthe (1984; Salthe 1985, 1993; Eldredge 1985, 1986, 1995) developed a framework based on viewing evolution as the result of the interaction between what they termed the ecological and the genealogical hierarchies. The ecological hierarchy is an economic system, manifested by organisms as interactors, characterized by patterns of matter and energy exchanges between organisms and their surroundings (*Nature of the conditions*). The genealogical hierarchy is an informational system, manifested by organisms as replicators, characterized by patterns of ancestral relationships among organisms and species (*Nature of the Organism*). Prebiotic environmental conditions established the boundary conditions within which life could originate; this was the origin of the ecological hierarchy. Conversely, genealogical processes that characterize life arose from abiotic origins but became autonomous enough from environmental conditions to be capable of overrunning available resources and of changing the environmental conditions substantially; the emergence of inheritance systems was the origin of the genealogical hierarchy. The longer life exists on this planet, the more it shapes the environment of the planet (*the nature of the organism is more important than the nature of the conditions*). Today, most of the environment relevant to biological systems consists of products of genealogical processes (Maynard Smith 1976; Brooks and Wiley 1986, 1988), suggesting that evolution driven by genealogical processes has resulted in an ever more intimate relationship between organisms and their surroundings (*phylogeny is part of the causal explanation for evolution*). The two hierarchies interact with each other causally, producing conflict caused by the production of organisms by rules that are autonomous, or at least insensitive to the surroundings.

Conflict resolution produces new levels of organization, each with its own form of selection (this has come to be known as hierarchical selection theory).

5.5.2 Brooks and Wiley (1986, 1988)

Living systems embody inherited information stored and transmitted by rules that are autonomous from the surroundings. Feedback between living systems imposing themselves on their surroundings using their inherited capabilities (*Nature of the Organism*) degrades environmental resources. This leads to conflict with the surroundings through reproductive overrun (*the nature of the organism is more important than the nature of the conditions*), and natural selection is an inevitable emergent property. Inherently irreversible processes (reproduction, ontogeny, speciation) increase biological information, constrained by the dynamics of the inheritance information system (*Nature of the Organism*) and selection (*Nature of the Conditions*). Evolution is characterized by irreversible increases in informational complexity (*phylogeny is part of the causal explanation for evolution*), including the emergence of new levels of organization, each of which is accompanied by the emergence of a novel form of cohesion and a novel form of selection focused on that level of organization. All conflict resolution enhances the organism/environment interface.

5.5.3 Maynard Smith and Szathmary (1995)

The origin of evolvable life involved analog replicators producing digital replicators with indefinite variation (*Nature of the Organism*). This allowed replication without regard to environmental resources (*nature of the organism is more important than the nature of the conditions*), leading to natural selection as an emergent property. This led to conflicts that were resolved by evolutionary transitions. Evolution is not about conflict, it is about conflict resolution. And conflict resolution often involves division of labor and cooperation. Evolutionary transitions arise as innovations in the inheritance system that increase the efficiency of storing and transmitting information (*Nature of the Organism*), thereby enhancing interactions between organisms and their surroundings (*Nature of the Conditions*). Transitions are characterized by the irreversible emergence of increased complexity and new levels of organization, each characterized by a novel form of cohesion, which sets it apart as a new level of organization, which is then reinforced by the novel form of selection generated which focuses on that level. Major transitions persist for indefinite periods of time, affecting evolutionary options (*phylogeny is part of the causal explanation for evolution*).

> Even hindsight is a kind of wisdom. It can be wise to look back. After all, we're our own past more than we are our future.—Jostein Gaarder (1999)

All three of the proposals discussed above are radical in the sense that they return to the roots of Darwinism. Viewed in the light of Darwin's major metaphors, we can see that their core assertions are virtually identical, and their divergent foci of attention are complementary. If Darwinism was encompassing enough to include all three of those perspectives, we are faced with a tantalizing answer to a question we proposed earlier—Darwinism *was* the first complex systems theory.

5.6 Summary

> When the past returns to us with all its glory and pain, we don't know whether to embrace it or flee—Brian Herbert and Kevin J. Anderson (2006)

Reading *Darwinism Today* at the beginning of the third decade of the twenty-first century is simultaneously illuminating and distressing. We have indeed accumulated mountains of new data, including sophisticated mathematical/statistical treatments and experimental studies providing a deep understanding of the chemical and biological basis of inheritance. But this has not led to a newer, grander, unified evolutionary framework for biology. Most textbooks, and most gatekeeper editors of journals of evolutionary biology promote and defend a relatively simple view of evolution, rarely discussing its shortcomings or proposed modifications or alternatives. As well—and this is the distressing part of Kellogg's book—the conceptual issues dividing biologists in 1907 are mostly still with us, some virtually unchanged in more than a century. The origins of the grand questions of Kellogg's time have been forgotten, so that few people engaged in academic conflicts today realize they are refighting old battles. We believe evolutionary biology is mired in a conceptual framework that is incapable of resolving conflicts arising from it.

In the transition from Darwin through Spencer to the Hardened Synthesis, we discarded the narrative, lost the metaphors, and found ourselves only with labels. We have to recover the lost metaphors before we move forward.

We are by no means original in asserting that there is a need for a fresh discussion, if not a fresh framework, of the core theory of Biology. And we are not claiming we have discovered any material aspects of biological systems that have been overlooked by researchers. Sometimes what is needed is looking at things that seem to be well-known, perhaps so well-known for so long that we take them for granted and no longer maintain them as integral parts of our explanations. Our perspective is fundamentally radical, to the extent that if anyone sees our discussion as a call for a replacement for the *Extended Hardened Synthesis*, we would nominate *Darwinism* for that role. We think simplifying Darwinism from Spencer to Kellogg to the New Synthesis to the Hardened Synthesis to the Extended Hardened Synthesis created paradoxes and inconsistencies that prevent it from being useful at a critical time for humanity. What is needed is a comprehensive framework based on an understanding of complex systems that provides grounds for effective action to cope with accelerating global climate change. All of life on this planet is evolved life, so evolutionary biology needs to be at the forefront offering useful advice and insights about how to cope with the myriad challenges facing humanity, most notably accelerating global climate change.

The *Extended Hardened Synthesis* is the elephant in the room—the current paradigm is so pervasive socially that there seems to be no way forward except through it, and it itself is so full of inconsistencies that it seems impossible to find a path ahead. It is bloated with an enormous amount of disorganized information and an accumulation of paradoxes. Only the most committed orthodox believers think that the existing paradigm can make sense of all of this—they argue that all we need is time to generate explanations that fit the current orthodoxy. But that raises the specter of scientists simply modifying their central theory on an ad hoc basis in an attempt to divert attention from the dry rot at the core.

> You rationalize...You defend. You reject unpalatable truths, and if you can't reject them outright you trivialize them.... "evidence is never enough for you...You turn incomprehension into mathematics..." "It served me well enough."—Peter Watts (2006)

We hope that what follows is helpful in overcoming that specter.

References

Alberch P, Gould SJ, Oster GF, Wake DB (1979) Size and shape in ontogeny and phylogeny. Paleobiology 5:296–315

Armbruster WS (1994) Early evolution of *Dalechampia* (Euphorbiaceae): insights from phylogeny, biogeography, and comparative biology. Ann Mo Bot Garden 81:302–316

Avise JC (1989) Gene trees and organismal histories: a phylogenetic approach to population biology. Evolution 43:1192–1208

Avise JC (2000) Phylogeography: the history and formation of species. Harvard University Press, Cambridge

Bak P (1996) How nature works: the science of self-organized criticality. Copernicus, New York

Barrowclough GF, Gutierrez RJ, Groth GG (1999) Phylogeography of spotted owl (*Strix occidentalis*) populations based on mitochondrial DNA sequences: gene flow, genetic structure, and a novel biogeographic pattern. Evolution 53:919–931

Beatty J (1994) Theoretical pluralism in biology. In: Grande L, Rieppel O (eds) Interpreting the hierarchy of nature: from systematic patterns to evolutionary process theories. Academic, London, pp 33–57

Boas F (1896) The limitations of the comparative method in anthropology. Science 4:901–908

Boas F (1898) A precise criterion of species. Science 7:860–861

Boucot AJ (1975a) Standing diversity of fossil groups in successive intervals of geologic time viewed in the light of changing levels of provincialism. J Paleontol 49:1105–1111

Boucot AJ (1975b) Evolution and extinction rate controls. Elsevier, New York

Boucot AJ (1981) Principles of benthic marine paleoecology. Academic, New York

Boucot AJ (1982) Paleobiologic evidence of behavioral evolution and coevolution. By the author, Corvallis

Boucot AJ (1983) Does evolution take place in an ecological vacuum? J Paleontol 57:1–30

Boucot AJ (1990) Community evolution: its evolutionary and biostratigraphic significance. In: Miller W III (ed) Paleocommunity temporal dynamics: the long-term development of multispecies assemblies. The Paleontological Society Special Publication No 5, pp 48–70

Brooks DR (1979) Testing the context and extent of host-parasite coevolution. Syst Zool 28:299–307

Brooks DR (1985) Historical ecology: a new approach to studying the evolution of ecological associations. Ann Mo Bot Garden 72:660–680

Brooks DR (1990) Parsimony analysis in historical biogeography and coevolution: methodological and theoretical update. Syst Zool 39:14–30

Brooks DR (1992) Incorporating origins into evolutionary theory. In: Varela F, Dupuy JP (eds) Understanding origins: contemporary ideas on the genesis of life, mind and society. Reidel/Kluwer Associates, Amsterdam, pp 191–212

Brooks DR (1994) Entropy, information and evolving biological systems. Theor Hist Sci 4:31–49

Brooks DR (1997) Biological evolution as a microcosm of cosmological evolution. Bridges 4:9–35

Brooks DR (1998) The unified theory of evolution and selection processes. In: van de Vijver G, Salthe SN, Delpos M (eds) Evolutionary systems: biological and epistemological perspectives on selection and self-organization. Kluwer Academic, Dordrecht, pp 113–128

Brooks DR (2000) The nature of the organism: life has a life of its own. Proc NY Acad Sci 901:257–265

Brooks DR, Agosta SJ (2012) Children of time: the extended synthesis and major metaphors of evolution. Fortschr Zool 29:497–514

Brooks DR, McLennan DA (1990) Searching for a general theory of biological evolution. J Ideas 1:35–46

Brooks DR, McLennan DA (1991) Phylogeny, ecology and behavior: a research program in comparative biology. University of Chicago Press, Chicago

Brooks DR, McLennan DA (1993a) Historical ecology: examining phylogenetic components of community evolution. In: Ricklefs RE, Schluter D (eds) Species diversity in ecological communities. University of Chicago Press, Chicago, pp 267–280

Brooks DR, McLennan DA (1993b) Parascript: parasites and the language of evolution. Smithsonian Institution Press, Washington, DC

Brooks DR, McLennan DA (1994) Historical ecology as a research programme: scope, limitations and the future. In: Eggleton P, Vane-Wright R (eds) Phylogenetics and ecology. Linnaean society symposium series no. 17. Academic Press, London, pp 1–27

Brooks DR, McLennan DA (1997) Biological signals as material phenomena. Rev pensee d'aujord d'hui 25:118–127. [in Japanese]

Brooks DR, McLennan DA (2002) The nature of diversity: an evolutionary voyage of discovery. University of Chicago Press, Chicago

Brooks DR, Wiley EO (1986) Evolution as entropy: toward a unified theory of biology, 1st edn. University of Chicago Press, Chicago

Brooks DR, Wiley EO (1988) Evolution as entropy: toward a unified theory of biology, 2nd edn. University of Chicago Press, Chicago

Brooks DR, Bandoni SM, Macdonald CM, O'Grady RT (1989) Aspects of the phylogeny of the Trematoda Rudolphi, 1808 (Platyhelminthes: Cercomeria). Can J Zool 67:2609–2624

Brooks DR, Hoberg EP, Boeger WA (2019) The Stockholm paradigm: climate change and emerging disease. University of Chicago Press, Chicago

Brown JH (1995) Macroecology. University of Chicago Press, Chicago

Brown JH, Maurer BA (1989) Macroecology: the division of food and space among species on continents. Science 243:1145–1150

Brundin L (1972) Evolution, causal biology and classification. Zool Scripta 1:107–120

Callebaut W, Müller GB, Newman SA (2007) The organismic systems approach: Evo-Devo and the streamlining of the naturalistic agenda. In: Sansom RE, Brandon B (eds) Integrating evolution and development. From theory to practice. MIT Press, Cambridge, MA, pp 25–92

Charlesworth B, Lande R, Slatkin M (1982) A neo-darwinian commentary on macroevolution. Evolution 36:474–498

Chernoff B (1982) Character variation among populations and the analysis of biogeography. Am Zool 22:425–439

Cheverud JM, Dow MM, Leutenegger W (1985) The quantitative assessment of phylogenetic constraints in comparative analyses: sexual dimorphism in body weight among primates. Evolution 39:1335–1351

Clutton-Brock TH, Harvey PH (1977) Primate ecology and social organization. J Zool Lond 183:1–39

Clutton-Brock TH, Harvey PH (1984) Comparative approaches to investigating adaptation. In: Krebs JR, Davies NB (eds) Behavioural ecology: an evolutionary approach, 2nd edn. Sinauer, Sunderland, pp 7–29

Coddington JA (1988) Cladistic tests of adaptational hypotheses. Cladistics 4:3–22

Coddington JA (1990) Bridges between evolutionary pattern and process. Cladistics 6:379–386

Coddington JA (1992) Avoiding phylogenetic bias. Trends Ecol Evol 7:68–69

Coddington JA (1994) The roles of homology and convergence in studies of adaptation. In: Eggleton P, Vane-Wright R (eds) Phylogenetics and ecology. Academic, London, pp 53–78

Collier J (1986) Entropy in evolution. Biol Philos 1:5–24

Collier J (1988) The dynamics of biological order. In: Weber BH, Depew DJ, Smith JD (eds) Information, entropy and evolution: new perspectives on physical and biological evolution. MIT Press, Cambridge, MA, pp 227–242

Collier J (1990) Two faces of Maxwell's demon reveal the nature of irreversibility. Stud Hist Phil Sci 21:257–268

Collier J (1998) Information increase in biological systems: how does adaptation fit? In: van de Vijver G, Salthe SN, Delpos M (eds) Evolutionary systems: biological and epistemological perspectives on selection and self-organization. Kluwer Academic, Dordrecht, pp 129–140

Collier J (2000) The dynamical basis of information and the origins of semiosis. In: Taborsky E (ed) Semiotics, evolution, energy. Shaker, Aachen, pp 111–138

Collier J, Hooker C (1999) Complexly organised dynamical systems. Open Syst Inf Dyn 6:241–302

Corning PA (1995) Synergy and self-organization in the evolution of complex systems. Syst Res 12:89–121

Cowan G, Pines D, Melzner D (eds) (1994) Complexity: metaphors, models and reality. Addison-Wesley, Reading

Crespi BJ (1996) Comparative analysis of the origins and losses of eusociality: causal mosaics and historical uniqueness. In: Martins EP (ed) Phylogenies and the comparative method in animal behavior. Oxford University Press, New York, pp 253–287

Croizat L, Nelson G, Rosen DE (1974) Centers of origin and related concepts. Syst Zool 23:265–287

Csanyi V (1989) Evolutionary systems and society: a general theory. Duke University Press, Durham

Darwin C (1872) Origin of species. John Murray, London

Day RL, Laland KN, Odling-Schmee FJ (2003) Rethinking adaptation: the niche-constructive perspective. Perspect Biol Med 46:80–95

Depew D, Weber B (1995) Darwinism evolving. Bradford Books, Cambridge

Dietz RS, Holden JC (1966) Miogeoclines (Miogeosynclines) in space and time. J Geol 74:566–583

Donoghue MJ (1990) Why parsimony? Evolution 44:1121–1123

Dunham AE, Miles DB (1985) Patterns of covariation in the life history traits of squamate reptiles: the effects of size and phylogeny reconsidered. Am Nat 126:231–257

Eldredge N (1979) Alternative approaches to evolutionary theory. In: Schwartz JH, Rollins HB (eds) Models and methodologies in evolutionary theory. Bull Carnegie Mus Nat Hist 13:7–19

Eldredge N (1985) The ontology of species. In: Vrba E (ed) Species and speciation, Transvaal Mus. Monogr. No., vol 4, pp 17–20

Eldredge N (1986) Information, economics and evolution. Ann Rev Ecol Syst 17:351–369

Eldredge N (1995) Reinventing Darwin: the great debate at the high table of evolutionary theory. Wiley, New York
Eldredge N, Gould SJ (1972) Punctuated equilibria: an alternative to phyletic gradualism. In: Schopf TJM (ed) Models in paleobiology. W.H. Freeman, San Francisco, pp 82–115
Eldredge N, Salthe SN (1984) Hierarchy and evolution. In: Dawkins R, Ridley M (eds) Oxford surveys in evolutionary biology, vol 1. Oxford University Press, Oxford, pp 182–206
Ellsworth DL, Honeycutt LR, Silvy NJ, Bickham JW, Klimstra WD (1994) Historical biogeography and contemporary patterns of mitochondrial DNA variation in white-tailed deer from the southeastern United States. Evolution 48:122–136
Endler JA (1977) Geographic variation, speciation, and clines. Monographs in population biology #10. Princeton University Press, Princeton
Febregas-Tejeda A, Vergara-Silva F (2018) The emerging structure of the extended evolutionary synthesis: where does evo-devo fit in? Theory Biosci 137:169–184
Felsenstein J (1982) Numerical methods for inferring phylogenetic trees. Q Rev Biol 57:379–404
Felsenstein J (1984) The statistical approach to inferring phylogeny and what it tells us about parsimony and compatibility. In: Duncan T, Stuessy TF (eds) Cladistics: perspectives on the reconstruction of evolutionary history. Columbia University Press, New York, pp 169–191
Felsenstein J (1985) Phylogenies and the comparative method. Am Nat 125:1–15
Felsenstein J (1988) The detection of phylogeny. In: Hawksworth DL (ed) Prospects in systematics. Systematics association. Clarendon, Oxford, pp 112–127
Fitter AH (1995) Interpreting quantitative and qualitative characteristics in comparative analyses. J Ecol 83:730
Frost DR, Kluge AG (1994) A consideration of epistemology in systematic biology, with special reference to species. Cladistics 10:259–293
Futuyma DJ (1989) Speciational trends and the role of species in macroevolution. Am Nat 134:318–321
Gaarder J (1999) Maya. H. Aschehoug (W. Nygaard), Oslo
Garland T Jr, Ives AR (2000) Using the past to predict the present: confidence intervals for regression equations in phylogenetic comparative methods. Am Nat 155:346–364
Garland T Jr, Midford PE, Ives AR (1999) An introduction to phylogenetically based statistical methods, with a new method for confidence intervals on ancestral values. Am Zool 39:374–388
Gittleman JL (1981) The phylogeny of parental care in fishes. Anim Behav 29:936–941
Gladyshev GP (1996) Thermodynamic direction of biological evolution: model and reality. Izvestiya Akad Nauk Ser Biol 4:389–397
Gladyshev GP, Kitaeva DK (1995) On thermodynamic direction of evolutionary processes. Izvestiya Rosk Akad Nauk Ser Biol 6:645–649
Goodwin BC (1982) Development and evolution. J Theor Biol 97:43–55
Goodwin BC, Trainor LEH (1983) The ontogeny and phylogeny of the pentadactyl limb. In: Goodwin BC, Holder N, Wylie CG (eds) Development and evolution. Cambridge University Press, Cambridge, pp 75–98
Gould SJ (1977) Ontogeny and phylogeny. Harvard University Press, Cambridge
Gould SJ (1980) Is a new and general theory of evolution emerging? Paleobiology 6:119–120
Gould SJ (1983) The hardening of the modern synthesis. In: Grene M (ed) Dimensions of Darwinism. Cambridge University Press, Cambridge, pp 71–93
Gould SJ (1986) Evolution and the triumph of homology, or why history matters. Am Sci 74:60–69
Gould SJ, Lewontin RC (1979) The spandrels of San Marco and the Panglossian paradigm: a critique of the adaptationist programme. Proc R Soc Lond B 205:581–598
Haldane JBS (1927) Possible worlds and other essays. Chatto and Windus, London
Hart MW, Byrne M, Smith MJ (1997) Molecular phylogenetic analysis of life-history evolution in asterinid starfish. Evolution 51:1848–1861
Harvey PH, Clutton-Brock T (1985) Life history variation in primates. Evolution 39:559–581

Harvey PH, Mace GM (1982) Comparisons between taxa and adaptive trends. In: King's College Sociobiology Group (ed) Current problems in sociobiology. Cambridge University Press, Cambridge, pp 343–361
Harvey PH, Pagel M (1991) The comparative method in evolutionary biology. Oxford University Press, Oxford
Harvey PH, Read AF, Nee S (1995a) Why ecologists need to be phylogenetically challenged. J Ecol 83:535–536
Harvey PH, Read AF, Nee S (1995b) Further remarks on the role of phylogeny in comparative ecology. J Ecol 83:733–734
Hedin MC (1997) Speciational history in a diverse clade of habitat-specialized spiders (Araneae: Nesticidae: *Nesticus*): inferences from geographic-based sampling. Evolution 51:1929–1945
Hennig W (1950) Grundzüge einer theory der phylogenetischen Systematik. Deutscher Zentralverlag, Berlin
Hennig W (1966) Phylogenetic systematics. University of Illinois Press, Urbana
Herbert B, Anderson KJ (2006) Hunters of Dune. Tor, New York
Hewzulla D, Boulter MC, Benton MJ, Halley JM (1999) Evolutionary patterns from mass originations and mass extinctions. Philos Trans R Soc Lond B 354:463–469
Holland J (1995) Hidden order: how adaptation builds complexity. Addison-Wesley, Reading
Huang S (2011) The molecular and mathematical basis of Waddington's epigenetic landscape: a framework for post-Darwinian biology? BioEssays 34:149–157. https://doi.org/10.1002/bies.201100031
Huey R, Garland T Jr, Turelli M (2019) Revisiting a key innovation in evolutionary biology: Felsenstein's "phylogenies and the comparative method". Am Nat 193:744–772
Hull DL (1988) Science as a process. University of Chicago Press, Chicago
Hunter E (1953) Blackboard jungle. Simon and Schuster, New York
Hutchinson GE (1957) Concluding remarks. Cold Spring Harb Symp Quant Biol 22:415–427
Huxley JS (ed) (1942) Evolution, the modern synthesis. Allen and Unwin, London
Juarrero A (1999) Dynamics in action: intentional behavior as a complex system, 1st edn. MIT, Boston
Juarrero A (2002) Dynamics in action: intentional behavior as a complex system, 2nd edn. MIT, Boston
Kampis G (1991) Self-modifying systems in biology and cognitive science: a new framework for dynamics, information and complexity. Pergamon, Oxford
Kampis G (1998) Evolution as its own cause and effect. In: van de Vijver G, Salthe SN, Delpos M (eds) Evolutionary systems: biological and epistemological perspectives on selection and self-organization. Kluwer Academic, Dordrecht, pp 255–265
Kauffman SA (1986) Autocatalytic sets of proteins. J Theor Biol 119:1–24
Kauffman S (1993) The origins of order: self-organization and selection in evolution. Oxford University Press, Oxford
Kjellstrom G (1996) Evolution as a statistical optimization algorithm. Evol Theory 11:105–117
Kjellstrom G, Taxen L (1981) Stochastic optimization in system design. IEEE Trans Circuit Syst CAS-28:702–715
Klein NK, Brown WM (1994) Intraspecific molecular phylogeny in the yellow warbler (*Dendroica petechia*), and implications for avian biogeography in the West Indies. Evolution 48:1914–1932
Kluge AG (1990) Species as historical individuals. Biol Philos 5:417–431
Korb KB, Dorin A (2011) Evolution unbound: releasing the arrow of complexity. Biol Philos 26:317–338
Kornet DJ (1993a) Permanent splits as speciation events: a formal reconstruction of the internodal species concept. J Theor Biol 164:407–435
Kornet DJ (1993b) Reconstructing species: demarcations in genealogical networks (Ph.D. Dissertation). Leiden University, Leiden

Kornet DJ, McAllister JW (1993) The composite species concept. In: Kornet DJ (ed) Reconstructing species: demarcations in genealogical networks. Ph.D. Dissertation. Leiden University, Leiden, pp 61–89
Kornet DJ, Metz AJ, Schellinx HAJM (1995) Internodons as equivalence classes in the genealogical network: building-blocks for a rigorous species concept. J Math Biol 34:110–122
Laland KN, Odling-Smee J, Hoppitt W, Uller T (2013) More on how and why: a response to commentaries. Biol Philos 28:793–810
Laland K, Uller T, Feldman M, Sterelny K, Müller GB, Moczek A, Jablonka E, Odling-Smee J (2014) Does evolutionary theory need a re-think? Yes, urgently. Nature 514:161–164
Laland KN, Uller T, Feldman MW, Sterelny K, Muller GB, Moczek A, Jablonka E, Odling-Smee J (2015) The extended evolutionary synthesis: its structure, assumptions and predictions. Proc R Soc B 282:20151019. https://doi.org/10.1098/rspb.2015.1019
Landweber LF, Simon PJ, Wagner TA (1998) Ribozyme engineering and early evolution. Bioscience 48:94–103
Layzer D (1978) A macroscopic approach to population genetics. J Theor Biol 73:769–788
Layzer D (1980) Genetic variation and progressive evolution. Am Nat 115:809–826
Lessios HA, Kessing BD, Roberston DR, Paulay G (1999) Phylogeography of the pantropical sea urchin *Eucidaris* in relation to land barriers and ocean currents. Evolution 53:806–817
Lewontin RC (1966) The principle of historicity in evolution. In: Moorhead PS, Kaplan MM (eds) Mathematical challenges to the neo-Darwinian interpretation of evolution. Alan R. Liss, New York, pp 81–94
Lewontin RC (1978) Adapt Sci Am 239:212–230
Lewontin RC (1983) Gene, organism, and environment. In: Bendall DS (ed) Evolution from molecules to men. Cambridge University Press, Cambridge, pp 273–285
Losos JB, Arnold SJ, Bejerano G, Brodie ED III, Hibbett D, Hoekstra HE, Mindell DP, Monteiro A, Moritz C, Orr HA, Petrov DA, Renner SS, Ricklefs RE, Soltis PS, Turner TL (2013) Evolutionary biology for the 21st century. PLoS Biol 11:e1001466. https://doi.org/10.1371/journal.pbio.1001466
Löther R (1990) Species and monophyletic taxa as individual substantial systems. In: Baas P, Kalkman K, Geesink R (eds) The plant diversity of Malesia. Kluwer Academic, The Hague, pp 371–378
Mabee PM (1993) Phylogenetic interpretation of ontogenetic change: sorting out the actual and artefactual in an empirical case study of centrarchid fishes. Biol J Linn Soc 107:175–291
Mabee PM (2000) Developmental data and phylogenetic systematics: evolution of the vertebrate limb. Am Zool 40:789–800
Mabee PM, Humphries J (1993) Coding polymorphic data: examples from allozymes and ontogeny. Syst Biol 42:166–181
Maddison WP (1990) A method for testing the correlated evolution of two binary characters: are gains and losses concentrated on certain branches of a phylogenetic tree? Evolution 44:539–557
Maddison DR (1994) Phylogenetic methods for inferring the evolutionary history and processes of change in discretely valued characters. Annu Rev Entomol 39:267–292
Maddison WP, Maddison DR (2000) MacClade. Analysis of hylogeny and character evolution. Version 4. Sinauer Association, Sunderland
Matsuno K (1989) Protobiology: physical basis of biology. CRC, Boca Raton
Matsuno K (1995) Consumer power as the major evolutionary force. J Theor Biol 173:137–145
Matsuno K (1998) Competence of natural languages for describing the physical origin of life. In: van de Vijver G, Salthe SN, Delpos M (eds) Evolutionary systems: biological and epistemological perspectives on selection and self-organization. Kluwer Academic, Dordrecht, pp 295–306
Maurer BA (1999) Untangling ecological complexity: the macroscopic perspective. University of Chicago Press, Chicago
Maurer BA, Brooks DR (1991) Energy flow and entropy production in biological systems. J Ideas 2:48–53

Maurer BA, Brown JH, Rusler RD (1992) The micro and macro in body size evolution. Evolution 46:939–953
Maynard Smith J (1968) Mathematical ideas in biology. Cambridge University Press, Cambridge
Maynard Smith J (1970) Time in the evolutionary process. Stud Gen 23:266–272
Maynard Smith J (1972) On evolution. Edinburgh University Press, Edinburgh
Maynard Smith J (1974) Models in ecology. Cambridge University Press, Cambridge
Maynard Smith J (1976) What determines the rate of evolution? Am Nat 110:331–338
Maynard Smith J (1978) The evolution of sex. Cambridge University Press, Cambridge
Maynard Smith J (ed) (1981) Evolution now. Macmillan, London
Maynard Smith J (1986) The problems of biology. Oxford University Press, Oxford
Maynard Smith J (1988) Did Darwin get it right?: essays on games, sex and evolution. Chapman & Hall, London
Maynard Smith J (1989) Evolutionary genetics. Oxford University Press, Oxford
Maynard Smith J (1993) The theory of evolution, 2nd edn. Penguin Books, London
Maynard Smith J, Price GR (1973) The logic of animal conflict. Nature 246:15–18
Maynard Smith J, Szathmary E (1995) The major transitions in evolution. W.H. Freeman, Oxford
Maynard Smith J, Burian R, Kauffman S, Alberch P, Campbell J, Goodwin B, Lande R, Raup D, Wolpert L (1985) Developmental constraints and evolution. Q Rev Biol 60:265–287
McKitrick MC (1994) On homology and ontological relationship of parts. Syst Biol 43:1–10
McLennan DA (1991) Integrating phylogeny and experimental ethology: from pattern to process. Evolution 45:1773–1178
McLennan DA (1993) Phylogenetic relationships in the Gasterosteidae: an updated tree based on behavioral characters with a discussion of homoplasy. Copeia 1993:318–326
McLennan DA (1994) A phylogenetic approach to the evolution of fish behaviour. Fish Biol Fish 4:430–460
McLennan DA (1996) Integrating phylogenetic and experimental analyses: the evolution of male and female nuptial coloration in the Gasterosteidae. Syst Biol 45:261–277
McLennan DA (2000) The macroevolutionary diversification of female and male components of the stickleback breeding system. Behaviour 137:1029–1045
McLennan DA, Brooks DR, McPhail JD (1988) The benefits of communication between comparative ethology and phylogenetic systematics: a case study using gasterosteid fishes. Can J Zool 66:2177–2190
McShea DW, Changizi MA (2003) Three puzzles of hierarchical evolution. Integr Comp Biol 43:74–81
Mesoudi A, Blanchet AS, Charmantier A, Danchin E, Fogarty L, Jablonka E, Laland KN, Morgan TJH, Muller GB, Odling-Smee FJ, Pujol B (2013) Is non-genetic inheritance just a proximate mechanism? A corroboration of the extended evolutionary synthesis. Biol Theory 7:189–195
Morgan TH (1932) The scientific basis of evolution. W.W. Norton and Co, New York
Moyers B (1987) The power of myth. Public Broadcasting System television series
Muller GB (2017) Why an extended evolutionary synthesis is necessary. Interface Focus 7:20170065. https://doi.org/10.1098/rsfs.2017.0065
Myers CE, Saupe EE (2013) A macroevolutionary expansion of the modern synthesis and the importance of extrinsic abiotic factors. Palaeontology 2013:1–20
Nelson G, Platnick N (1981) Systematics and biogeography: cladistics and vicariance. Columbia University Press, New York
Nelson G, Rosen DE (eds) (1980) Vicariance biogeography: a critique. Columbia University Press, New York
Newman SA (1970) Note on complex systems. J Theor Biol 28:411–413
Niklas KJ (1999) Evolutionary walks through a land plant morphospace. J Exp Bot 50:39–52
Niklas K (2004) Computer models of early plant evolution. Annu Rev Earth Planet Sci 32:47–66
O'Hara RJ (1993) Systematic generalization, historical fate, and the species problem. Syst Biol 42:231–246
O'Hara RJ (1994) Evolutionary history and the species problem. Am Zool 34:12–22

Odling-Schmee FJ (1988) Niche constructing phenotypes. In: Plotkin HC (ed) The role of behavior in evolution. MIT Press, Cambridge, MA, pp 73–132
Odling-Schmee FJ, Laland KN, Feldman MW (1996) Niche construction. Am Nat 147:641–648
Pagel MD (1994) Detecting correlated evolution on phylogenies: a general method for the comparative analysis of discrete characters. Proc R Soc Lond B 255:37–45
Patterson C (1982) Morphological characters and homology. In: Joysey KA, Friday AE (eds) Problems of phylogeny reconstruction. Academic, London, pp 21–74
Phillips CA (1994) Geographic distribution of mitochondrial DNA variants and the historical biogeography of the spotted salamander, *Ambystoma maculatum*. Evolution 48:597–607
Pigliucci M (2007) Do we need an extended evolutionary synthesis? Evolution 61:2743–2749
Pigliucci M (2009) An extended synthesis for evolutionary biology. Ann NY Acad Sci 1168:218–228
Pigliucci M, Muller GB (eds) (2010) Evolution – the extended synthesis. MIT, Cambridge
Platnick NI, Nelson G (1978) A method of analysis for historical biogeography. Syst Zool 27:1–16
Raff RA (1996) The shape of life: genes, development, and the evolution of animal form. University of Chicago Press, Chicago
Raup DM, Gould SJ (1974) Stochastic simulation and evolution of morphology – towards a nomothetic paleontology. Syst Zool 23:305–322
Read AF, Nee S (1995) Inference from binary comparative data. J Theor Biol 173:99–108
Reilly SM, Wiley EO, Meinhardt DJ (1997) An integrative approach to heterochrony: the distinction between interspecific and intraspecific phenomena. Biol J Linn Soc 60:119–143
Ricklefs RE (1987) Community diversity: relative roles of local and regional processes. Science 235:167–171
Riddle BR (1996) The molecular phylogeographic bridge between deep and shallow history in continental biotas. Trends Ecol Evol 11:207–211
Ridley M (1983) The explanation of organic diversity: the comparative method and adaptations for mating. Clarendon, Oxford
Riedl R (1978) Order in living organisms. Wiley, New York
Rieppel O (1992) Homology and logical fallacy. J Evol Biol 5:701–715
Rocha LM (1998) Selected self-organization and the semiotics of evolutionary systems. In: Van de Vijver G, Salthe SN, Delpos M (eds) Evolutionary systems: biological and epistemological perspectives on selection and self-organization. Kluwer Academic, Dordrecht, pp 341–358
Rose MR, Oakley TH (2007) The new biology: beyond the modern synthesis. Biol Direct 2:30. https://doi.org/10.1186/1745-6150-2-30
Rosen DE (1975) A vicariance model of Caribbean biogeography. Syst Zool 24:431–464
Rosen DE (1978) Vicariant patterns and historical explanation in biogeography. Syst Zool 27:159–188
Rosen DE (1979) Fishes from the uplands and intermontane basins of Guatemala: revisionary studies and comparative biogeography. Bull Am Mus Nat Hist 162:267–376
Rosen DE (1985) Geological hierarchies and biogeographic congruence in the Caribbean. Ann Mo Bot Garden 72:636–659
Ross HH (1972a) The origin of species diversity in ecological communities. Taxon 21:253–259
Ross HH (1972b) An uncertainty principle in ecological evolution. In: Allen RT, James FC (eds) A symposium on ecosystematics. University Arkansas Mus. occ. paper, vol 4, pp 133–157
Roth VL (1984) On homology. Biol J Linn Soc 22:13–29
Roth VL (1988) The biological basis of homology. In: Humphries CJ (ed) Ontogeny and systematics. Columbia University Press, New York, pp 1–26
Roth VL (1991) Homology and hierarchies: problems solved and unresolved. J Evol Biol 4:167–194
Roth VL (1994) Within and between organisms: replicators, lineages, and homologues. In: Hall BK (ed) Homology: the hierarchical basis of comparative biology. New Academic, New York, pp 301–337

Salthe SN (1985) Evolving hierarchical systems: their structure and representation. Columbia University Press, New York

Salthe SN (1993) Development and evolution: complexity and change in biology. MIT, Boston

Salthe SN (1998) The role of natural selection theory in understanding evolutionary systems. In: Van de Vijver G, Salthe SN, Delpos M (eds) Evolutionary systems: biological and epistemological perspectives on selection and self-organization. Kluwer Academic, Dordrecht, pp 13–20

Schneider TD (1988) Information and entropy of patterns in genetic switches. In: Erickson GJ, Smith CR (eds) Maximum entropy and Bayesian methods in science and engineering, vol 2. Kluwer, Brussels, pp 147–154

Schwenk K (1995) A utilitarian approach to evolutionary constraint. Zoology 98:251–262

Schwenk K, Wagner GP (2001) Function and the evolution of phenotypic stability: connecting pattern with process. Am Zool 41:552–563

Seutin G, Brawm J, Ricklefs RE, Bermingham E (1993) Genetic divergence among populations of a tropical passerine, the streaked saltator (*Saltator albicolllis*). Auk 110:117–126

Shaffer HB, McKnight ML (1996) The polytypic species revisited: genetic differentiation and molecular phylogenetics of the tiger salamander *Ambystoma tigrinum* (Amphibia: Caudata) complex. Evolution 50:417–433

Sillén-Tullberg B (1993) The effect of biased inclusion of taxa on the correlation between discrete characters in phylogenetic trees. Evolution 47:1182–1191

Simpson GG (1944) Tempo and mode in evolution. Columbia University Press, New York

Simpson GG (1953) The major features of evolution. Columbia University Press, New York

Smith JDH (1988) A class of mathematical models for evolution and hierarchical information theory. Inst Math Appl Preprint Series 396:1–13

Smith JDH (1998) Canonical ensembles, competing species, and the arrow of time. In: Van de Vijver G, Salthe SN, Delpos M (eds) Evolutionary systems: biological and epistemological perspectives on selection and self-organization. Kluwer Academic, Dordrecht, pp 141–154

Sober E (1988) Reconstructing the past: parsimony, evolution and inference. MIT, Cambridge

Sokal RR, Sneath PHA (1963) The principles of numerical taxonomy. W. H. Freeman, San Francisco, CA

Soto A, Longo G, Miquel PA, Montevil M, Mossio M, Perret N, Pocheville A, Sonnenschein C (2016) Toward a theory of organisms: three founding principles in search of a useful integration. Prog Biophys Mol Biol 122:77–82

Stenseth NC (1984) Why mathematical models in evolutionary ecology? In: Cooley JH, Golley FB (eds) Trends for ecological research for the 1980's. Plenum, New York, pp 239–287

Stocking GW Jr (1968) Race, culture and evolution. University of Chicago Press, Chicago, IL

Swofford DL, Olsen GJ (1990) Phylogeny reconstruction. In: Hillis DM, Moritz C (eds) Molecular systematics. Sinauer Assoc, Sunderland, pp 411–501

Trivers RL (1972) Parental investment and sexual selection. In: Campbell B (ed) Sexual selection and the descent of man, 1871–1971. Aldine, Chicago, pp 136–179

Tuomi J, Vuorisalo T, Laihonen P (1988) Components of selection: an expanded theory of natural selection. In: de Jong G (ed) Population genetics and evolution. Springer, Berlin, pp 109–118

Ulanowicz RE (1986) Growth and development: ecosystems phenomenology. Springer, New York

Ulanowicz RE (1997) Ecology: the ascendent perspective. Columbia University Press, New York

van de Vijver G, Salthe SN, Delpos M (eds) (1998) Evolutionary systems: biological and epistemological perspectives on selection and self-organization. Kluwer Academic, Dordrecht

Vogel G (1998) Tracking the history of the genetic code. Science 281:329–331

Wagner GP (1984) Coevolution of functionally constrained characters: prerequisites for adaptive versatility. Biosystems 17:51–55

Wagner GP (1985) The adaptive significance of developmental constraints. In: Proceedings of international symposium evolution and morphogenesis. Academia, Prague, pp 97–103

Wagner GP, Schwenk K (2000) Evolutionary stable configurations: functional integration and the evolution of phenotypic stability. Evol Biol 31:155–217

Wake DB, Roth G (eds) (1989) Complex organismal functions: integration and evolution in vertebrates. Wiley, New York
Wake DB, Roth G, Wake MH (1983) On the problem of stasis in organismal evolution. J Theor Biol 101:211–224
Wanntorp H-E (1983) Historical constraints in adaptation theory: traits and non-traits. Oikos 41:157–160
Wanntorp H-E, Brooks DR, Nilsson T, Nylin S, Ronqvist F, Stearns SC, Weddell N (1990) Phylogenetic approaches in ecology. Oikos 57:119–132
Watts P (2006) Blindsight. Tor, New York
Weber BH (2011) Extending and expanding the Darwinian synthesis: the role of complex systems dynamics. Stud Hist Phil Biol Biomed Sci 42:75–81
Weber BH, Depew D, Smith JD (1988) Entropy, information and evolution. MIT, Cambridge
Werdelin L, Sillen-Tullberg B (1995) A comparison of two methods to study correlated discrete characters on phylogenetic trees. Cladistics 11:265–277
Westoby M, Leishman MR, Lord JM (1995a) On misinterpreting the "phylogenetic correction". J Ecol 83:531–534
Westoby M, Leishman MR, Lord JM (1995b) Further remarks on phylogenetic correction. J Ecol 83:727–730
Wicken JS (1987) Evolution, thermodynamics and information: extending the Darwinian paradigm. Oxford University Press, Oxford
Wiley EO (1978) The evolutionary species concept reconsidered. Syst Zool 27:17–26
Wiley EO (1980) Is the evolutionary species concept fiction? A consideration of classes, individuals, and historical entities. Syst Zool 29:76–80
Wiley EO (1981) Phylogenetics: the theory and practice of phylogenetic systematics. Wiley, New York
Wiley EO (1986) The evolutionary basis for phylogenetic classification. In: Hovenkamp P (ed) Systematics and evolution: a matter of diversity. Utrecht University Press, Utrecht, pp 55–64
Wiley EO, Brooks DR (1982) Victims of history – a nonequilibrium approach to evolution. Syst Zool 31:1–24
Wiley EO, Siegel-Causey DJ, Brooks DR, Funk VA (1991) The compleat cladist: a primer of phylogenetic procedures. Special Publ. Mus. Nat. Hist. Univ. Kansas, Lawrence
Wilson EO (1965) A consistency test for phylogenies based on contemporaneous species. Syst Zool 14:214–220
Wimsatt WC, Schanck JC (1988) Adaptations and the means of their avoidance. In: Nitecki MH (ed) Evolutionary progress. University of Chicago Press, Chicago, pp 000–000
Windley BF (1986) The evolving continents. Wiley, New York
Wray GA, Lowe CJ (2000) Developmental regulatory genes and echinoderm evolution. Syst Biol 49:28–51
Wray GA, Hoekstra HE, Futuyma DJ, Lenski RE, Mackay TFC, Schluter D, Strassman JE (2014) [Does evolutionary theory need a rethink?] No, all is well. Nature 514:161–164
Zink RM (1994) The geography of mitochondrial DNA variation, population structure, hybridization, and species limits in the fox sparrow (*Passerella iliaca*). Evolution 48:96–111
Zink RM (1996) Comparative phylogeography in North American birds. Evolution 50:308–317
Zotin AI, Zotina RS (1978) Experimental basis for qualitative phenomenological theory of development. In: Lamprecht I, Zotin AI (eds) Thermodynamics of biological processes. de Gruyter, Berlin, pp 61–84

Chapter 6
Buying Time

Abstract What is life? How can life, which is highly organized, arise spontaneously and persist in an entropic universe, where order naturally decays to disorder? This was a major puzzle for twentieth-century chemists and physicists who eventually identified two phenomena, *autocatalysis* and *open thermodynamic systems*, as key to understanding this conundrum. Life uses *metabolism* (autocatalytic reactions) to bring high-grade energy into the system, which it stores and degrades to do work in the system, and then exports back out of the system to the surroundings as heat and other forms of low-grade energy. Metabolism is how organisms *exploit* their surroundings, creating a sense of Circular Time. This "buys" living systems the time to be "alive," but to what end? The goal that Darwin identified is indefinite persistence, which life achieves through its *inheritance systems*. The essence of inheritance is the storage and transmission of information that can be passed on to offspring in a way that specifies the development of a new organism, including the metabolic system. Inheritance allows organisms to *explore* their surroundings, creating a sense of Linear Time through the irreversible processes (growth, ontogeny, reproduction, evolution, speciation) that produce a distinction between past and future. To persist indefinitely, life needs to be *evolvable*, and this requires constantly exploiting—through metabolism and the production of Circular Time—without losing the ability to explore—through inheritance and the production of Linear Time. The Nature of the Organism is life as a combined metabolic-inheritance system that becomes embedded in its own time; therefore, any *unified theory of evolvable life* requires integrating the two.

To be, or not to be, that is the question.—William Shakespeare (1609), *Hamlet*

Bending the rules of time, we take a seat around a prehistoric campfire, joining our ancestors and other fearful, storytelling, dreaming people. The fire crackles and glows. The night sky is in full bloom, undiluted by the light, air, and noise pollution of modern times, a theater of mysteries that sets the human mind ablaze with wonder. The discussions among those people begin with an exchange of information—what happened today? Having established that no new and immediate threat exists,

S. J. Agosta, D. R. Brooks, *The Major Metaphors of Evolution*, Evolutionary Biology – New Perspectives on Its Development 2,
https://doi.org/10.1007/978-3-030-52086-1_6

conversations turn to discussions of the meaning of their lives and the world around them. The sophistication of those discussions might surprise some, but our ancestors lived in a complex and challenging world. The fragility of early hominids—largely lacking natural weapons—placed a premium on accurately perceiving and generalizing the complexity of their surroundings. *Underestimate it and the leopard eats you, overestimate it and you huddle terrified in a cave until you starve.* Their ability to accurately assess complexity in their surroundings allowed our ancestors to survive by finding recurring patterns in nature and encapsulating them into action plans, without reducing what was happening to individual cause and effect interactions. Successful generalizing gave them a sense of control over their lives. And yet, there were many features of our ancestors' world that could not be generalized; many aspects of life are unpredictable. These contingencies of life were frightening because they were where the predators and pestilences lurk. Eventually, our ancestors came to associate "good" with the sense of security that came with successful generalizing and "evil" with the sense of danger that came with the complexity that defied generalizing.

Early on, in this highly complex and only somewhat predictable world, humans recognized a duality about the time they experienced. On the one hand, our early ancestors recognized much of time as a cyclical phenomenon. Each day the sun rose and set. Each night the moon rose and proceeded through a series of nightly changes that repeated each month. Each year, there were seasonal changes in weather patterns—dry and wet, hot and cold—with each kind of change following another in a predictable way. And associated with those seasonal climate changes were changes in the availability of food and water. Many elements of their lives were thus organized around the recurring challenges of climate and the availability of water and food. Early humans learned that annual animal migrations and seasonal transitions in weather and food were paralleled by changes in the phases of the moon and the patterns of the stars in the nighttime sky, even changes in wind patterns denoting a coming storm.

Our ancestors also experienced linear time. Each baby born is a unique human being, who will—barring accident—grow through childhood to adulthood, to old age and death. This process never reverses. Once an adult, a person never becomes a child again. Once dead, a person never returns. But there is continuation in this kind of time—each person will die but life will go on because before they die, many people will leave offspring behind, ensuring the persistence of their piece of the line of time stretching into the future. The recognition that linear time consisted of the old always passing away but making way for the new was a profound insight that helped humans not give in to despair about their own mortality. This was also the initial stirrings of our understanding of the evolutionary nature of life.

Classical Greek philosophers recognized that the dual nature of time in everyday life presented a conundrum. An extreme version of one perspective, most often associated with Parmenides, teaches that the fundamental nature of the world is stasis. Any appearance of change is an illusion or accident, an indication that something has gone wrong. A static world is highly predictable and thus fundamentally "good," giving us hope that we can eliminate contingencies that plague us. An extreme version of the alternative, most often associated with Heraclitus, teaches the opposite—the fundamental nature of the world is constant change. Any appearance of stasis is an illusion or accident. We can never hope to eliminate contingencies that plague us, so the world appears fundamentally "evil" (Neiman 2002).

Aristotle refined the Parmenidean view into the assertion that "nature operates in the shortest way possible." It is human nature to find comfort in simplicity and to distrust complexity, so the development of Western science drew inspiration and comfort from Aristotle's message. By the time of the Enlightenment, therefore, imaginings about the nature of time and the duality of temporal patterns in nature had largely gone out of style. The age of modernism and naturalism began to emerge with a focus on time-independent laws. The universe was structured by a few simple but powerful laws, and science was the search for those laws, and the promulgation of explanations about the world based on them. This was accomplished by *reductionism*—breaking complex systems into smaller parts, each expected to represent a specific case of one of those general laws. Any appearance of complexity in nature is the result of incomplete information. Simple is good. Simple reinforces our belief that we can control the universe to our benefit, eliminating every natural evil that plagues us. From this perspective, we can easily outrun the past and control the future. Time, it seemed, was no longer a problem.

By the nineteenth century, post-Enlightenment physics had largely abandoned the concept of time except as a measuring device. Ironically, at the same time, a new view was emerging, based on the notion that everything that happens in the universe has a material cost. Or, as the saying goes, "there is no free lunch in the universe." Physics treats this reality as an accounting system, in which everything that "happens" (e.g., movement of particles) can be measured in terms of the transformation of energy from one form to another. These transformations are governed by the canonical Laws of Thermodynamics:

First Law—Energy/matter cannot be created or destroyed, only transformed from one form to another.

Second Law—While the *quantity* of energy/matter remains the same (First Law), the *quality* of energy/matter deteriorates with each transformation.

Everything that happens in the universe involves a transformation of energy and according to the second Law, there is always a net cost. The universe is considered to be a self-contained or "closed" system with respect to the matter and energy within it. The initial bank account of "usable energy" is the enormous but finite amount of energy released by the Big Bang. For each event in the universe, energy is transformed into a less useful state, depleting the bank account of usable energy. *Entropy* is an abstract accounting system for taking note of those costs. Entropy increases in proportion to usable energy diminished as a result of doing work within the system, so we always expect entropy to increase over time. And because we can only assess changes in entropy through temporal sequences, thermodynamics resurrected a role for time in explanations in the natural sciences. Thermodynamics in fact dictates an "arrow of time" and entropy records its passage (Blum 1968).

By placing limits on what could be achieved, thermodynamics produced a view of the nature of the universe that offended the same socioeconomic and philosophical preferences of nineteenth-century western society that found Darwinism offensive. When the usable energy inside any system, even the universe, is spent there can be no more net activity because the initial organization of energy into gradients, which permitted the flow of energy from one form to another, has been decayed to maximum disorganization where all energy and matter is equally distributed. While the same quantity of energy remains (First Law), the quality has been transformed from *useful* to *useless* (Second Law). "Equilibrium" has been reached, no more work can be done, and the system has achieved *maximum entropy*. This formulation of thermodynamics led physicists to the vision of "heat death of the universe"—eventually, all the usable thermodynamic free energy (i.e., heat differentials) released by the Big Bang will be spent and the universe will cool to a single uniform temperature.

6.1 Becoming Alive: From Non-life to Life

The Second Law seems to dictate that any structure (order) in the universe spontaneously decays (becomes disordered) and equilibrates with its surroundings. How can life, which is highly organized, arise spontaneously and persist under such conditions? If order naturally decays to disorder, then life seems highly improbable, if not supernatural. This was a major puzzle for physicists and chemists in the twentieth century (e.g., Lotka 1913, 1925; Schrödinger 1945; Prigogine and Wiame 1946; Blum 1968; Prigogine 1980), most of whom argued that life was so improbable that it demanded an explanation based on extremely low probabilities. Physical scientists working throughout the twentieth century eventually focused on two phenomena as key to understanding the origins and nature of life.

The first of these is a chemical phenomenon called *autocatalysis*. An *autocatalytic reaction* is one that literally catalyzes itself—one of the products of the reaction is also a catalyst for the same reaction or another coupled reaction (Fig. 6.1).

An *autocatalytic set of reactions* is one in which enough of the reaction products are catalysts for enough of the other reactions so that, as a whole, the entire set is able to catalyze itself. Autocatalysis has been a central concept in attempts to explain the origins, maintenance, and dynamics of biological systems (e.g., Eigen 1971; Eigen

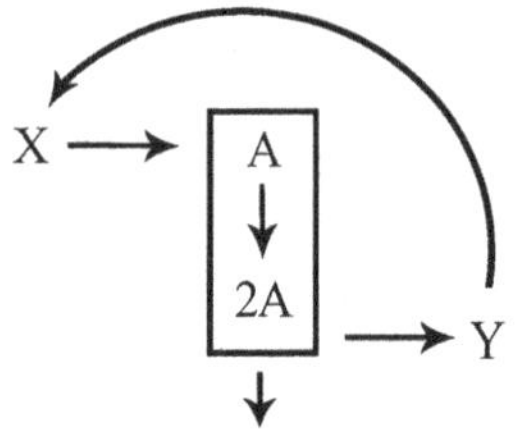

Fig. 6.1 Diagram of a simple autocatalytic reaction where the product *y* is a catalyst for the same reaction that produces *y*. Redrawn and modified from Brooks and Wiley (1988)

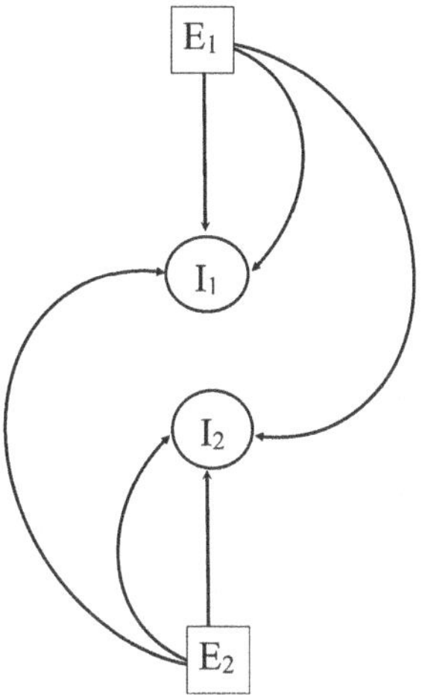

Fig. 6.2 Diagram of the simplest form of hypercycle. The "messenger" molecules I_1 and I_2 produce the enzymes E_1 and E, respectively, each of which is capable of catalyzing the production of both messengers. Redrawn and modified from Brooks and Wiley (1988)

and Shuster 1978; Ulanowicz 1986, 1997; Brooks and Wiley 1988; Kauffman 1995, 2000; Maynard Smith and Szathmàry 1995). Nobel laureate chemist Manfred Eigen (1971) famously envisioned life as a complex of autocatalytic sets joined together with other autocatalytic sets to form what he called *hypercycles*. He suggested the possibility that nonliving autocatalytic systems could give rise to hypercycles from which could emerge living systems. Two key characteristics that distinguish living systems from most nonliving systems are self-replication and a physical separation between "inside" and "outside." Eigen and Shuster (1978) suggested that before life arose, there was a diversity of nonliving autocatalytic systems that they called "replicators." Within this collection of replicators, molecules acting as catalysts for the central product of one cycle could, by serendipity, have the ability to catalyze the formation of the central product in another cycle. If two mutually augmenting autocatalytic systems happened to meet, then they could pair spontaneously to form a *hypercycle* (Fig. 6.2).

The simplest version of a hypercycle that could be a precursor to life would involve the coupling of four separate autocatalytic cycles, one for each of the bases of RNA. To the extent there was variation in the efficiency of replication, some hypercycles would come to predominate over others, an emergent form of natural selection. Although central to mainstream origins of life theory for decades, the first empirical support for the idea of a spontaneously emerging RNA hypercycle came only recently. In a pioneering experiment, Lincoln and Joyce (2009) were able to create self-sustaining autocatalytic reactions in which two RNA enzymes acted as catalysts for each other's synthesis, using only four simple oligonucleotide substrates. Much like Darwin intuited the existence of a material basis for the inheritance system long before the discovery of DNA, it seems that students of the origins of life problem had the notion that autocatalysis must play a role long before any empirical evidence was found.

The idea of autocatalysis—chemical systems recycling themselves—is the first inkling of an understanding of the nature of the organism. And biological systems are full of autocatalytic cycles. But autocatalysis per se does not address the presumptive conundrum of the Second Law and life. In the early 1900s, the famous biophysicist Alfred Lotka (1913, 1925) made a critical insight when he characterized biological systems as *metabolic systems.* He suggested that the way in which organisms maintain themselves in highly organized, low entropy states is by exchanging matter and energy with their surroundings. As open systems with a clear distinction between the inside and outside, organisms could bring energy into the system to do the work needed to maintain their own organization. In this way, Lotka reasoned, organisms could delay the inevitable consequences of the Second Law. By bringing in energy and matter from the surroundings, and by depositing waste (entropy) in the form of heat and other forms of degraded energy back to their surroundings, organisms could slow their inevitable decay. From this perspective, organisms do not simply exist in time but instead use metabolism to "make" or "buy" enough time to "be alive." Metabolism has a direction in time, step by step; but because it is autocatalytic, after a certain number of steps you arrive back where you started. In aerobic metabolism, one begins with citric acid and eventually ends up with. . .citric acid. Entropy increases at each step but the process as a whole is characterized by circular time.

Understanding how organisms "buy enough time to be alive" gained traction with the emergence of the concept of open thermodynamics systems. For closed thermodynamic systems, once the energy inside the system is maximally dispersed, equilibrium has been reached and no more work can be done in the system. But in open systems, new matter and energy can flow through the system, which allows it to function (stay alive) so long as the flows continue. To be sure, open systems are still bound by the Second Law. But their ability to cycle in energy from the outside and do work with it on the inside opens the possibility for a *lag* in the transformed energy that will *eventually* be released from the system. Some of the energy taken in stays inside the system long enough to do work internally to keep the system operational. The Nobel laureate Ilya Prigogine summarized the open-systems perspective using the following metaphorical equation:

$$\mathrm{d}S = \mathrm{d_e}S + \mathrm{d_i}S, \mathrm{d_i}S > 0$$

where $\mathrm{d}S$ refers to the total change in entropy in the surroundings and the system, $\mathrm{d_e}S$ refers to entropy changes occurring as a result of exchanges of matter and energy between the system and its surroundings, and $\mathrm{d_i}S$ refers to changes in entropy resulting from energy transformations within the system. Entropy in closed systems is something like a state of the system, whereas in open systems, entropy refers to ongoing activities, and thus we speak not of *entropy*, but of *entropy production*. As well, the notion of *entropy production* suggests that entropy need not be conserved locally, i.e., for any given open system (Denbigh 1975), though it may be conserved universally, i.e., in the universe as a whole (as asserted by the First Law). Open systems, including living ones, thus may exist indefinitely. To be thermodynamically open there must be localized flows of matter and energy (a classic exemplar is a convection current), but to make a thermodynamically open *system*, there has to be a material boundary between the surroundings and something internal that becomes d_iS, which can be largely insensitive to the surroundings. This characterization of life originating through the emergence of membrane-enclosed, autocatalytic, thermodynamically open systems thus brings with it the Darwinian distinction between the nature of the organism and the nature of the conditions.

In his essay *What is Life* the physicist Erwin Schrödinger, another Nobel laureate, famously described living systems metaphorically as systems that feed on "negative entropy" or *negentropy* (Schrödinger 1945). In a footnote, he indicated that he believed *free energy* to be the more appropriate technical term, but felt it obscured his main point: what is "usable energy" to the organism is essentially "missing entropy" from the universe. Despite popular misunderstandings, Schrödinger did not suggest that living systems disobey the Second Law even temporarily. No system can use low entropy energy to do work that *decreases its own entropy* (this violates the First Law); moreover, no system can use low entropy energy to do work that even *maintains an initial low entropy state* (this violates the Second Law). If you feed on negentropy, therefore, you are in a higher entropy state than what you are feeding upon. Furthermore, once the system uses energy from the surroundings internally, the only way even to maintain its original state would be to "export" 100% of the degraded energy and matter from the system. In the case of organisms, this would involve the complete loss of metabolic heat necessary to catalyze reactions essential for life. It is true that *most* of the degraded energy is lost from living systems, with the net result that organisms degrade the ability of the surroundings to continue to provide low-entropy energy and matter. But it is also true that some of the energy transformations *remain within the living system*. This means living systems are indeed systems of increasing entropy in accordance with the Second Law, even though they dissipate more of their entropy than they retain.

One man's rubbish is another man's treasure—Herbert Urquhart (1860)

Relative to each organism, the surroundings are made up of usable energy (negentropy) that can be taken up and used by the organism *plus* degraded energy dissipated from the organism that it can never use again. Plants are not organized photons, they use photonic energy to organize themselves according to inherited information; and the resulting plant biomass is rubbish compared to the photons, which were rubbish compared to the thermonuclear system—the sun—from which they emanated. Schrodinger's concept of negentropy must be considered in this subjective context if it is to be physically and biologically realistic. Whether something is called *negentropy* or *entropy* depends on which end of the process we focus on. All the negentropy of the universe could be seen as entropy produced by the Big Bang. In biological systems, photons are negentropy to plants but are also part of the entropy produced by the sun. Some of the negentropy taken up by a living system is transformed into entropy relative to an initial event, but that entropy may still be useful negentropy for a different transformation within the organism.

Whether we call it negentropy or entropy is subjective, but each energy transformation proceeds over time in the direction of decreasing usefulness of the energy (increasing entropy). Perhaps this is what Boltzmann had in mind when he said, in a lecture in 1905 in Leipzig, that he felt his only disagreement with Darwin was that in the future we would find that life was a struggle for entropy rather than for survival (Broda 1983). We think Darwin and Boltzmann may both have been right.

In sum, if we can surround autocatalytic operations with a material boundary, making an open thermodynamic system capable of resisting the inevitabilities of the Second Law for an indefinite period of time, we have the minimal elements necessary for some form of life (Maynard Smith and Szathmàry 1995). For organisms, the most fundamental boundary is a hydrophobic cell membrane. Perhaps the most promising possibility for both creating a boundary and for getting autocatalytic cycles inside has been the discovery of certain clay substrates with some fascinating properties: not only do they help catalyze autocatalytic cycles on their surfaces, but they are capable of peeling off in thin layers and curling up in a manner suggestive of a membrane. This "primordial pizza" metaphor for the origin of life represents a major competitor to the more traditional culinary metaphor of a "primordial soup" (for excellent discussions, see Maynard Smith and Szathmàry 1995; Szathmàry 2015), but food is the unifying theme.

6.2 Staying Alive: The First Rule of Life

To this point, we have established that organisms are autocatalytic open thermodynamic systems bounded by membranes. They take up high-grade energy and materials from their surroundings, use the energy to do work inside the system, and in the process, they degrade the energy and eventually release it back to the surroundings. So long as unused (high grade) energy flows in and used (degraded) energy flows out of the system, life persists—the organism is *alive*. There is a diverse nomenclature describing this state of affairs—dissipation of energy and matter; entropy production; entropy increases; cost of living; work done; exploitation of resource gradients; exploitation of entropy gradients—and for our purposes they can all be used interchangeably.

Time to be alive in an entropic universe is, however, not cheap. Organisms produce a great deal more entropy in exchanging matter and energy with their surroundings (d_eS; mostly in the form of heat) than they do internally (d_iS) so that Prigogine's summary equation should be a little more elaborate than he imagined:

$$dS = d_eS + d_iS, \text{ where not only is } d_iS > 0 \text{ but } d_eS >>> d_iS$$

In other words, internal entropy production (d_iS) represents a small amount of an organism's overall energy budget, yet it is that tiny fraction of total work done by the organism that is essential for life. To understand further, we need to take a closer look at what is happening inside the organism and two components of internal entropy production (Brooks and Wiley 1986, 1988; Brooks et al. 1989; Brooks and McLennan 1990; Maurer and Brooks 1991). The first component accounts for entropy produced as the result of exchanges between the system and its surroundings (d_eS) or the *heat-generating transformations*. Exchanges between organisms and their surroundings produce an enormous amount of waste, mostly in the form of heat dissipated from the organism, and thus d_eS is very large. Put another way, it is extremely expensive for organisms to exchange materials with their surroundings, but this is the cost of "doing business" in an entropic universe. From this perspective, it is wrong to view the heat and other forms of degraded energy given off by organisms as "waste." Organisms do not "waste" anything to be alive; instead, they pay a hefty price to stay in the game.

The second component of total entropy changes in open systems pertains to production within the system (d_iS) or *conservative transformations* (Maurer and Brooks 1991). These transformations happen inside the system and are conservative in that they convert energy from the outside into states that can be stored and utilized in subsequent transformations. In other words, conservative transformations produce the "stuff" of life. This stuff or *biomass* exists fundamentally in the form of large complex molecules, the bonds of which represent potential energy that can be used in the future. Once used, the entropy produced internally (transformed negentropy) is stored or *bound* inside the system until it becomes unusable, and then it will be released into the surroundings. Thus, it is not so much the notion of *feeding* on negentropy that is key, as Schrödinger (1945) posited, but *retaining* negentropy within the system where it can continue to do work, becoming transformed into entropy that allows life to persist.

Production rules govern the conservative transformations inside living systems. Brooks and Wiley (1986, 1988) followed earlier workers (e.g., Prigogine and Wiame 1946; Zotin and Zotina 1978) by denoting such allocations using the symbol ψ, which can be used to specify two classes of internal production processes. First are those processes involved in dissipation from the system, or the *external dissipation function* (ψ_α) (Collier 1990 called this *intropy*; Ulanowicz 1986 called it *overhead*). External dissipation involves mostly heat generated by production within the organism and lost to the surroundings at a slower rate than the heat-generating transformations denoting the costs of transporting matter and energy into and out of the system (d_eS). The second class of energy transformations includes processes involved in maintaining structure *within* the system, the *bound dissipation function* (ψ_μ) (Collier 1990 called this *enformation*). Total internal entropy production by an organism (d_iS) can thus be viewed heuristically as:

$$d_iS = \psi_\alpha + \psi_\mu$$

Models of the origin of life for which the production rules specify only metabolic processes embody only cyclical time. These processes are essential for allowing organisms to be alive, but they are not enough in our quest for an understanding of evolution, which is characterized by linear time. As Vasas et al. (2010) so eloquently showed, metabolism-only life forms are destroyed by the introduction of variation, the very thing that allows life on this planet to evolve (see also Maynard Smith and Szathmàry 1995). There must be a particular aspect of the metabolic system that allows the system to cope with disruptive variation and with the thermodynamic inevitability of the system's breakdown. Evolvable life requires something like a *stochastic corrector* so that catalytic templates occurring in two cellular compartments—replication and metabolism—can provide protection against selfish template mutants if the template with the higher genetic information constrains the template with the lower genetic information (Szathmàry and Demeter 1987; Zintzaras et al. 2002) and a way to perpetuate correctable cyclical time (metabolic) machines. For that to be the case, there must a subsystem within an organism capable of producing an indefinite sequence of correctable cyclical systems. That subsystem is inheritance, the means by which information specifying particular metabolic systems are perpetuated linearly in time.

Among all the chemical and physical models of the origin of life, only one—Tibor Ganti's (1971, 1974, 1979, 1987, 2003) *chemoton* model—adds the second major ingredient needed to get closer to evolvable life. Ganti was a Hungarian biochemist who conceived that life could not be "metabolism-only," but instead must be *metabolism plus information* inside a membrane. Maynard Smith and Szathmàry (1995) provide an in-depth discussion of Ganti's chemoton model, which we only briefly summarize here. In the model, a chemoton has three main features, or subsystems, that collectively permit growth, reproduction, and variation within the hereditary system. These subsystems include: (1) an autocatalytic network that drives metabolism, (2) a molecule that carries information that can be replicated, and (3) a semipermeable membrane to separate "inside" from "outside." Maynard Smith and Szathmàry (1995) referred to this system collectively as Ganti's microsphere. The essential point here is that for Ganti's microsphere to grow and reproduce, both metabolism *and* information are required. In contrast to "metabolism-only" theories, the chemoton model makes a critical leap forward in the pursuit of a comprehensive understanding of life. However, while it may explain how metabolism can be linked with heredity to explain replication and reproduction, it is still not sufficient for a full understanding of *evolvable* life.

6.3 Being Evolvable: The Second Rule of Life

For Ganti, as for Eigen, the issue of the nature of life was a chemistry problem. And yet, evolvable life comprises organisms that are autocatalytic open thermodynamic systems. Metabolism buys circular time that keeps organisms alive, *but to what end*? The goal Darwin identified is *indefinite persistence*, which living systems achieve through their *inheritance systems*. The essence of inheritance is the storage and transmission of information that can be passed on to offspring in such a way that specifies the development of a new organism, including information on how to cope with external changes that could threaten metabolic function. In addition to protecting the integrity of the metabolic system, the inheritance system is what allows life to persist in time and space through reproduction; this is what allows living systems to explore possibilities for survival. It is what makes living systems *evolvable*. While existence may be a chemistry problem, buying time for persistence is a thermodynamic problem. Another Hungarian, Ervin Bauer, provided the first post-Boltzmann inkling of a solution to that problem (Bauer 1920a, b).

> All living organisms are characterized by being a system that is not in equilibrium in its environment and is so organized that it transforms the sources and forms of energy taken up from its environment in such state that acts against the establishment of equilibrium in the given environment.—Ervin Bauer (1920a)

The key to this is recognizing that the functional autonomy of living organisms stems from their ability to produce their own internal time, that must be paid for by the system's *own* activities that are fundamentally autonomous from the surroundings (Arnellos and Moreno 2012; Bich et al. 2012; Moreno and Mossio 2015; Ruiz-Mirazo et al. 2017).

6.3.1 *Slow Down and Live: It Is the Fluxes (d_iS), Not the Flows (d_eS)*

As mentioned earlier, if energy and entropy simply flowed in and out of an organism at the same rate, there would be no persistent organism (organized structure) to speak of. Thus, a key to being alive is to control the rate and direction of the flow of energy/entropy through the system to slow the inevitable decay. The heat-generating processes (d_eS plus ψ_α) discussed above involve energy and entropy flowing in opposite directions: energy comes into the organism which produces a large amount of entropy (heat) that drives the system toward more disorganized states. If all the heat generated by processes associated with exchanging materials with the surroundings stayed in the organism, it would rapidly die. Organisms mitigate this problem by "exporting" a lot of entropy back to the surroundings; this is easily observed and measured as heat radiated across the surface of any organism. Conservative transformations, on the other hand, are characterized by energy and entropy flowing in the *same* direction: entropy production that results from energy transformations within the organism is retained in the system, driving it toward more organized states. And now here is the key point: as entropy and energy flow through biological systems in different directions and at different rates, structure accumulates at different levels of organization (Maurer and Brooks 1991). And, as Lotka (1913) suggested, the inevitable trend toward structural decay in biological systems can be delayed, although not reversed, by this accumulation of bound dissipation in the system.

Elek and Müller (2013) emphasized that Bauer's perspective differed fundamentally from what became the prevailing view in the latter half of the twentieth century, namely that metabolism produces an internal equilibrium state. Bauer believed that the flow of energy through an organism in some manner maintained a nonequilibrium state that was in some way stable with respect to the surroundings. He suggested the key to understanding that was in understanding how organisms used the energy they took up from the environment.

> ...All the energy taken up by the organism from the environment must be fully used to deviate from the equilibrium state—Ervin Bauer (1920a)

This produces the maximum amount of entropy (ergo the maximum amount of useful work) but does so by maximizing not the flows but the fluxes—constraints or resistance—to the flows. Ganti recognized that organisms must contain within their boundaries the capacity to produce the fluxes that allow them to use the matter and energy taken up from the surroundings to make and maintain themselves. He called that capacity *information*, which Maynard Smith and Szathmàry (1995) linked to inheritance systems, i.e., to the *Nature of the Organism*. Some recent studies follow in the footsteps of the Hungarian triad of Bauer, Ganti, and Szathmary (e.g. Tessera 2011; Bai et al. 2018; Vitas and Dobovišek 2019; Tetz and Tetz 2020).

Many have assumed that the entropy of living systems must be decreasing because the entropy of the surroundings is increasing due to dissipation from the system. This is only true if *Onsager reciprocity* (Onsager 1931) holds; if not, then entropy density internally can stay the same or even increase (Denbigh 1975). This is the case for living systems in which external dissipation > internal dissipation ($d_eS > d_iS$). Because energy goes into internal cycling, not immediate dissipation, the rate of entropy production is minimized the more times energy is used internally, though entropy increases at each step. Onsager (1931) was following Bauer when he wrote that constraints (fluxes) are impediments to entropy increases (see Andrade 2000; Lucia 2016; Grmela et al. 2019; Roach et al. 2018). Metabolic cycles relying on enzymes to power them, which Liu (1989, 1990) characterized as *flow-resistance cycles*, are a wonderful example of this concept. Imprecise language has led to the appearance of a disagreement between a maximum entropy production and a minimum entropy production perspective. In reality, maximizing the quantity of entropy produced (the total amount of work done) is achieved in living systems by minimizing the *rate* of entropy production. In the end, the tortoise beats the hare.

Bound dissipation represents all structure maintained within the organism and can be further subdivided into two components that account for allocations to (1) making and maintaining biomass (ψ_μ^b) and (2) information that specifies the forms of matter and energy which any given organism can take in and use to maintain its life, and also can be passed on through inheritance (ψ_μ^i). Putting this all together, total internal entropy production by an organism can be viewed heuristically as:

$$d_iS = \psi_\alpha + \psi_\mu^b + \psi_\mu^i$$

Maurer and Brooks (1991) presented a visualization of the fluxes and flows of entropy production within a single multicellular sexual organism (Fig. 6.3). It begins with energy and matter flowing from the organism's environment, *E*, into the organism. There is a cost to bringing matter and energy into the organism, so energy

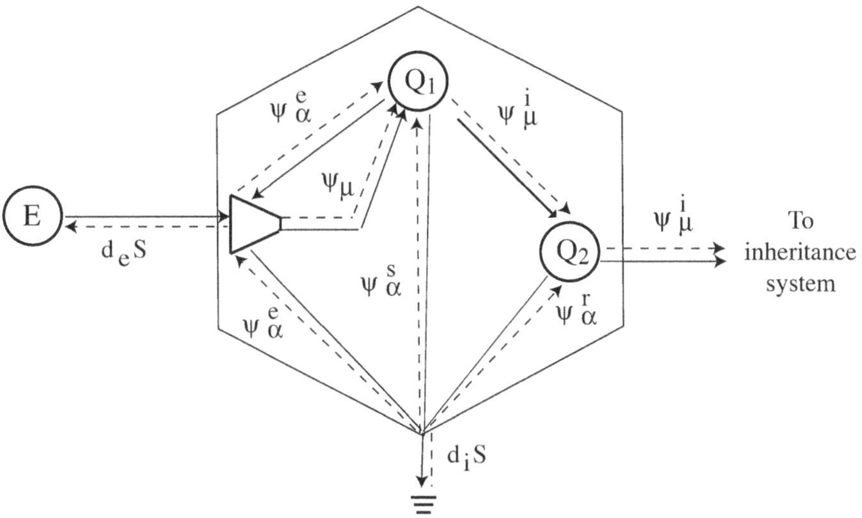

Fig. 6.3 Cartoon diagram of an organism as an open thermodynamic system in which internal fluxes constrain the flows. Taking energy and matter from the surroundings (E) has a cost (d_eS), represented by the keystone icon (representing a filter: Odum 1983). That energy is used to do work internally, characterized by entropy production associated with heat-generating transformations (flows: dashed lines) and conservative transformations (fluxes: solid lines) in somatic (Q_1) and inheritance system (Q_2) cells. Entropy production due to flows within the system (ψ_α) associated with modulating exchanges with the surroundings (ψ^e_α), somatic activities (ψ^s_α) and inheritance system activities (ψ^r_α) is quickly lost from the system, becoming part of the overall entropy production resulting from exchanges between the system and its surroundings (d_eS). Entropy production resulting from fluxes (ψ_μ) is dissipated into growth, differentiation, and maintenance of somatic (ψ_μ) and germ cells (ψ^r_μ), until it is eventually released back to the surroundings, where it later becomes part of the overall entropy production resulting from exchanges between the system and its surroundings. Redrawn and modified from Maurer and Brooks (1991)

exchanges between the organism and its surroundings are accompanied by an increase in the entropy of the energy source (d_eS). The inherited capacities for taking in matter and energy in particular forms are depicted by a filter symbol (using the formalism of Odum 1983). Once inside the system, everything that happens, every bit of work done by that energy, also costs something. And all of that work done internally is, initially, part of the entropy production of the system (d_iS). That work involves both heat-generating transformations (flows) and conservative transformations (fluxes). Flows refer to the cost of obtaining energy at less than perfect efficiency and are associated with direct loss from the system (ψ_α). This is the part of internal entropy production (d_iS) that is lost (d_eS) as the cost of doing business. Fluxes refer to allocations to growth and maintenance of structure (ψ_μ).

Some of the energy taken in is used immediately as flows modulating exchanges with the surroundings (ψ^e_α), and some is stored for later use. Most fluxes (ψ_μ) are in the form of structure that persists in various forms for various periods of time. That which is stored is degraded relative to the energy that was taken in initially but is still of high enough grade to perform additional work within the organism. Eventually, the energy taken up to maintain that structure is degraded to the point that the organism can do no more work with it, and it is lost from the system. Multicellular organisms store and use energy in somatic cells (denoted by Q_1 in Fig. 6.3), through a variety of means, including cell division and growth (ψ_μ). A myriad of biochemical pathways produces fluxes in the flows of matter and energy. Storage of energy in these various biochemical structures requires the expenditure of energy. Thus, a second type of heat loss is attributable to the costs of storage in somatic cells (ψ^s_α). Finally, a very small, but critical source of entropy production occurs in the generation, maintenance, and perpetuation of germ cells (denoted by Q_2 in Fig. 6.3) (ψ^r_μ) and associated heat loss costs (ψ^r_α).

The story of entropy production and circular time in living systems is thus about how slowly the energy is lost from the system, and that is associated with how many times it is used to do work, which in turn is associated with the number of fluxes in the system. Flows are always coupled with fluxes, and that slows the flow of energy through the system, so more work can be done using it. Slowing the rate of entropy production increases the total amount of entropy production. How much work is done, how much entropy is produced, how much energy is dissipated, how much structure is built and maintained and ultimately lost—these are simply different ways to say the same thing, to envision the entropic dynamic process of maintaining life.

While living in Russia, Bauer became interested in the process of aging and death (see discussion in Elek and Müller 2013), which was later taken up and elaborated on by Russians possibly inspired by Bauer's work. Zotin and Zotina (1978), for example, showed that early in ontogeny, organisms exhibit high metabolic rates associated with rapid growth and development. However, as an organism's life continues this "immature" phase of early ontogeny is replaced by a "mature" phase in which metabolic rate and rate of entropy production *decrease*, then remain in a steady-state (Bauer's stable nonequilibrium) until the organism begins to senesce. Senescent organisms exhibit decreasing rates of entropy production, indicating reduced ability to maintain life functions, ultimately to the point that life can no longer sustain itself and death ensues. The longer an organism processes matter and energy internally, the longer it lives and the more chances it has for extending itself in time through reproduction and inheritance. The accumulation of bound dissipation (the fluxes) aids life's own cause. Moreover, the ways and rates at which organisms pass through the stages of decreasing entropy production are determined by an *interaction* between the surroundings and a "sense of self" the organism inherits from its parents. Life from life, with death as an unavoidable by-product. This underscores the idea that inheritance is a critical piece of being "self-organized"—again, the nature of the organism takes precedence over the nature of the conditions.

6.3.2 *Keeping It Affordable*

The keys to making life and its evolution affordable are mechanisms for making it cheap and slow. It is not a matter of how fast you can go, or how powerful you can be. It is a matter of how long you can persist. This is true for both the metabolic system—buying circular time—and for the inheritance system—buying linear time. Metaphorically, the longer a system keeps its secrets the more chances it has to survive. And in a Darwinian world, survival is the only thing that matters.

> We're just recycled history machines, cavemen in faded blue jeans—Jimmy Buffet (1995), "Don't Chu-Know"

The extended hardened synthesis, with its emphasis on power and efficiency rather than persistence, has put itself in a position of assuming that life is expensive and evolution requires massive and rapid energy processing (Dewar 2005; Sharma and Annila 2007; Annila and Annila 2008; Jaakkola et al. 2008; Kaila and Annial 2008; Annila and Kuismanen 2009; Tuisku et al. 2009; Mahulikar and Herwig 2009; Annila and Salthe 2010; Salthe 2010; Annila and Salthe 2012a, b; Hartonen and Annila 2012; Giangaspero and Sciubba 2013; Cheng and Liang 2014; Annila 2015; Zheng et al. 2019). In reality, the metabolism of any given organism is no more expensive than the organism can afford. Autocatalysis reduces the costs by recycling matter, while using the same energy for multiple transformations within the organism reduces the costs even more. Inheritance is even cheaper than metabolism, being a tiny fraction of the bound dissipation of the system. Costs are reduced even more by another form of autocatalytic recycling, the repeated use of preexisting templates of DNA-based information that makes replication a largely "rinse and repeat" affair. And finally, inheritance originates as part of internal entropy production and finishes by expelling its products into the surroundings where they become metabolic and inheritance systems of their own. The inheritance system reduces costs to the organism by virtue of being part of the dissipated entropy loss from the system—through cell division, gamete release, or zygote release (Rastegar-Sedehi et al. 2018). Given that the cost of producing new cells, gametes or zygotes is cheap, producing many of them is not nearly as wasteful a process as is commonly believed.

> Homologous parts tend to vary in the same manner, and homologous parts tend to cohere—Darwin (1872)

Using preexisting templates lowers the cost of operating the inheritance system but is also the means by which the whole process is slowed down. Darwin recognized that inheritance produces an extremely high level of historical cohesion in biological systems. And though the Second law implies that perfectly faithful copying cannot continue indefinitely, the nature of the inheritance system means that while the system is open-ended, all innovations must integrate functionally with the rest of the system in order to produce living offspring characterized by the innovation. Even though the inheritance system is open to change, the criteria of functional integration ensures it is conservative enough to be stable. Conservative inheritance is the primary flux slowing the evolutionary flow. As living systems grow and develop, the accumulated products of their previous activities manifest themselves as correlations among the parts of the system, which decrease the rate of entropy production and "buy time" for different components of the system to interact in innovative ways and also for reproduction and evolution to occur. This ensures that innovations produce new but nonrandom correlations among the preexisting parts of the system. These nonrandom correlations produce inherited *lagload* (Maynard Smith 1976: also called historical constraints or inertia, and phylogenetic constraints or inertia; for a discussion, see Brooks and McLennan 2002), slowing parts of the system and limiting the ways parts can interact, reinforcing the linearity of the time inheritance buys. A basic example is self-reflexive nucleotide copying for replicating DNA. The more faithful the replication, the higher the degree of historical correlation, and therefore the slower the rate of evolutionary change. Many mechanisms associated with inheritance produce correlations among parts, including molecular affinities, cell–cell adhesion, genetic correlations within the inheritance system and between the inheritance system and the surroundings, genetic compatibility, mate recognition, and symbiosis.

Conservative inheritance lowers the cost of innovation because recycling old information is less expensive than making new information. As Jacob (1977) recognized in his essay *Evolution and Tinkering*, evolution is more about modifying and recombining existing pieces than about inventing new pieces. Conservative inheritance lowers the cost of coping with the changing conditions of life because old information can always be recycled and potentially used in new ways to cope with changes in the surroundings, rather than inventing solutions de novo. When faced with new challenges, biological systems do not need to wait for new solutions to arise, but instead can co-opt preexisting information in new ways (Darwin 1872; see McLennan 2008 for an excellent review).

Maynard Smith and Szathmàry (1995) suggested that the conjunction of two or more events, each of which is improbable on its own, is sufficient to give life its distinctively linear time. Lagload increases the chance of such conjunctions by maintaining improbable events long enough to increase the likelihood that two or more of them will become causally intertwined. Inheritance systems accumulate historical correlations that slow down the expansion of the system, ensuring that "what's realized" always lags behind "what's possible." By binding negentropy in the system long enough to create a lag in total entropy production, life "slows the flow," creates new opportunities for itself, and generates its own arrow of linear time. This allows life to "move forward" and "explore the future" as irreversible changes in structure/information/diversity accumulate in the system. In a sense, living systems become embedded in their own time, which is the reason history plays such a predominant role in their evolution. Life tells time, evolvable life tells history. This is the sense in which the nature of the organism buys linear time and extends itself into a future largely of its own making.

> The only thing that makes life possible is permanent, intolerable uncertainty: not knowing what comes next—Ursula Le Guin (1969)

The historical nature of inheritance turns linear time into a *causal agent* (Juarrero 1999) in the system, whose primary function is keeping the process cheap and slow. The inheritance system, therefore, is the core of the *nature of the organism*. Armed with information transmitted from their ancestors through conservative inheritance, organisms are capable of deploying this information to impose themselves on the surroundings in whatever ways they can to stay alive and reproduce. As well, while lagload makes evolution more affordable, accumulated history slows the system down enough to limit its options and reduce responsiveness to changes in the nature of the conditions. Lagload limits the expression of genetic diversity, which determines how populations can respond to their surroundings; in this way, history makes space between accident and necessity (Neiman 2002). History links inheritance to the emergence of natural selection and to heritability in the sense of Darwin's "inherited tendencies," so that

If it's causal, it's heritable
If it's heritable, it's historical
Therefore, if it's causal, it's historical

6.3.3 *Intimate Details of Inheritance Dynamics*

To this point, we have envisioned both circular time and linear time in terms of the flow of energy. Metabolism and other physiological processes that keep organisms alive are commonly expressed in energetic terms. Because the total energy budget used for storing and transmitting information is so small, and yet the impact is so large, it is easier to examine the dynamics of inheritance systems in terms of product—information—rather than cost. This "information view" will be harder to understand than the metabolic view because inheritance is harder to understand in energetic terms. What the system "dissipates" is copies of itself, and these are not easily understood as "degraded energy."

In order to accomplish this, we need an abstract language that allows us to see flows and transformations of energy in terms of changes in structure. Fortunately, there is a generalization or abstraction of thermodynamics, called *statistical mechanics*, that is more inclusive than just measuring the flow of energy and which will allow us to draw parallels between the energy and information views of life using the same formalism. At the core of statistical mechanics is the idea that energy can be measured statistically as the movement, configuration, and distribution of particles. The origins of this idea can be traced backed to the Swiss physicist and mathematician Daniel Bernoulli and his seminal work *Hydrodynamica* published in 1738. In it, he laid the groundwork for a *kinetic theory of gases* by postulating that gases are comprised of particles and that the motion of these particles can be felt as the kinetic energy we call *heat.* A century later, the Scottish physicist James Clerk Maxwell presented a theory of the distribution of the motion of these particles (Maxwell 1860a, b, 1871) in what is considered to be the first statistical law in physics. These ideas were developed further by others in the decades that followed, most notably

Austrian physicist Ludwig Boltzmann (1877) and American physicist Josiah Gibbs (1902), who coined the term *statistical mechanics*.

In a public lecture in Leipzig in 1905, Boltzmann made a point to link his ideas with Darwin's, proclaiming Darwin's theory of evolution as the first step toward a statistical mechanical theory of biology (see Broda 1983). But for biology, the statistical movement of particles is not a very useful concept for understanding the inheritance system. Fortunately, there is a more general and abstract language of statistical mechanics, called *information theory*, in which the fundamental units can be something other than particles or energy. Going from thermodynamics to statistical mechanics in its most general form will show us that energy and information are interconvertible: the combined metabolic-inheritance system is a functional whole where the information has an energetic basis and the energetics are specified by inherited information. Energy and information complement each other, but it is in the informational dynamics that we most clearly see how life generates the ability to live and evolve.

6.3.4 An Information View of Evolvable Life

To understand evolvable life from the standpoint of information, we first need to define "information" as a material phenomenon with a physical basis. At the beginning of formal information theory, information was defined statistically as deviations from an equiprobable distribution of variables making up any system (Shannon and Weaver 1949). If all letters of an alphabet occur at random with equal frequency in a long string, there can be no higher-order organization of letters into words, sentences, paragraphs, and so on. In this case, the system (the string of letters) is said to be at maximum entropy and there is no "information" in the system. Information is thus inversely related to the entropy (disorder) of the system. From this basis, *statistical entropy functions* were developed to measure the amount of entropy, H, in a system based on the number of different components (parts, pieces, states, etc.) and the relative distribution of these components. They take on the following generalized form:

$$H = -\sum_{i=1}^{N} p_i \log_b p_i.$$

where p_i is the proportion of type i in the system. This equation has various guises. Many biologists will recognize it as the Simpson Diversity Index or Shannon Diversity Index. In all variations, the equation is *maximized at equality*, when all types occur in equal proportions, which is the state of maximum entropy ($H_{\max}$). A maximally diverse system expresses all possibilities at once, without constraint, and this by definition means there is no order in the system. When all parts of a system are expressed at once in near-equal proportions, entropy (disorder) is high and information content (order) is low. Consider the following strings of six letters: ABAAAA versus ABBABA. In which string is it more difficult to predict the seventh letter? Answer: the second string because the diversity of letters is higher (3 A's and 3 B's vs. 5 A's and 1 B in the first string). A higher diversity of "things" means a higher uncertainty of predicting the next "thing," which is the same as saying the second string of letters has less information content than the first string.

From this foundation, information theory has developed from two general perspectives, "communications theory" and "measurement theory." In communications theory, initiated by Shannon and Weaver (1949), the amount of information sent from a "source" through a "transmission channel" to a "receiver" is calculated using a statistical entropy function, as described above. Errors in transmission can result from poor encoding at the source or from noise in the transmission channel. Meaningful information emerges as the subset of information sent through the transmission channel that is actually recorded by the receiver. In this way of thinking, all of the processes that affect transmission and reception of the information cause a *decrease* in the entropy (disorderliness) of the message from its maximal value at the source. But physically, entropy is expected to *increase* as a result of work done on a system, so either the transmission of information is not a physical process, or the communications view is nonphysical. Since biological information has a clear physical basis, this cannot be an adequate description of information for living systems.

Measurement theory provides a second way of thinking about information that gets us closer, but not all the way, to biology. The famous French physicist and information theorist Léon Brillouin (1962) proposed the idea of *bound information* to describe material properties of systems with a clear distinction between microstates and macrostates. Bound information is calculated with respect to the number and distribution of microstates (what Brillouin called "complexions") in the system, again using a statistical entropy function like described above. But contrary to communications theory, bound information emerges only in systems for which there is a clear nonarbitrary distinction between microstates and macrostates. Bound information is defined heuristically as:

$$I = H_{\max} - H_{\text{obs}}$$

where $H_{\max}$ (maximum entropy) refers to the state of the system in which all components are randomized and H_{obs} (observed entropy) refers to the actual state of the system calculated based on the observed configuration of components. Brillouin defined I—the difference between maximum possible entropy and actual realized entropy—as "negentropy," which then gets converted into bound information by measurement devices (i.e., receivers). Thus, Brillouin equated the amount of information bounded within a system with what can be measured as the difference between "what's possible" ($H_{\max}$) and "what's realized" (H_{obs}). If all possible states of a system are expressed at the same time, then $I = H_{\max} - H_{\text{obs}} = 0$ (i.e., the system is at equilibrium) and from the perspective of an observer (measuring device) the system lacks information content. As the difference between "what's possible" and

Fig. 6.4 Biological information flows as a process of self-communication. Organisms transmit information about themselves (*Source*) to themselves in the future (*Receiver*) through channels of information flow determined by reproduction and ontogeny. Redrawn and modified from Brooks and Agosta (2012)

"what's realized" in an information system grows, the amount of missing entropy ("negentropy") grows and, from a measuring device's perspective, so does the amount of information.

Like communications theory, Brillouin's measurement theory of "negentropy" leaves room for information to be an abstraction, not a material part of the system with a real physical basis. It is also not necessarily dynamical—it does not change or evolve. Both the communications and measurement perspectives move us closer to an understanding of information in biological systems, but as Brooks and Wiley (1988) recognized, the bridge to "evolvable life as an inheritance system" is incomplete. Both communications and measurement theory propose that information is (1) anything transmitted from a "source" through a "channel" to a "receiver" and (2) not necessarily a physical part of the system. But biological information has a material basis, DNA. And the information encoded in DNA flows through a form of *self-communication* system, where organisms "talk to themselves about themselves in the future" through the mechanisms of ontogeny and reproduction. In this sense, organisms are replicators that act as both source and receiver (Fig. 6.4) of information sent from themselves in the present to themselves in the future (Csanyi 1989; Kampis 1991, 1998; Brooks and McLennan 1997; Collier and Hooker 1999; Brooks and Agosta 2012). Information is meaningful in such systems to the extent that it is organized in such a way to specify the development of an organism and is transmitted to the next generation of replicators.

Life sends its past into the future through a static-filled present. Simple brute force redundancy is a feature of the system because it is so cheap, but efficient coding wins the day, and the favored algorithm is "maximize historical correlations."

Collier (1986, 1988, 1990, 1996, 1998, 2000, 2003, 2010, 2011) related information to the causal capacity of a system. Causal capacity is the ability of a system to impose distinctions on its surroundings (the nature of the organism in biology). Information is whatever converts the meaningless surroundings into something meaningful to the organism, and which allows organisms to continue to impose themselves on their surroundings. This implies that organisms are complex wholes, not just decorated replicators. They do not inherit their surroundings; they inherit the ability to interact with their surroundings in particular ways. As we said earlier, plants are not organized photons, they are organized by photons; and plant biomass is rubbish compared to the photons, which were rubbish compared to the thermonuclear system from which they emanated. This is how organisms can be causally dependent on their surroundings without being determined by them.

Collier envisioned physical information systems occurring as arrays, or multidimensional messages, where the information has (1) a material basis, (2) an energetic cost, and (3) a real nonarbitrary difference between microstates and macrostates. The material basis for biological information has been clear since the discovery of DNA, satisfying the first condition (Collier and Hooker 1999; Brooks et al. 1989; Smith 1988, 1998, 2000, 2001). The second condition is satisfied by showing that energy and information are interconvertible, which provides a direct link between the informational and physical views of entropy. Recall that the *conservative transformation* processes within biological systems are coupled with *heat-generating transformation* processes. In other words, all conservative processes that convert energy from the outside into states that can be stored and utilized in subsequent transformations produce entropy, mostly in the form of heat. It takes energy and produces entropy to make, maintain, and transmit biological information. Energy and biological information are therefore interconvertible.

Collier's third condition for physical information systems is satisfied by showing that microstate/macrostate distinctions can be determined objectively by part/whole associations. In living systems, nonarbitrary part/whole associations occur across all levels of the biological hierarchy. The number of accessible microstates is increased by the production of new components, which increases diversity at a given level and can open up the possibility of new levels of organization. For example, autocatalytic processes producing monomers make "monomer space" a level of organization. Some monomers have high chemical affinities for each other and may spontaneously clump (self-organize) into dimers and higher-order polymers. Once polymers form, "polymer space" emerges as a new level of organization in the evolving system. At this level, polymers are macrostates, and monomers and dimers are microstates. New levels of organization emerge from interactions among polymers; now polymers are the microstates that combine to make new macrostates, and so on. Protein coding units could be macrostates, while all actual sequences that code for proteins would be microstates. Loci could be macrostates, while all alleles that correspond to loci would be microstates. Phenotypes could be macrostates, while all genotypes that correspond to phenotypes would be microstates, and so on. The linked phenomena of the emergence of new levels of organization and expansion of the realm of possibilities within a growing biological "space" is what Kauffman (1993, 1995, 2000) called the "adjacent possible."

6.3.5 Temporal Dynamics of Biological Information

Brooks and Wiley (1988) proposed a general theory of biological information that accounts for its material basis, flows through a self-communication system of replicators, and dynamical nature. Their theory has its roots in the physics of nonequilibrium thermodynamics used to explain cosmological evolution and the emergence of structure in the expanding universe (e.g., Layzer 1975; Frautschi 1982, 1988; Landsberg 1984a, b). It is an extension of classical statistical mechanics used to compute *configurational entropy*, called *algorithmic entropy* or Kolmogorov Complexity (Kolmogorov 1965, 1968) in which the fundamental units, instead of particles of energy and matter, are biological "parts." A key feature that emerges from their theory is linear or sequential time marked by the type of changes that are the hallmark of inheritance systems and provide our sense of past and future. To start, they used Chaitin–Komolgorov Algorithmic Information Theory (Chaitin 1975), in which computational complexity is directly proportional to entropy and information, to summarize biological information based on three key elements of diversity: the number of kinds of parts, the relative frequency of the different kinds of parts, and changes in the number and relative frequency of kinds of parts. This simple entropy function has four key formulations:

1. The *observed informational entropy* (H_{obs}) that is actually expressed in the system. This is calculated based on the observed distribution of components in the system at any given time. It is synonymous with "what's realized" or the *Expressed Information Content* of the system.
2. The *maximum possible informational entropy* (H_{max}) that could be expressed in the system in its totally relaxed state (i.e., without constraints). This is calculated based on all components of the system being distributed equiprobably throughout the system at any given time. It is synonymous with "what's possible" or the *Potential Information Capacity* of the system.
3. The amount of *information* (I) in the system. This is calculated as the absolute difference between the maximum and observed entropy ($I = H_{max} - H_{obs}$). It is analogous to Brillouin's "negentropy" and can be taken as a macroscopic measure of the degree of organization in the system (Brillouin 1962; Gatlin 1972; Layzer 1975; Landsberg 1984a, b).
4. Two conceptually related relative differences between maximum entropy and observed entropy. The first is *order* (Landsberg 1984a, b) calculated as the ratio of observed entropy to maximum entropy (H_{obs}/H_{max}). The second is *redundancy* (Gatlin 1972) calculated as the ratio of *information density* to maximum entropy (I_d/H_{max}), where $I_d = H_{max} - H_{obs}$.

Brooks et al. (1984) modeled organisms as linear inheritance systems—temporal strings—and simulated evolution based on two simple rules: (1) the rate of replication is higher than the rate of mutation and (2) whatever happens first has an impact on what happens later. Using the statistical entropy function to keep track of changes in information content in the system through time, they discovered three generalities about the temporal dynamics of biological inheritance illustrated in Fig. 6.5:

1. Both Expressed Information Content (observed entropy; H_{obs}) and Potential Information Capacity (maximum entropy; H_{max}) increase as a function time. As linear temporal processes unfold the system evolves and diversifies (entropy increases), and the realms of "what's possible" and "what's realized" grow in tandem.

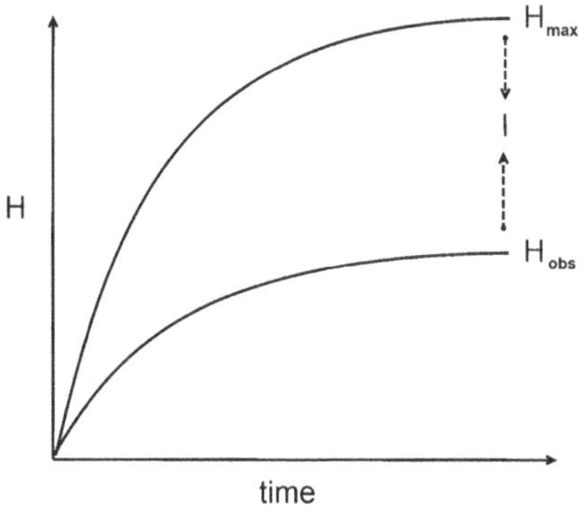

Fig. 6.5 Temporal dynamics of biological information. The entropic growth of potential informational capacity (H_{max}) in living systems produces information (I) proportional to the constraints on realized expressed information content (H_{obs}). Redrawn and modified from Brooks and Wiley (1988)

2. Both Expressed Information Content (H_{obs}) and Potential Information Capacity (H_{max}) are *concave* functions of time, as suggested mathematically by Landsberg (1984a, b; see also Danielli et al. 2003). Although entropy production continues to increase as the system unfolds, the rate at which this occurs slows down as *historical constraints* build up in the system. In other words, diversity keeps increasing, but the accumulation of constraints in the ways the parts of the system can be assembled slows down the accumulation of new diversity. For example, additive genetic variation can be viewed as population-level entropy (genetic diversity), while genetic correlations can be viewed as organizing principles that constrain the expression of that variance (e.g., gene A requires gene B to function, and so on). Likewise, species diversity can be viewed as higher-level entropy in which interspecific associations act as organizing principles (e.g., plant A needs insect B to reproduce, metazoan X needs microbiome Y to survive) that constrain the expression of food webs, ecosystems, and on up to the biosphere. At all levels of biological organization, the accumulation of historical correlations is the "cost" of greater integration among parts of the system and a major constraint on both "what's realized" and "what's possible" at any given time.
3. The difference between Expressed Information Content (H_{obs}) and Potential Information Capacity (H_{max}) is an increasing function of time. This difference is a measure of the amount of Information ($I = H_{max} - H_{obs}$) in the system and is indicative of the emergence and growth of organized complexity (see also Collier and Hooker 1999; Roach et al. 2018). As a corollary, the relative difference between H_{max} and H_{obs}—known as "order" (Landsberg 1984a, b) or "redundancy" (Gatlin 1972)—is also an increasing function of time.

These generalities reveal a fundamental insight into the dual nature of time in biological systems: entropy/diversity *and* negentropy/information/organization/order *all* increase over time but at a decreasing rate as emerging historical correlations act as fluxes constraining the system. This perspective was discussed in the context of evolutionary biology in the latter part of the twentieth century (e.g., Brooks and Wiley 1986, 1988; Smith 1988, 1998, 2000, 2001; Brooks et al. 1984, 1988, 1989; Brooks and McLennan 1990, 1991, 2000; Brooks 1992, 1994, 1997,

1998, 2000, 2001, 2010, 2011a, b, c). Similar ideas have emerged independently in a broader context during the early part of the twenty-first century (e.g., Danielli et al. 2003; Verlinde 2011; Wolpert 2013; Cheng and Liang 2014; Mitrokhin 2014; Bartolotta et al. 2016; Van den Broeck and Esposito 2015; Annila and Baverstock 2016; Adams et al. 2017; Sciubba and Zullo 2017; Wang 2017; Alves and Pumariño 2018; Kolchinsky and Wolpert 2018; Dikranjan and Bruno 2019; Omidvarnia et al. 2018; Palazzo 2018; Roberts et al. 2019; Tozzi and Peters 2019).

The simulations thus corroborated Komolgorov's perspective that entropy and complexity are directly, not inversely, proportional. Increasing complexity is an expected outcome of the second law of thermodynamics, and biological evolution, driven by inheritance, is fundamentally a system of increasing complexity driven by increasing entropy. And it is the historical nature of biological inheritance that creates the lag between "what's realized" and "what's possible" ($H_{max} - H_{obs} > 0$) that permits the growth of structure and organization, with the difference getting larger as the system evolves. Or, as Darwin recognized, "diversity begets diversity."

6.4 Summary

We're all children of Time.—Ursula Le Guin (1974)

Darwin identified the essence of life as a drive for indefinite persistence through continued reproduction. The rules of physics set the stage for the challenges faced by this "drive to stay alive," and living systems provided the solution to those challenges. When you have metabolism + information surrounded by a barrier (Ganti's *Chemoton* model), the system can be far from equilibrium with no necessity of a local equilibrium [*Onsager reciprocity* (Onsager 1931) no longer holds] and material information systems can emerge in which the autocatalytic arrow, the thermodynamic arrow and the information arrow all point in the same direction. Evolvable life is contingent, existing between the improbable and the inevitable, between memory and imagination. It is beautiful and wonderful, all the more so because it is not magical.

In an entropic universe governed by the laws of thermodynamics, biological systems make the time in which they exist, and have to pay for it. Metabolism buys living systems circular time to *exploit* their surroundings, taking advantage of energy or "resource" gradients in the environment to fuel the maintenance of their own organization (Eldredge 1986; Wicken 1987; Ulanowicz 1986, 1997; Matsuno 1989, 1995, 1996, 1998, 2000; Hirata 1993; Salthe 1993; Schneider and Kay 1994; Depew and Weber 1995).

Inheritance buys living systems linear time to *explore* their surroundings. These are the processes that carry life forward and produce our sense of a distinction between past and future—ontogeny, growth, inheritance, reproduction, evolution. Being a good exploiter is about being well suited to the current conditions, while being a good explorer is about being prepared for coping with future contingencies.

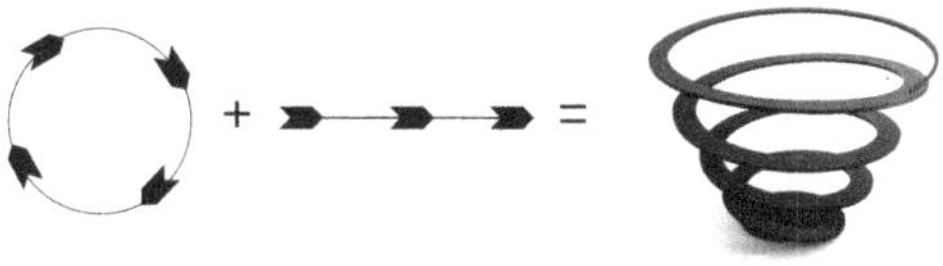

Fig. 6.6 Visual metaphor for the temporal dualism in living systems. Life creates circular time (through metabolism) and linear time (through inheritance) as the result of its own entropic behavior. These combine to produce evolutionary time, represented as an expanding helix (Brooks and Agosta 2012)

If life were only good at exploiting the surroundings, then perhaps it could maintain itself in a spatially-localized but timeless existence so long as the conditions do not change. But once the conditions changed, there would be no capacity to respond. To be responsive to changes in the surroundings, life also needs to be a good explorer (Kováč 2007; Popadiuk 2011). The capacity for indefinite exploration emerges and grows naturally in living systems as a result of the nature of the inheritance system. Inheritance systems accumulate historical correlations that slow down the expansion of the system, ensuring that "what's realized" always lags behind "what's possible." By binding negentropy in the system long enough to create a lag in the rate of entropy production, life "slows the flow," creates new opportunities for itself, and generates its own arrow of linear time. Organisms are truly children of time.

Whether organisms are buying time that is already immanent in the universe or buying new time is a philosophical question beyond our capabilities. Fortunately, we do not believe this is an impediment to providing a general metaphorical framework for understanding evolvable life. A useful visual metaphor for the temporal duality of evolvable life is what is known as the Great Circle of Life or "helix" in classical Greek philosophy (Fig. 6.6). This symbol occurs carved and painted on rocks and in caves throughout the world, showing that humans understood the duality of time long before the classical Greek philosophers wrote about it. For us, it is symbolic of the unfolding biosphere embedded in its own time from its inception at a possible Biological Big Bang (Koonin 2007; David and Alm 2011; Drosera and Gehling 2015; Slater 2015; Nutman et al. 2016; Morris et al. 2018; Betts et al. 2018; Garcia-Pichel et al. 2019; Staps et al. 2019) to its expansion into a future largely of its own making.

A somewhat poetic way to sum up the temporal metaphors of life might be

If there is no circle of life, there cannot be life.
But a circle of life alone cannot persist.
The circle fights against the second law, the line draws it out.
And the struggle for existence is the outcome.

The fundamental point of this part of our story is, as Maynard Smith and Szathmàry (1995) recognized, that any theory of "evolvable life" requires integrating both metabolism and inheritance. Both are essential for life to persist. It is far cheaper for an organism to reproduce than continue to maintain itself, and therefore cheaper for life to evolve than remain static. So much cheaper that once life originated as open thermodynamic autocatalytic systems of metabolism and inheritance inside a membrane, it became overwhelmingly probable that evolution would occur.

Life is astonishing, but not magical; it is not improbable, it is contingent. The Information Age began more than 3.5 billion years ago with the Big Bang origin of evolvable life. And there was time then even though we were not there to measure it. But the temporal dimensions of life are only part of the story. Buying cyclical time (metabolism) is largely about *exploiting* the surroundings; buying linear time (inheritance) is largely about *exploring* the surroundings. In order to persist indefinitely, life needs to be evolvable, and to be evolvable, life needs to constantly exploit without losing the ability to explore (see also Page 2011). A good visual metaphor of the exploitation–exploration duality is a bicycle, made of nothing but circular components, none of which "go anywhere" but without which the bicycle would not be capable of moving directionally and exploring.

But what, exactly do living systems explore? Like an expedition to discover new lands, life must extract sufficient energy from its surroundings all along the way to propel itself forward into an uncertain future. Upon life's inception, at the moment of the Biological Big Bang, the combined metabolic-inheritance system opened up temporal production which immediately catalyzed the expansion of the *space* that life continues to explore. Ulanowicz (1997) referred to that space as life's "window of vitality." The material basis for the window of vitality, and for its growth, will be addressed in the next chapter.

References

Adams AM, Berner A, Davies PCW, Walker SI (2017) Physical universality, state-dependent dynamical laws and open-ended novelty. Entropy 19:461

Alves JF, Pumariño A (2018) Entropy formula and continuity of entropy for piecewise expanding maps. arXiv preprint arXiv:1806.01095

Andrade E (2000) From external to internal measurement: a form theory approach to evolution. Biosystems 57:49–62

Annila A (2015) Natural thermodynamics. Entropy 17:6995–7020. https://doi.org/10.3390/e170x000x

Annila A, Annila E (2008) Why did life emerge? Int J Astrobiol 7:293–300

Annila A, Baverstock K (2016) Discourse on order vs. disorder. Commun Integr Biol 9:e1187348–e1187342

Annila A, Kuismanen E (2009) Natural hierarchy emerges from energy dispersal. Biosystems 95:227–233

Annila A, Salthe S (2010) Physical foundations of evolutionary theory. J Non-Equilib Thermodyn 35:301–321

Annila A, Salthe S (2012a) Threads of time. Int Sch Res Netw Thermodyn 2012:850957

Annila A, Salthe S (2012b) On intractable tracks. Phys Essays 25:233–238

Arnellos A, Moreno A (2012) How functional differentiation originated in prebiotic evolution. Ludus Vitalis 20:1–23

Bai S, Ge H, Qian H (2018) Structure for energy cycle: a unique status of the second law of thermodynamics for living systems. Sci China Life Sci 61:1266–1273

Bartolotta A, Carroll SM, Leichenauer S, Pollack J (2016) Bayesian second law of thermodynamics. Phys Rev E 94:022102

Bauer E (1920a) Die definition des Lebewesens auf Grund seiner thermodynamischen Eigenschaften und die daraus folgenden biologischen Grundprinzipien [Definition of living organisms on the ground of their thermodynamic features and the ensuing basic biological principles]. Naturwissenschaften 8:338–340

Bauer E (1920b) Zur Bemerkungen von L. Ebert über meinen Artikel in Heft 18 d.J. [On the comments to my paper by L. Ebert in number 18 of this year]. Naturwissenschaften 8:582

Betts HC, Puttick MN, Clark JW, Willliams TA, Donoghue PCJ, Pisani D (2018) Integrated genomic and fossil evidence illuminates life's early evolution and eukaryote origin. Nat Ecol Evol 2:1556–1562

Bich L, Mossio M, Ruiz-Mirazo K, Moreno A (2012) Biological regulation: controlling the system from within. Biol Philos 31:237–265

Blum HF (1968) Time's arrow and evolution, 3rd edn. Princeton University Press, Princeton

Boltzmann L (1877) Uber die Beziehung eines allgemeine mechanischen Satzes zum zweiten Haupsatzes der Warmtheorie. Sitzungsber Akad Wiss Wien, Math-Nat Kl 75:67–73

Brillouin L (1962) Science and information theory, 2nd edn. Academic, New York

Broda E (1983) Darwin and Boltzmann. In: Geissler E, Scheler W (eds) Darwin today: the 8th Kühlungsborn colloquium on philosophical and ethical problems of biosciences. Abhandlungen der Akademien der Wissenschaften der DDR. Akademie, Berlin, pp 61–70

Brooks DR (1992) Incorporating origins into evolutionary theory. In: Varela F, Dupuy JP (eds) Understanding origins: contemporary ideas on the genesis of life, mind and society. Reidel/ Kluwer Associates, Amsterdam, pp 191–212

Brooks DR (1994) Entropy, information and evolving biological systems. Theor Hist Scient 4:31–49

Brooks DR (1997) Biological evolution as a microcosm of cosmological evolution. Bridges 4:9–35

Brooks DR (1998) The unified theory of evolution and selection processes. In: van de Vijver G, Salthe SN, Delpos M (eds) Evolutionary systems: biological and epistemological perspectives on selection and self-organization. Kluwer Academic, Dordrecht, pp 113–128

Brooks DR (2000) The nature of the organism: life takes on a life of its own. Proc NY Acad Sci 901:257–265

Brooks DR (2001) Evolution in the information age: rediscovering the nature of the organism. Semotics Evol Energy Dev 1:1–26. Available at: http://www.library.utoronto.ca/see

Brooks DR (2010) The mastodon in the room: how Darwinian is neo-Darwinism? Stud Hist Phil Biol Biomed Sci 42:82–88

Brooks DR (2011a) The extended synthesis: something old, something new. Evol Edu Outreach 4:3–7

Brooks DR (2011b) The extended synthesis: the law of the conditions of existence. Evol Edu Outreach 4:254–261

Brooks DR (2011c) The major transitions of evolution: visualizing the extended synthesis. Evol Edu Outreach 4:446–452

Brooks DR, Agosta SJ (2012) Children of time: the extended synthesis and major metaphors of evolution. Fortschr Zool 29:497–514

Brooks DR, McLennan DA (1990) Searching for a general theory of biological evolution. J Ideas 1:35–46

Brooks DR, McLennan DA (1991) Phylogeny, ecology and behavior: a research program in comparative biology. University of Chicago Press, Chicago

Brooks DR, McLennan DA (1997) Biological signals as material phenomena. Rev pensee d'aujord d'hui 25:118–127. [in Japanese]

Brooks DR, McLennan DA (2000) The nature of the organism and the emergence of selection processes and biological signals. In: Taborsky E (ed) Semiotics, evolution, energy. Shaker, Aachen, pp 185–218

Brooks DR, McLennan DA (2002) The nature of diversity: an evolutionary voyage of discovery. University of Chicago Press, Chicago
Brooks DR, Wiley EO (1986) Evolution as entropy: toward a unified theory of biology, 1st edn. University of Chicago Press, Chicago
Brooks DR, Wiley EO (1988) Evolution as entropy: toward a unified theory of biology, 2nd edn. University of Chicago Press, Chicago
Brooks DR, LeBlond PH, Cumming DD (1984) Information and entropy in a simple evolution model. J Theor Biol 109:77–93
Brooks DR, Cumming DD, LeBlond PH (1988) Dollo's law and the second law of thermodynamics: analogy or extension? In: Weber BH, Depew DJ, Smith JD (eds) Information, entropy and evolution: new perspectives on physical and biological evolution. MIT, Cambridge, pp 189–224
Brooks DR, Collier J, Maurer BA, Smith JDH, Wiley EO (1989) Entropy and information in evolving biological systems. Biol Philos 4:407–432
Buffet J (1995) "Don't Chu-know"
Chaitin GJ (1975) A theory of program size formally identical to information theory. J ACM 22:329–340
Cheng XT, Liang XG (2014) Entransy: its physical basis, applications and limitations. Chin Sci Bull 59:5309–5323
Collier J (1986) Entropy in evolution. Biol Philos 1:5–24
Collier J (1988) The dynamics of biological order. In: Weber BH, Depew DJ, Smith JD (eds) Information, entropy and evolution: new perspectives on physical and biological evolution. MIT, Cambridge, pp 227–242
Collier J (1990) Two faces of Maxwell's demon reveal the nature of irreversibility. Stud Hist Philos Sci 21:257–268
Collier J (1996) Information originates in symmetry breaking. Symmetry: Sci Cult 7:247–256
Collier J (1998) Information increase in biological systems: how does adaptation fit? In: van de Vijver G, Salthe SN, Delpos M (eds) Evolutionary systems: biological and epistemological perspectives on selection and self-organization. Kluwer Academic, Dordrecht, pp 129–140
Collier J (2000) The dynamical basis of information and the origins of semiosis. In: Taborsky E (ed) Semiotics, evolution, energy. Shaker, Aachen, pp 111–138
Collier J (2003) Hierarchical dynamical information systems with a focus on biology. Entropy 5:100–124
Collier J (2010) A dynamical approach to identity and diversity in complex systems. In: Cilliers P, Preiser R (eds) Complexity, difference and identity: an ethical perspective. Springer, Berlin, pp 83–97
Collier J (2011) Information, causation and computation. In: Crnkovic GD, Burgin M (eds) Information and computation: essays on scientific and philosophical understanding of foundations of information and computation (World scientific series in information studies). World Scientific, Singapore, pp 89–105
Collier J, Hooker C (1999) Complexly organised dynamical systems. Open Syst Inf Dyn 6:241–302
Csanyi V (1989) Evolutionary systems and society: a general theory. Duke University Press, Durham
Danielli D, Garofalo N, Nhieu D-M (2003) Notions of convexity in Carnot groups. Commun Anal Geom 11:263–341
Darwin C (1872) Origin of species. John Murray, London
David LA, Alm EJ (2011) Rapid evolutionary innovation during and Archaean genetic expansion. Nature 469:93–96
Denbigh KG (1975) A non-conserved function for organized systems. In: Kubat L, Zemen J (eds) Entropy and information in science and philosophy. American Elsevier, New York, pp 83–92
Depew D, Weber B (1995) Darwinism evolving. Bradford Books, Cambridge
Dewar RC (2005) Maximum entropy production and the fluctuation theorem. J Phys A Math Gen 38:L371

Dikranjan D, Bruno AG (2019) Entropy on normed semigroups (A unifying approach to entropy). Dissertationes Mathematicae 542. https://doi.org/10.4064/dm791-2-2019
Drosera ML, Gehling JG (2015) The advent of animals: the view from the Ediacaran. Proc Natl Acad Sci USA 112:4865–4870
Eigen M (1971) Self-organization of matter and the evolution of biological macromolecules. Naturwissenschaften 48:465–522
Eigen M, Shuster P (1978) The hyper-cycle: a principle of natural self-organization. Part C. The realistic hyper-cycle. Naturwissenschaften 65:341–369
Eldredge N (1986) Information, economics and evolution. Ann Rev Ecol Syst 17:351–369
Elek G, Müller M (2013) The living matter according to Ervin Bauer (1890–1938) (on the 75th anniversary of his tragic death) (History). Acta Physiol Hung 100:124–132
Frautschi S (1982) Entropy in an expanding universe. Science 217:593–599
Frautschi S (1988) Entropy in an expanding universe. In: Weber B, Depew DJ, Smith JD (eds) Entropy, information and evolution: new perspectives on physical and biological evolution. MIT, Cambridge, pp 11–22
Gánti T (1971) Az élet principuma (The principle of life). Gondolat, Budapest. (in Hungarian)
Gánti T (1974) Theoretical deduction of the function and structure of the genetic material. Biologia 22:17–35
Gánti T (1979) A theory of biochemical supersystems and its application to problems of natural and artificial biogenesis. University Park Press, Baltimore
Gánti T (1987) The principle of life, 6th edn. OMIKK, Budapest
Gánti T (2003) The principles of life. Oxford University Press, Oxford
Garcia-Pichel F, Lombard J, Soule T, Dunaj S, Wu SH, Wojciechowski MF (2019) Timing the evolutionary advent of cyanobacteria and the later great oxidation event using gene phylogenies of a sunscreen. mBio 10:e00561-19. https://doi.org/10.1128/mBio.00561-19
Gatlin LL (1972) Information theory and the living system. Columbia University Press, New York
Giangaspero G, Sciubba E (2013) Application of the entropy generation minimization method to a solar heat exchanger: a pseudo-optimization design process based on the analysis of the local entropy generation maps. Energy 58:52–65
Gibbs JW (1902) Elementary principles in statistical mechanics, developed with especial reference to the rational foundation of thermodynamics. Dover, New York
Grmela M, Pavelka M, Klika V, Cao BY, Bendian N (2019) Entropy and entropy production in multiscale dynamics. J Non-Equilib Thermodyn 44:217–233
Hartonen T, Annila A (2012) Natural networks as thermodynamic systems. Complexity 18:53–62
Hirata H (1993) Information of organization in ecological systems: nutrient > energy > carbon. J Theor Biol 162:187–194
Jaakkola S, El-Showk S, Annila A (2008) The driving force behind genomic diversity. Biophys Chem 134:232–238
Jacob (1977) Evolution and tinkering. Science 196:1161–1166
Juarrero (1999) Dynamics in action. MIT, Boston
Kaila VRI, Annial A (2008) Natural selection for least action. Proc R Soc Ser A 464:3055–3070
Kampis G (1991) Self-modifying systems in biology and cognitive science: a new framework for dynamics, information and complexity. Pergamon, Oxford
Kampis G (1998) Evolution as its own cause and effect. In: van de Vijver G, Salthe SN, Delpos M (eds) Evolutionary systems: biological and epistemological perspectives on selection and self-organization. Kluwer Academic, Dordrecht, pp 255–265
Kauffman SA (1993) The origins of order: self-organization and selection in evolution. Oxford University Press, Oxford
Kauffman SA (1995) At home in the universe: the search for the laws of self-organization and complexity. Oxford University Press, Oxford
Kauffman SA (2000) Investigations. Oxford University Press, Oxford
Kolchinsky A, Wolpert DH (2018) Semantic information, autonomous agency and non-equilibrium statistical physics. Interface Focus 8:20180041. https://doi.org/10.1098/rsfs

Kolmogorov AN (1965) Three approaches to the quantitative definition of information. Problems of inform. Transmission 1:1–7
Kolmogorov AN (1968) Logical basis for information theory and probability theory. IEEE Trans Inf Theory 14:662–664
Koonin EV (2007) The biological big bang model for the major transitions in evolution. Biol Direct 2:21. https://doi.org/10.1186/1745-6150-2-21
Kováč L (2007) Information and knowledge in biology: time for reappraisal. Plant Signal Behav 2:65–73
Landsberg PT (1984a) Is equilibrium always an entropy maximum? J Stat Phys 35:159–169
Landsberg PT (1984b) Can entropy and "order" increase together? Phys Lett 102A:171–173
Layzer D (1975) The arrow of time. Sci Am 233:56–69
Le Guin U (1969) The left hand of darkness. Ace Books, New York
Le Guin U (1974) The dispossessed. Harper Collins, New York
Lincoln TA, Joyce GF (2009) Self-sustained replication of an RNA enzyme. Science 323:1229–1232
Liu D (1989) Evolution mechanism: flow-resistance cycles. Proc Int Symp Pac Neogene Cont Mar Events 223–231
Liu D (1990) Outline of systems geology and its flow-resistance outlook on time-space. Acta Micropalaeontol Sin 7:83–90
Lotka AJ (1913) Evolution from the standpoint of physics, the principle of the persistence of stable forms. Sci Am Supp 75:345-6, 354, 379
Lotka AJ (1925) Elements of physical biology. Williams and Wilkins, Baltimore
Lucia U (2016) Considerations on nonequilibrium thermodynamics of interactions. Phys A: Stat Mech Appl 447:314–331
Mahulikar SP, Herwig H (2009) Exact thermodynamic principles for dynamic order existence and evolution in chaos. Chaos, Solitons Fractals 41:1939–1948
Matsuno K (1989) Protobiology: physical basis of biology. CRC, Boca Raton
Matsuno K (1995) Consumer power as the major evolutionary force. J Theor Biol 173:137–145
Matsuno K (1996) How many trophic levels are there? J Theor Biol 180:105–109
Matsuno K (1998) Competence of natural languages for describing the physical origin of life. In: van de Vijver G, Salthe SN, Delpos M (eds) Evolutionary systems: biological and epistemological perspectives on selection and self-organization. Kluwer Academic, Dordrecht, pp 295–306
Matsuno K (2000) Material contextualization in time. In: Taborsky E (ed) Semiotics, evolution, energy. Shaker, Aachen, pp 219–230
Maurer BA, Brooks DR (1991) Energy flow and entropy production in biological systems. J Ideas 2:48–53
Maxwell JC (1860a) Illustrations of the dynamical theory of gases. Part I. On the motions and collisions of perfectly elastic spheres. Philos Mag 19:19–32
Maxwell JC (1860b) Illustrations of the dynamical theory of gases. Part II. On the process of diffusion of two or more kinds of moving particles among one another. Philos Mag 20:21–37
Maxwell JC (1871) Theory of heat. Longmans, Green, London
Maynard Smith J (1976) What determines the rate of evolution? Am Nat 110:331–338
Maynard Smith J, Szathmàry E (1995) The major transitions in evolution. W.H. Freeman, Oxford
McLennan DA (2008) The concept of co-option: why evolution often looks miraculous. Evol Edu Outreach 1:246–258
Mitrokhin Y (2014) Two faces of entropy and information in biological systems. J Theor Biol 359:192–198
Moreno A, Mossio M (2015) Biological autonomy: a philosophical and theoretical enquiry. Springer, Dordrecht
Morris JL, Puttick MN, Clark JW, Edwards D, Kenrick P, Pressel S, Wellmane CH, Yang Z, Harald Schneider H, Donoghue PCJ (2018) The timescale of early land plant evolution. Proc Nat Acad Sci 115. https://doi.org/10.1073/pnas.1719588115

Neiman S (2002) Evil in modern thought: an alternative history of philosophy. Princeton University Press, Princeton

Nutman AP, Bennett VC, Friend CLR, Van Kranendonk MJ, Chivas AR (2016) Rapid emergence of life shown by 3,700-million-year-old microbial structures. Nature 537:535–538

Odum HT (1983) Systems ecology. Wiley, New York

Omidvarnia A, Mesbah M, Pedersen M, Jackson G (2018) Range entropy: a bridge between signal complexity and self-similarity. Entropy 20:962. https://doi.org/10.3390/e20120962

Onsager L (1931) Reciprocal relations in irreversible processes. I. Phys Rev 37:405–426

Page SE (2011) Diversity and complexity. Princeton University Press, Princeton

Palazzo P (2018) Hierarchical structure of generalized thermodynamic and informational entropy. Entropy 20:553. https://doi.org/10.3390/e20080553

Popadiuk S (2011) Scale for classifying organizations as explorers, exploiters or ambidextrous. Int J Inf Manag 32:75–87

Prigogine I (1980) From being to becoming. W. H. Freeman, San Francisco

Prigogine I, Wiame JM (1946) Biologie et thermodynamique des phénomènes irréversìbles. Experientia 2:451–453

Rastegar-Sedehi HR, Radhakrishnan C, Nehme SI, Birman L, Lau PMV, Byrnes T (2018) Nonequilibrium time dynamics of genetic evolution. Phys Rev E 98:022403

Roach TNF, Salomon P, Nulton J, Andresen B, Felts B, Haas A, Calhoun S, Robinett N, Rohwer F (2018) Application of finite time and control thermodynamics to biological processes at multiple scales. J Non-Equilib Thermodyn 43:193–210

Roberts E, Sindi S, Smith SA, Mitchell KA (2019) Ensemble-based topological entropy calculation (E-tec). Chaos 29:013124

Ruiz-Mirazo K, Briones C, de la Escosura A (2017) Chemical roots of biological evolution: the origins of life as a process of development of autonomous functional systems. Open Biol 7:170050

Salthe SN (1993) Development and evolution: complexity and change in biology. MIT, Boston

Salthe S (2010) Maximum power and maximum entropy production: finalities in nature. Cosmos Hist: J Nat Soc Philos 6:114–121

Schneider ED, Kay JJ (1994) Life as a manifestation of the second law of thermodynamics. Math Comput Model 19:25–48

Schrödinger E (1945) What is life? Cambridge University Press, Cambridge

Sciubba E, Zullo F (2017) A novel derivation of the time evolution of the entropy for macroscopic systems in thermal non-equilibrium. Entropy 19:594. https://doi.org/10.3390/e19110594

Shakespeare W (written between 1599–1602, first performed 1609) The tragedy of hamlet, Prince of Denmark

Shannon CE, Weaver WJ (1949) The mathematical theory of communication. University of Illinois Press, Urbana

Sharma V, Annila A (2007) Natural process – natural selection. Biophys Chem 127:123–128

Slater GJ (2015) Not-so-early bursts and the dynamic nature of morphological diversification. Proc Nat Acad Sci USA 112:3595–3596

Smith JDH (1988) A class of mathematical models for evolution and hierarchical information theory. Inst Math Appl Preprint Series 396:1–13

Smith JDH (1998) Canonical ensembles, competing species, and the arrow of time. In: Van de Vijver G, Salthe SN, Delpos M (eds) Evolutionary systems: biological and epistemological perspectives on selection and self-organization. Kluwer Academic, Dordrecht, pp 141–154

Smith JDH (2000) On the evolution of semiotic capacity. In: Taborsky E (ed) Semiotics, evolution, energy. Shaker, Aachen, pp 283–309

Smith JDH (2001) Some observations on the concepts of information-theoretic entropy and randomness. Entropy 3:1–11. https://doi.org/10.3390/e3010001

Staps M, van Gestel J, Tarnita CE (2019) Emergence of diverse life cycles and life histories at the origin of multicellularity. Nat Ecol Evol 3:1197–1208. https://doi.org/10.1038/s41559-019-0940-0

Szathmàry E (2015) Toward major transition theory 2.0. Proc Natl Acad Sci USA 112:10104–10111
Szathmàry E, Demeter L (1987) Group selection of early replicators and the origin of life. J Theor Biol 128:463–486
Tessera M (2011) Origin of evolution *versus* origin of life: a shift of paradigm. Int J Mol Sci 12:3445–3458
Tetz VV, Tetz GV (2020) A new biological definition of life. Biomol Concepts 11:1–6
Tozzi A, Peters JF (2019) Entropy balance in the expanding universe: a novel perspective. Entropy 21:406. https://doi.org/10.3390/e21040406
Tuisku P, Pernu TK, Annila A (2009) In the light of time. Proc R Soc Ser A 465:1173–1198
Ulanowicz RE (1986) Growth & development: ecosystems phenomenology. Springer, New York
Ulanowicz RE (1997) Ecology: the ascendent perspective. Columbia University Press, New York
Urquhart H (1860) Introduction to popular tales of the west highlands: orally collected with a translation by J.F. Campbell. Edmonston and Douglas, Edinburgh
Van den Broeck C, Esposito M (2015) Ensemble and trajectory thermodynamics: a brief introduction. Phys A: Stat Mech Appl 418:6–16
Vasas V, Szathmáry E, Santos M (2010) Lack of evolvability in self-sustaining autocatalytic networks constrains metabolism-first scenarios for the origin of life. Proc Natl Acad Sci USA 107:1470–1475
Verlinde E (2011) On the origin of gravity and the laws of Newton. J High Energ Phys 29. https://doi.org/10.1007/JHEP04(2011)029
Vitas M, Dobovišek A (2019) Towards a general definition of life. Orig Life Evol Biosph 49:77–88
Wang LS (2017) The second law: from Carnot to Thomson-Clausius, to the theory of exergy, and to the entropy-growth potential principle. Entropy 19:57. https://doi.org/10.3390/e19020057
Wicken JS (1987) Evolution, thermodynamics and information: extending the Darwinian paradigm. Oxford University Press, Oxford
Wolpert DH (2013) Information width: a way for the second law to increase complexity. In: Lineweaver CH, Davies PCW, Ruse M (eds) . Cambridge University Press, Complexity and the arrow of time, pp 246–275
Zheng Y, Yu H, Du J (2019) The dual formalisms of nonextensive thermodynamics for open systems with maximum entropy principle. Contin Mech Thermodyn 31:1503–1516
Zintzaras E, Santos M, Szathmáry E (2002) "Living" under the challenge of information decay: the stochastic corrector vs. hypercycles. J Theor Biol 217:167–181
Zotin AI, Zotina RS (1978) Experimental basis for qualitative phenomenological theory of development. In: Lamprecht I, Zotin AI (eds) Thermodynamics of biological processes. de Gruyter, Berlin, pp 61–84

Chapter 7
Making Space

Abstract In the process of "buying" time, life not only disturbs material space but creates an abstract space specifying the capacity to engage functionally with the surroundings. This *capacity space* emerges from the inheritance system, which provides the material information and therefore causal capacity for organisms to impose themselves on their surroundings for survival and reproduction. The realm of all possible inheritances represents *potential capacity*; the realm of actual inheritances is *realized capacity*. In the evolving expanding space, realized capacity grows but potential capacity grows even more, ensuring an ever-present realm of possibilities—an "adjacent possible"—for the inheritance system to explore given opportunities presented by the nature of the conditions. *Opportunity space* emerges from the conditions, but capacity space (nature of the organism) limits how these opportunities may be used. *Fitness space* is the intersection between capacity and opportunity, the subset of realized opportunity space that supports survival and reproduction. The portion of *fundamental fitness space* accessed by organisms at any given time and place is *realized fitness space*; the difference between these is proportional to how "sloppy" fitness space is, i.e., how much capacity there is to do something new when conditions change. Within *sloppy fitness space*, the historically conservative and largely autonomous nature of inheritance produces reproductive over-run, creating natural selection in proportion to the degree of *conflict* between organisms and the environment (i.e., Darwin's Necessary Misfit). The primary mechanism for life to resolve this conflict is *ecological fitting*, an umbrella term for a fundamental phenomenon that allows living systems to cope with the conditions by exploring novel portions of fitness space using preexisting information they inherited from their ancestors.

> ...life...a mad and futile ferment of substances meant originally to occupy space without disturbing it—Rex Stout (1935)

In the process of buying time, living systems also disturb space. They stay alive by *exploiting* resources in their material surroundings, degrading them as a result. Their offspring, therefore, must *explore* the same or different surroundings in search

S. J. Agosta, D. R. Brooks, *The Major Metaphors of Evolution*, Evolutionary Biology – New Perspectives on Its Development 2,
https://doi.org/10.1007/978-3-030-52086-1_7

of resources of their own. Biologists have observed the interactions between living systems and their surroundings closely for more than a century. This has led to four generalities: (1) each living system is capable of using only a small part of the environmental resources capable of supporting life; (2) those restricted set of abilities are inherited from previous generations, not endowed by the surroundings; (3) the inheritance of abilities to use portions of the surroundings are, like all inheritance, conservative, so the more closely related living systems are, the more likely it is they have the same or similar requirements and abilities; and (4) despite a high degree of conservatism in the inheritance of abilities, evolutionary history—phylogeny—shows a pattern of diversification in the abilities to use parts of the surroundings. Inheritance thus produces not only organisms that vary in their appearance, but also organisms that vary in the functions they can perform, and in their ability to perform a given set of functions.

Inheritance systems, therefore, do not simply disturb the material surroundings in which they reside and upon which they are dependent. Their activities produce an abstract space within the material space of basic existence that is responsible for the production, persistence, and proliferation of the nature of the organism. That space—or at least the rules of operation within that space—must in some way be independent of, or insensitive to, the details of the material space in which life occurs. This is the space of inherited capacities.

7.1 The Nature of the Organism: Capacity Space

Living systems have the capacity to impose themselves on their surroundings (Collier 1986, 1988a, b, 1990a, b, 1992, 1993, 1996a, b, 1998, 1999a, b, 2000a, b, c, 2002, 2003, 2004, 2008, 2010, 2011, 2013, 2015). That capacity, residing in inherited information dominated by, but not limited to, genetic inheritance (Jablonka and Lamb 1995, 2014; Maynard Smith and Szathmáry 1995, 1999; Szathmary 2000) determines the kinds of matter and energy needed to cope with the conditions of life and the ways they are extracted from the environment. A squirrel and an oak tree have very different needs and they meet these needs in very different ways, allowing them to live in the same environment and even inadvertently provide mutual support (the oak tree provides food and protected habitat for the squirrel and the squirrel buries acorns, some of which germinate before they are dug up and eaten).

Each organism is unique, but each bears a strong "family resemblance" to its relatives. Some of these family resemblances are truly ancient—all eukaryotic cells have a nucleus surrounded by a membrane, which is a resemblance so old it encompasses a billion years of "family history." This is why Darwin asserted that similarities among species are mostly due to common ancestry, not to living in common environments.

> The bond is simply inheritance, that cause which alone, so far as we positively know, produces organisms quite like each other...—Darwin (1872)

This capacity for life is distributed globally in material space and extensively through time, giving living systems their distinctly historical nature. But material space provides only the raw materials for life and does not determine how particular living systems go about the business of living. As well, the extension of living systems in time is a product of their imposition on the surroundings, not the reverse.

Gauging the overall capacity for organisms to impose themselves on the surroundings requires that we assess how much information is contained in inheritances. We begin by recognizing that, while cohesive wholes, organisms are *digital replicators*; the inheritance system is fundamentally digital and combinatorial (Maynard Smith and Szathmáry 1995; Szathmary 2000; Smith 1988, 1998, 2000, 2001). Imagine an organism whose inheritance system comprises a string of DNA 1000 base pairs long. Recall, each base can take one of four potential "states," the nucleobases adenine (A), cytosine (C), guanine (G), and thymine (T). To describe a binary system (e.g., on/off) requires a single bit of information. Each DNA site will be one of four bases, so two bits of information would be required to describe them. Now, if organisms were only holistic replicators (sensu Maynard Smith and Szathmáry 1995), the string of DNA would function strictly as a single unit. The sequence of DNA would contain a maximum of 2000 bits of potential information (1000 base pair sites times 2 bits of information per site). The sequence could only be read as a complete whole and no matter how many times it was read, it would always contain the same 2000 bits of potential information. But now imagine that our DNA string 1000 base pairs long can be read at multiple levels, from single bases to the entire sequence (Brooks and Wiley 1988; Smith 1988, 1998, 2000; Collier 2003). The ability to assess the same system in different ways at different times allows the same information to be additive since at each time it is a different set of the total potential information being expressed. Each potential reading would contain a maximum information capacity of 2000 bits. If each reading is of equal length, and assuming no interactions among the bases, the total set of all possible readings would have a maximum information capacity of 2,002,000 bits—a thousand times larger than if the string could only be read as a whole. If the bases were interactive, then these self-interactions could act as organizing principles that constrain the total information content of the system. The total amount of information that could be expressed at any point in time (potential capacity) would be constrained by the fact that bases (and genes, tissues, organisms, etc.) are causally linked. These linkages mean that expressing some information limits the expression of other information, which produces an upper bound on the amount of potential information that could be expressed at any one point in time. For example, if gene B were expressed only when gene A was expressed, then the bases that comprise these genes are linked and potential diversity is constrained by the fact that the expression of one gene is dependent on another. Alternatively, if the expression of gene A excludes the expression of gene B, then again potential diversity is limited by this interaction. Scaling up, if species B requires species A to survive, then entire genomic complexes

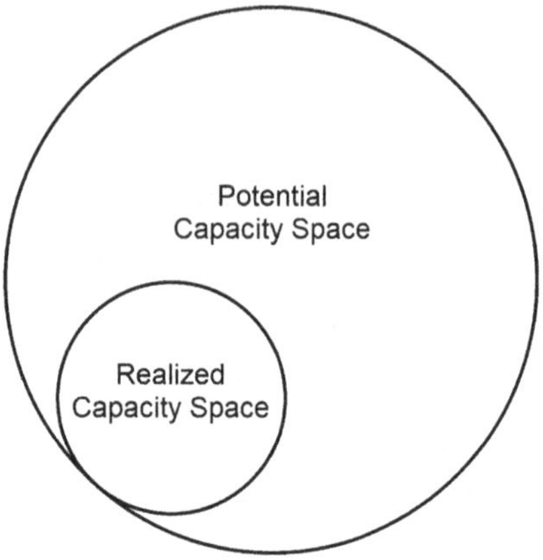

Fig. 7.1 Potential capacity space is always larger than realized capacity space, but they grow in tandem as the inheritance system evolves and diversifies

are linked, and again potential diversity is constrained. Despite these inherent constraints on potential information, the fact that organisms are digital combinatorial replicators gives them enormous potential for variation. The inheritance system is truly a system of indefinite variation (Maynard Smith and Szathmáry 1995, 1999; Szathmary 2000, 2015; de Vladar et al. 2017). In fact, there is so much potential information digitally encoded in DNA that producing a coherent analog product—an organism—must involve a distillation of enormous amounts of potential into a small amount of actuality. Ontogeny—the growth and development of an organism from its inception to maturity—is this distillation process. The inherited program followed during ontogeny is an irreversible energy-efficient algorithm for converting digitally encoded potential information into realized analog output. Metaphorically, we could say this is paid for by the dissipation of potential information (see also Andrade 1999, 2000).

The realm of all possible inheritances represents *potential capacity*, or the maximum diversity of all possible states of the inheritance system. The realm of actual inheritances represents *realized capacity*, or the states of the inheritance system that are actually expressed. We believe it is un-problematical to assert that while realized capacity at any point in time is enormous (Thomas et al. 2020) it is a small subset of potential capacity (Wolf et al. 2015; de Vladar et al. 2017; Hansen et al. 2019) (Fig. 7.1).

7.2 Evolvable Space–Time: An Integrated View of the Nature of the Organism

Inheritance drives both the production of historical time and the exploration of capacity space. When described in terms of the abstract formulation of information theory, we find that buying time and making space appear to be different manifestations of a common phenomenon. A critical part of that phenomenon is the distinction between potential and realized outcomes in space, both of which grow over time.

We think this is more than coincidence. In general, there are two classes of processes that generate entropy (Frautschi 1982, 1988; Landsberg 1984a, b). The first class is an energy-oriented view and involves the equilibration of temperatures between systems and their surroundings. Biological systems exhibit this type of entropic behavior through metabolism and other homeostatic processes involved in the maintenance of the organism that dissipate heat and other forms of degraded energy to the surroundings. The second class is an information-oriented view and involves the expansion of the realm of possibilities—the phase space—in which systems reside. Once again, there is nomenclatural diversity. Kolmogorov (1933) equated sample space, event space, and probability space; and many scientists use *frequency distribution*, all in a manner compatible with our metaphor of informational phase space. Systems increase their informational phase space by increasing the number of possible microstates and thereby the number of higher-order configurations (macrostates) that can be accessed. Recall from the previous chapter the insights from Brooks and Wiley (1986, 1988) into the temporal dynamics of biological information: as the realm of "what's possible" grows, so does the realm of "what's realized". As the number of possible states of the phase space expands, the amount of organization in the system can also expand so long as equilibration of the system (into an equiprobable distribution of microstates) happens at a slower rate than the rate of expansion of the possibilities. This permits a lag between the increase in realized entropy/expressed information and the increase in the maximum possible entropy/potential information of the phase space.

Thus, in a growing phase space, so long as all possible states of the system cannot be expressed at once (maximum entropy), the system can remain in a relatively low-entropy state, permitting the emergence of structure and organization. This happens whenever realized diversity lags behind maximum potential diversity (Soberón and Arroyo-Peña 2017), which is always the case in systems that generate their own conditional probabilities and therefore *adjacent possibles* (Kauffman 1993, 1995, 2000; see also de Vladar et al. 2017). In cosmology, this dynamic explains the spontaneous and irreversible emergence of structure in the universe. Gravity and other fundamental forces cause matter and energy to cohere, thereby slowing down their entropic expansion enough that organized structures (stars, planets, solar systems, galaxies, etc.) emerge even though the total entropy of the system (the universe) is increasing (Badiali 2005). In biology, linear time processes arising from the inheritance system (ontogeny, reproduction, speciation, etc.) are what produce these effects. Innovations in the inheritance system emerge from mutations and other higher-order genetic and epigenetic phenomena that expand capacity space (Jablonka and Lamb 1995, 2014; O'Dea et al. 2016; Lebedeva et al. 2017; de Vladar et al. 2017; Tikhodeyev 2018). Correlation structure within the inheritance system plays a role analogous to gravity and other fundamental forces in cosmology, providing cohesion among parts of the system and causing expressed informational capacity to lag behind potential informational capacity (Brooks and Wiley 1986, 1988). The autonomous and conservative nature of biological

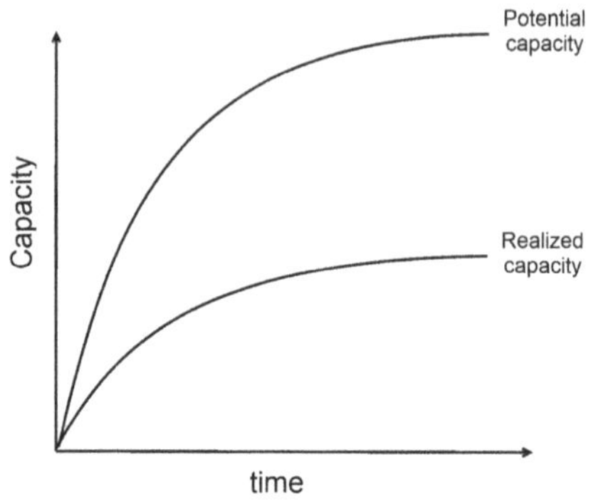

Fig. 7.2 Graphical representation of integrated evolvable space–time. Life's inception opened up temporal production by the system, catalyzing the production of the capacity space that it continues to explore. As the living system evolves and diversifies, realized capacity space grows but potential capacity grows even more. The buildup of historical correlations in the system slows the overall rate of growth of capacity space over time, but the difference between potential capacity and realized capacity (i.e., evolvability) continues to grow

inheritance systems allows life to combine memory and imagination. Realized evolutionary innovations always produce new potential innovations, so realized capacity space increases through time but potential capacity space grows even more.

Buying time and making space are thus the same thing—actualizations of some potential paid for by the dissipation of the rest, digital information converted into analog output (organisms), which then produce additional potential, and repeat. Organisms are a subset of the previous generation's potential, making living systems inherently historical. The production of new potential simultaneously with the actualization of old potential drives, pulls, pushes, or permits (depending on your nomenclatural preferences) living systems to constantly exploit their present circumstances while maintaining the ability to explore possible futures, both based on capacities rooted firmly in the past. In this way, evolvable life is able to extend itself and diversify in time and space, although at a less-than-maximal rate as historical conservatism slows the expansion of the system. The nature of the organism is thus integrated evolvable space–time (Fig. 7.2), an evolutionary version of Hutchinson's niche concept embedded in the flow of historical time. *Fundamental* inherited space–time is the realm of all possible capacities. *Realized* inherited space–time is the realm of achieved capacities; the difference between them is the realm of *evolutionary potenti*al, or *evolvability* (Sniegowski and Murphy 2006; Daniels et al. 2008; Vasas et al. 2015; Crother and Murray 2018, 2019; Hansen et al. 2019). This is the space in which evolutionary creativity occurs, and from which emerge modularities, hierarchical structure, and developmental stability domains (Szathmary 2015).

Through its own capacities—the nature of the organism—life creates and occupies evolvable surroundings. This is a manifestation of Darwin's notion that "diversity-begets-diversity"—in any system of expanding phase space "what's possible" grows faster than "what's realized," and the realization of one possibility always produces new possibilities. Life "buys" the time to "make" the space to stay alive by exploiting and exploring its space–time. In the case of living systems, however, we should never assume that *adjacent possible* necessarily means *adjacent actual-in-waiting*. Some, perhaps even most, potential will never be realized.

We have now reached a critical point of integration in which living systems can be viewed dualistically as explorers and exploiters in time and space, dependent on their surroundings for raw materials and upon their internal rules of inheritance for the capacities that determine which raw materials are put to use and how. This is the most general metaphorical notion of the nature of the organism we can offer. Now we can begin to ask how the tendencies and fates of those evolvable capacities play out in material space.

7.3 The Nature of the Conditions: Opportunity Space

Evolution is more than the capacity to exploit and explore. It is exploiting and exploring in the context of opportunities provided by the nature of the conditions. It is the nature of the conditions, therefore, that produces opportunity space. But the limitations on how opportunity space is used arise from the nature of the organism, and this is the major reason Darwin asserted that the nature of the organism, capacity space, was more important than the nature of the conditions. Individual living systems can survive only in a subset of the *potential opportunity space* that is available to all species. That subset of potential opportunity space that is accessible to any giving living system is *realized opportunity space*, more commonly known as *fitness space*.

7.3.1 Capacity Meets Opportunity: Fitness Space

Fitness space is more than the intersection of capacity space and opportunity space. It is the result of an asymmetrical interaction between two relatively autonomous spaces that are also intimately connected. While the two spaces are not coequal, neither are they independent of each other, so their interaction is complex. We have left the realm of tight integration between time and space and, paradoxical as it may seem, that lack of integration is critical to understanding the coherent nature of the organism–environment relationship in biological evolution. Capacity space imposes itself on opportunity space, not the reverse. Inherited capacities are produced according to inheritance rules that are largely insensitive to the nature of the conditions. And yet, they determine the ways in which and the extent to which living systems can cope with, and take advantage of, opportunities presented by their surroundings. In imposing themselves on their surroundings, living systems perturb opportunity space in a way that makes their existence and persistence possible, but at a cost—they reduce the availability of environmental resources necessary for survival. A tiny fraction of the resources needed to maintain the organism is used to produce new organisms, which are the products of inheritance systems that are extremely conservative, adding an inextricable historical element to the interaction between the two spaces. Darwin understood these fundamental aspects of the nature of the organism implied reproductive overrun, with more organisms being produced than there are resources to support them.

Fit and *fitness* are related to the notion that capacity and opportunity must complement each other but this does not mean that capacity is perfectly attuned to opportunity. This could only be possible if capacity and opportunity responded synchronously or if opportunity determined capacity. Before Darwin, it was widely believed that organisms were perfectly fit to their surroundings, either by a creative deity or due to romantic views associated with the concept of a "balance of nature." Indeed, Lamarck's evolutionary ideas were grounded in the notion that organisms were perfectly fit to their surroundings and, moreover, had the capacity to mold themselves to changes in the surroundings.

Darwin had a very different perspective. If it is in the nature of the organism (as inheritance systems) to be largely insensitive to the nature of the conditions, the kind of fit between capacity and opportunity (organism and environment) envisioned by the eighteenth- and nineteenth-century Romanticists and by Lamarckian evolutionists is unlikely to be achieved. Moreover, if it were achieved on rare occasions, it would be ephemeral, buffeted by the whims of changing conditions. What Darwin found remarkable, based on his extensive natural history observations, was that, while there was a correspondence between capacity and opportunity, it was not precise:

> ... for nearly similar variations sometimes arise under, as far as we can judge, dissimilar conditions; and, on the other hand, dissimilar variations arise under conditions which appear to be nearly uniform.—Charles Darwin (1872)

Living systems do not occur where they cannot survive, but this is not an explanation for how they live within their opportunity space. Living systems are suited, or fit, for any part of opportunity space where they have the capacity to survive and reproduce. The question is not how well do you fit the surroundings, but how fit are you for coping with the surroundings in which you find yourself? This returns the focus of understanding the dynamics of fitness space on the nature of the organism rather than on the nature of the conditions.

So, for each living system its imposition on the surroundings in order to survive changes opportunity space into fitness space that is contextual and contingent in space and time. Living systems are not likely to be uniformly distributed in fitness space. Implicit in Darwin's understanding of the primacy of the nature of the organism and the complexity of the interaction between capacity and opportunity is Darwin's Necessary Misfit (Brooks and Hoberg 2007) and the recognition that fitness space is inherently sloppy, in part because it can extend beyond the bounds of current local conditions at any given place and time (Agosta 2006; Agosta and Klemens 2008; Agosta et al. 2010) (Fig. 7.3). The "slop" is the difference between what organisms are doing (realized fitness space) and what organisms could be doing (fundamental fitness space) to achieve positive fitness—the adjacent possible of

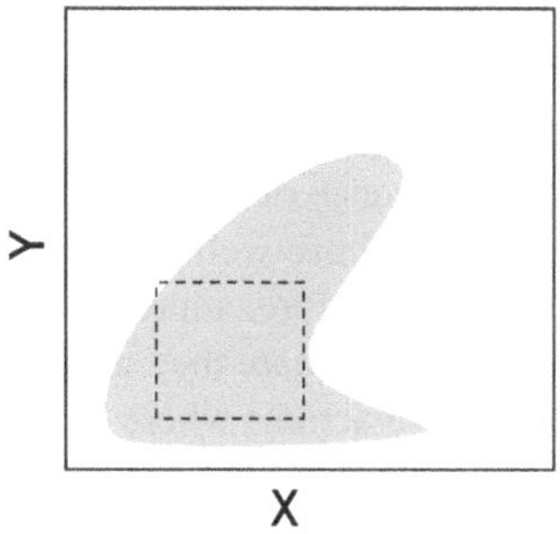

Fig. 7.3 Simple graphical representation of sloppy fitness space. The set of all possible conditions is defined by the environmental variables X and Y. The dashed box defines the realized conditions. Fitness space is defined by the shaded region which includes the currently realized conditions, but also some unrealized conditions. The degree of mismatch between fundamental fitness space (shaded region) and realized fitness space (shaded region inside the dashed box) is a measure of how sloppy fitness space is, which is proportional to the capacity to respond when conditions change

fitness space. For evolution to occur, fundamental fitness space needs to be larger than realized fitness space, and this seems to be a ubiquitous condition of life (Soberón and Arroyo-Peña 2017).

To persist, living systems need only to survive, not thrive. They only need to be adequate, not the best. Nonetheless, resources are not distributed uniformly in opportunity space (especially when the resources include other living systems) so we should expect to find both preferred and marginal parts of fitness space, where living systems perform better or worse depending on the distribution of resources essential for surviving and thriving. Furthermore, the historically contingent nature of inheritance means that the occupation of even preferred fitness space may be constrained. One penalty for being autonomous from the surroundings is that living systems cannot guarantee that they find themselves in the best possible conditions.

Four characteristics of the nature of the organism play a role in helping us understand the sloppy nature of fitness space. First, some portions of fitness space are inaccessible to the living system due either to physical barriers or to preemption by other living systems. Second, while it is expensive for an organism to survive, only a tiny amount of the total energy budget for an organism is necessary to reproduce—offspring are cheap to produce. Third, inheritance is highly conservative, especially for traits of functional capacity, so the vast majority of those cheaply produced offspring will share a common set of capacities. And finally, exploitation trumps exploration. So, there will be over-use and over-reproduction within the preferred fitness space, and that will be complemented by under-use and under-reproduction within marginal fitness space. The imposition by highly conservative and specific offspring may be so localized in fitness space that it appears that the conditions are structured in such a way as to provide a specific place for those offspring. But this is an illusion—evolutionary lagload (Maynard Smith 1976) (also known as genealogical conservatism, phylogenetic inertia, or historical constraint:

see Brooks and McLennan 2002 for a discussion of this nomenclature) leads to the relentless production of offspring regardless of the nature of the conditions. Every organism that inherits a common capacity for imposing itself on the surroundings will reconstitute that illusory appearance of a "niche" in opportunity space.

Fitness is a metaphor describing how well a living system is coping with the conditions of life. Fitness space, therefore, represents the realm of conditions in which organisms can "make a living," where they are fit to live given their inherited capabilities. It is the realm of the interactions where Darwin's Law of the Conditions of Existence and natural selection are emergent properties of the system in which the nature of the organism takes precedence over the nature of conditions. Natural selection emerges whenever there is a misfit. The degree of Darwin's Necessary Misfit creates *selection in proportion to the misfit*. Misfit is proportional to the sloppiness of fitness space, so the force of natural selection is also proportional to the sloppiness of fitness space. In both preferred and marginal portions of fitness space, every member of each living system will experience a struggle for survival. And in all cases, the historical contingency and the complexity of the interactions that characterize fitness space will result in multiple demands made upon the members of each living system, many of them conflicting. In more popular terminology, there will be many different fitness vectors affecting each inheritance system, with the result—as Darwin explicitly stated—that the overall impact of natural selection will be incremental.

Despite the inevitability of conflicts, the processes that make fitness space sloppy ensure realized fitness space will always be a subset of fundamental fitness space. *This means there will always be some under- or unoccupied fitness space, an ever-present adjacent possible to explore conditions sufficient for survival and reproduction*. Whatever living systems are doing to achieve non-zero fitness at any given point in space–time, it is only a subset of what they could be doing. This difference between "what's realized" and "what's possible" is the critical ingredient for persisting indefinitely in an unpredictable world—it provides fitness space to explore. *Fundamental fitness space* is equivalent to *realized opportunity space*, encompassing all possible places and conditions in which organisms could survive and produce offspring. The portion of fundamental fitness space accessed by organisms at any given time and place is *realized fitness space* (Fig. 7.4).

The "fittest" (those concentrated in preferred fitness space) will thus be a snapshot description of a given state of existence at a given place and time, while "the fit" (the total distribution of an inheritance system in fitness space) will be explanatory (i.e., the snapshot "present" emerges from historical processes and sets the stage for what may emerge in the future). Even if fitness space is sloppy, there will be conflicts of interest among offspring for preferred fitness space but also the potential for coping with conflict by surviving in marginal fitness space. There will be little room for innovation in the preferred fitness space due to the intensity of the competition and extensive room for innovation in marginal fitness space. The more organisms produced with the same fundamental capacities, the higher the cost of living, but never more than can be afforded. As long as the system can afford the conflict it will

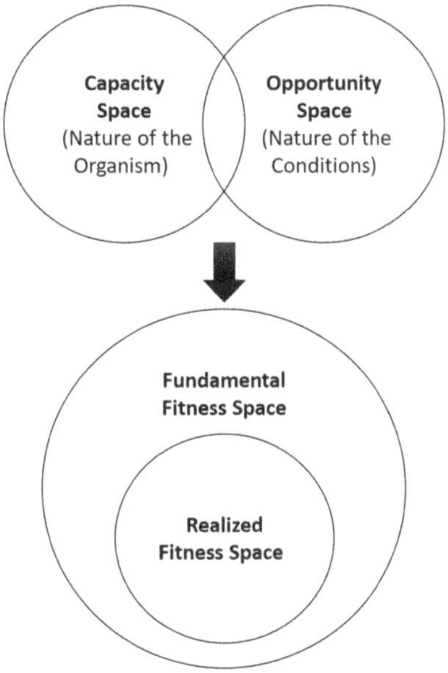

Fig. 7.4 Fitness space emerges from the interaction between capacity space and opportunity space. Realized fitness space is always a subset of fundamental fitness space

continue, and there is no guarantee that a resolution will ever emerge. But, there are coping mechanisms that make conflict more affordable.

7.4 Coping with Conflict

He that fights and runs away, May turn and fight another day; But he that is in battle slain, Will never rise to fight again.—Tacitus

Interactions within fitness space are complex. Capacity space includes the means by which living systems attempt to exploit and explore opportunity space. Opportunity space is largely indifferent to those attempts. Many of the traits important for exploiting environmental resources are both specific and conservative, leading to conflicts due to over-production in preferred fitness space. Inheritance also includes traits that allow living systems to lower the costs of those conflicts by shifting from exploitation-biased to exploration-biased behavior. Moving into unused or under-used portions of fitness space allows marginally fit organisms to survive and persist, and for marginal portions of fitness space to be occupied at least to some extent. As Tacitus knew, using marginal fitness space is better than dying.

7.4.1 The Means: Ecological Fitting

God helps those who help themselves—Algernon Sidney (1698)

The secret to success is to be ready when your opportunity comes—Benjamin Disraeli

Ecological fitting, formally proposed in 1985 (Janzen 1985) but anticipated more than a decade earlier (Janzen 1968, 1973a, b, 1980, 1981, 1983; Brooks 1979, 1985: for reviews, see Brooks and McLennan 2002; Agosta 2006; Agosta and Klemens 2008; Brooks et al. 2019), is the umbrella term for a fundamental phenomenon that lowers the costs of conflicts arising in fitness space. It emerges from three classes of inherited capacities—phylogenetic conservatism, co-option, and phenotypic plasticity—that together allow living systems to cope with conflict by exploring novel portions of fitness space.

7.4.1.1 Phylogenetic Conservatism: The Past Is Always with Us

Biological inheritance is not only conservative, it clings to its past tenaciously. What organisms inherit from their ancestors is a historically unique portion of potential capacity embodied in a larger inheritance system. Because it is inherited from ancestors that survived, it is a record of what worked in the past. In the face of an unknown future, conservative inheritance is equivalent to assuming that the future will be mostly like the past. This "assurance of adequate performance" is of course no guarantee; conditions experienced in the present may be sufficiently different from the past so that inherited capacity provides little or no opportunity to perform. *Evolutionary lagload* (Maynard Smith 1978) was originally coined as an indicator of how difficult it is for selection to make substantial changes in an inheritance system. For us, lagload indicates the totality of potential capacity to function adequately in both preferred and marginal portions of fitness space.

Just how conservative are the functional traits that provide species with the capacity to survive and persist? Ross (1972a) proposed that approximately one in every 30 speciation events in a variety of insect groups was correlated with geographic dispersal from the primitive climatic zone to a derived one, or with shifts from the ancestral condition to any subsequent state in reproductive behavior, ecology, or host preferences. He concluded that such shifts in "functional diversity" were consistent with, but much less frequent than, phylogenetic diversification. Furthermore, he felt there were no predictable patterns explaining the shifts that did occur and suggested such transitions constituted a biological "uncertainty principle" (Ross 1972b). By the beginning of the twenty-first century, phylogenetic studies had uncovered ample evidence of conservative and specific resource use supporting Ross's views in contemporaneous species (Brooks and McLennan 2002; Brooks et al. 2019) and in the fossil record (Boucot 1982, 1983, 1990). More recently, experimental studies have shown that phytophagous insects may retain the ecological fitting capacity to use novel hosts even after 40 million years of isolation with their ancestral hosts (Garcia-Robledo et al. 2017).

Janzen and Martin (1982) used the term *evolutionary anachronisms* to refer to the components of evolutionary lagload that were persistent ancestral traits that may have functioned in a particular way in the past but no longer serve that function.

Their exemplar was large thorns on some Neotropical trees, which they postulated had served as a defense against being eaten by Gomphotheres, members of an ancient and honorable, but extinct, group of large herbivores related to elephants.

The neo-Darwinian framework assumes all traits "cost" something, and selection will eliminate those that do not provide "benefits" as great as, or greater than, their costs. This sounds reasonable at first, but there is a problem. Just as new traits do not magically appear on command, old traits do not magically disappear if they are no longer in the limelight of selection. If a trait does not interfere with any function that is the focus of current selection, there is no reason to think it will disappear. Also, inheritance systems are complex, and virtually all traits are interconnected in some way during development. The maintenance of a not-so-special trait may be a necessary precursor for a different trait that is of critical importance at another point in time. The evolutionary "cost" of some old traits, like the thorns of those Latin American trees, may not be high enough to provoke selection against them—they may cost more to eliminate than to maintain. This illustrates one form of "survival of the adequate"; traits that once were critical for coping with the environment are now silent partners but could once again function in the selection spotlight. Anachronisms thus complement chance innovations that are only marginally fit with their surroundings at the time they arise but may become more fit later if the spotlight of selection shines on them.

This is the evolutionary significance of persistent ancestral capacities. Rather than representing a wasteful evolutionary cost, they have a particular value when it comes to ecological fitting. Because they are traits that functioned well in past environmental conditions, anachronisms allow species that retain them to survive if anything like those ancient environmental conditions recur as marginal fitness space (e.g., Weinstein et al. 2019).

7.4.1.2 Co-option: Old Wine in New Bottles

Darwin proposed co-option, using existing structures to perform novel functions, as one means of coping with conflicts in fitness space. This might occur if two structures performed the same function, one with increasing efficiency leaving the second, and now superfluous, structure free to perform a different function. Darwin (1872) believed that this type of dynamic was "an extremely important means of transition" in evolution. In other cases, an organ that served a major and minor function was modified to serve the latter at the expense of the former. In both of the preceding processes, the starting conditions might be obscured and the "transitional stage" might be missing. Darwin also discussed traits that served no apparent function because they arose as the byproduct of processes such as the developmental dynamics leading to the "complex laws of growth" and the "mysterious laws of the correlation of parts." He believed that such traits were an important part of organismal evolution because they might eventually acquire a function that aided survival in what was initially marginal fitness space: "But structures thus indirectly gained, although at first of no advantage to a species, may subsequently have been taken advantage of by its modified descendants, under new conditions of life and newly acquired habits." In each case imagined by Darwin, important evolutionary change was accomplished without creating new structures. Co-option is critical to Darwinism because, without it, it is often impossible to explain dramatic movements into novel portions of fitness space without invoking a magical "need" for the trait that led to its origin (see the comprehensive and highly readable analysis by McLennan 2008).

Co-option is a key element of ecological fitting. If preexisting traits are capable of performing new functions that allow new conditions to be utilized, we have a relatively cheap and nonmystical coping mechanism. Without it, we are left with a variety of almost magical scenarios, most of which involve waiting for a fortuitous mutation to arise. Deciduousness in southern hemisphere beeches (*Nothofagus*) originated as a mechanism for coping with seasonally dry habitats but turned out to also function well in seasonally cold habitats (Wanntorp 1983; Wanntorp et al. 1990). Gene co-option played a significant role in the evolution of the molluscan radula (Hilgers et al. 2018). Persistent traits co-opted to perform novel functions also are common in interspecific associations (Arnold 1994; Armbruster 1996, 1997; Kardon 1998). For example, co-option explains how dipterans, hymenopterans, and coleopterans have become parasitoids (Eggleton and Belshaw 1993) and co-options featured prominently in the evolution of plant–herbivore and plant–pollinator interactions among *Dalechampia* (Euphorbiaceae) species (Armbruster 1997; see also Brooks and McLennan 2002).

Some of the best examples of co-option involve some of the least attractive but most diverse members of the biosphere. Because pathogens require hosts, it seems reasonable that hosts existed before pathogens. It also seems reasonable that the ancestors from which each group of pathogens arose were themselves not pathogens. Did those ancestors evolve an entire suite of traits specific to being pathogens, or did preexisting traits allow them to become pathogens through co-option? The life history traits of parasitic flatworms, for example, do not differ from the life history traits of their free-living relatives, indicating that these species do not have a "parasitic mode of life" but rather a "platyhelminth mode of life" which was co-opted to function in a host-parasite context (Trouvé et al. 1998; Solà et al. 2015). The basic conditions of life with which a free-living flatworm had to contend, feeding on decomposing material in low-oxygen conditions in mudflats of a tidal marsh, might not be that different from living in the intestine of a vertebrate. If that free-living ancestor also had a novel thick kind of epidermis (a *neodermis*) that helped it fend off other decomposers in its surroundings, members of that ancestral species might well have been protected when eaten by a vertebrate, allowing it to enter a new portion of fitness space—the vertebrate intestine. This seemingly simple episode of co-option set the stage for the entire diversity of parasitic flatworms including trematodes and tapeworms now called the Neodermata.

7.4.1.3 Plasticity: Making the Best of Bad Situations

Living systems not only impose themselves on their surroundings, they can be flexible in the way they do it, doing different things in different situations. This is the built-in flexibility of organisms to respond to changing and possibly novel conditions postulated by Darwin, more commonly referred to today as *phenotypic plasticity* (Bradshaw 1965; Stearns 1989, 2015; West-Eberhard 1989, 2003; Schlichting and Pigliucci 1998; Yeh and Price 2004; Mason 2015). Plasticity becomes important in situations in which a living system—by chance or by choice—finds itself in a less than ideal situation. For example, it is not uncommon for living systems to have the capacity to cope with conflicts in fitness space by facultatively reducing the number of conflicting organisms; this includes plastic responses leading to reductions in fecundity, such as lowered fertility rates and skewed sex ratios favoring migration away from the conflict arena. Alternatively, there are crowding effects, in which plasticity leads to reductions in body size that allow a larger number of individuals to live in the same place with reduced conflict.

The surroundings, however, are largely indifferent to the aspirations of living systems. As a consequence, for many organisms finding preferred fitness space is something like participating in a lottery with a time limit. Organisms that are "locked-in" to preferred fitness space to the exclusion of all marginal portions die and are eliminated from evolution. This reduces evolution to a brute-strength approach: simply blast the environment with enormous numbers of offspring (buy *many* lottery tickets) and enough will find preferred resources to produce the next generation. All "non-winning lottery tickets" are presumed to be evolutionary waste. But they are not—they are the cost of doing business, including the cost of exploring marginal fitness space. Having the plasticity to explore marginal parts of fitness space changes the rules of the "find the resource" lottery in favor of the nature of the organism. For instance, drought tolerance in barley depends on the expression of plasticity prior to the occurrence of drought (Janiak et al. 2018).

Survival is paramount in evolution, but surviving does not require thriving. The capacity to survive in suboptimal fitness space provides living systems with the essential ability to "be fit" in their surroundings even if they are not ideal. Ecological fitting allows living systems to provide a positive answer to the evolutionary question, "what happens if your preferred resource isn't available?"

7.4.2 The Opportunity: Ecological Fitting in Sloppy Fitness Space

'Cause I can't give the best
Unless I got room to move.—John Mayall (1970)

In really good mystery novels, "who did it" is not as interesting as "how it was done." Thus it is with members of living systems coping with conflict in fitness space by moving from preferred to marginal fitness space. The answer, as with the denouement of a good mystery novel, seems obvious and simple in retrospect. Survival is paramount in evolution, and the ability to persist indefinitely is linked to how flexible living systems can be on short notice. The ability to move away from unsuitable conditions to find suitable conditions is more important than how well adapted the system is to any particular set of conditions. Conservative capacity provides the possibility of conflict, even if fitness space is sloppy, but ecological fitting provides the coping mechanisms, reducing the cost of conflict, by exploring marginal portions of fitness space, and in this regard, the sloppier the fitness space the more chances there are for survival.

Because it is sloppy, fitness space always includes some amount of non-zero fitness space to potentially explore. Ecological fitting allows inheritance systems to explore fitness space with some degree of autonomy and without sacrificing the ability to exploit local conditions as much as possible. This autotomy gives organisms the essential ability to minimize the risk of *moving away* from conditions that are deteriorating with respect to fitness. Sloppy fitness space provides an adjacent possible that provides biological systems options to remain fit when they explore marginal fitness space. This potential is a function of how many historical alternatives a species maintains in its inherited capacities, and of how many ways—old and new—those capacities can be used for survival and reproduction in novel conditions (Fig. 7.5).

Agosta and Klemens (2008) presented a graphical model depicting ecological fitting in sloppy fitness space (Fig. 7.6). For simplicity, they visualized a bivariate world where the conditions could be described by the two *operative* environmental variables *X* and *Y*. Fundamental fitness space is the suite of abiotic and biotic variables that affect fitness. At any given time and place the local conditions (the *operative environment*: Spomer 1973; Dunham et al. 1989; Dunham 1993; Dunham and Overall 1994) are defined by the subset of *X* and *Y* that exist. For a particular living system, *fundamental* fitness space is defined by all combinations of *X* and *Y* for which positive fitness can be achieved and *realized fitness space* is the portion of fundamental fitness space being accessed to meet the current local conditions. The misfit between fundamental and realized fitness space is proportional to how sloppy fitness space is. It is this fundamental process that allows living systems to cope with conflicts associated with exploiting preferred portions of fitness space by exploring and persisting in marginal portions of fitness space using preexisting capacities.

7.5 Summary

The ability to *survive in less than ideal conditions* is the source of life's capacity for indefinite persistence in the face of an uncertain future. Not everything gets to survive and those that survive do not all do equally well. Any organism that survives long enough to reproduce is fit. Think of fitness space as an arena; the players are

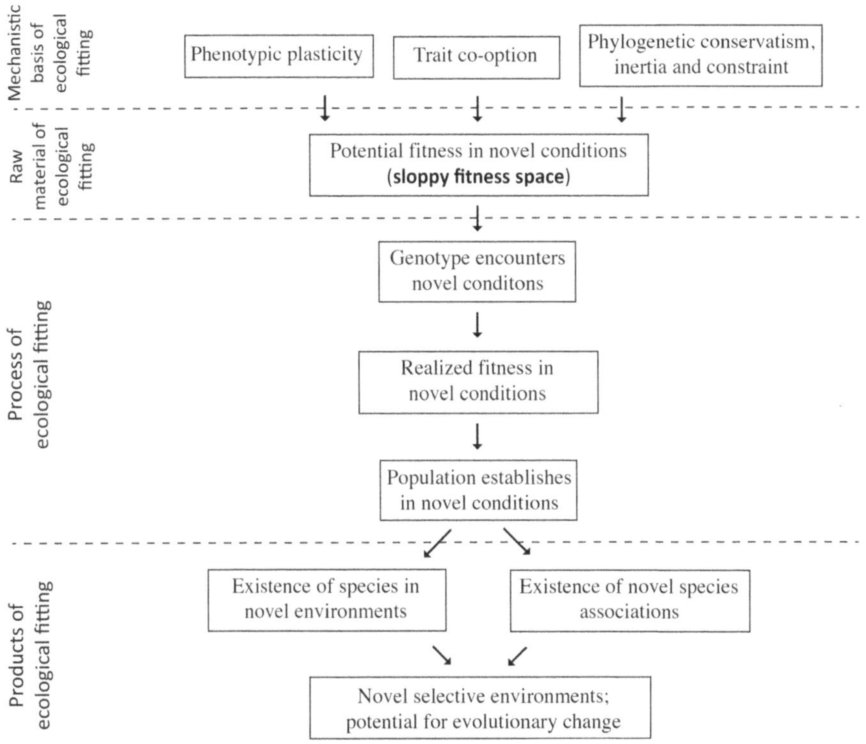

Fig. 7.5 Flow diagram of the major elements involved in the process of ecological fitting in sloppy fitness space, from its mechanistic basis arising from capacities in the inheritance system to its final outcome of organisms coping with environmental change using the information they inherited. Redrawn and modified from Agosta and Klemens (2008)

drawn from inheritance space, and the play they perform is an outcome of the opportunities provided by the arena, given their capacities. Every performance differs, though conservative inheritance produces a significant amount of continuity from one to another. Having the capacity and opportunity to perform do not guarantee a superb performance. Some performances are better than others, but even poor performances count. The point of the game of life is to stay alive; it is not a matter of "they who die with the most toys win," but of "they who live longest win." This is the *window of vitality* (Ulanowicz 1997) for evolvable life on this planet.

Ecological fitting in sloppy fitness space is Darwin's Law of the Conditions of Existence

The dynamics of ecological fitting in sloppy fitness space should be familiar to most biologists, although the Darwinian emphasis on the primacy of the nature of the organism rather than the nature of the conditions may seem odd. Despite all the capacity space and all the opportunity space, the conservative nature of inheritance

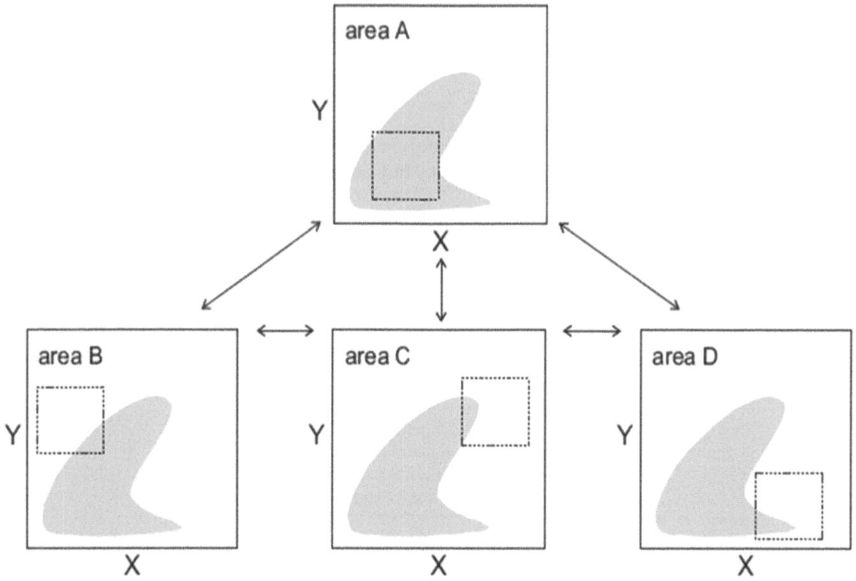

Fig. 7.6 Graphical model of ecological fitting in sloppy fitness space. Each panel represents a "population" separated geographically from other populations but connected by gene flow (arrows). Environmental conditions are defined by the variables *X* and *Y*. For each population, the realized set of conditions is defined by the dashed box and fundamental fitness space is defined by the shaded region. The difference between realized fitness space (shaded region inside the dashed box) and fundamental fitness space is proportional to how sloppy fitness space is. In the model, the ancestral population (area A) has fitness under the full set of realized conditions and then expands geographically to areas B, C, and D by accessing unused parts of fitness space (i.e., the slop) to establish in novel conditions. Redrawn and modified from Agosta and Klemens (2008)

means that mismatches and conflicts will occur (even in an expanding universe, asteroids sometimes collide with planets). So, what emerges at this point is evolutionary existence (exploitation) and persistence (exploration), but with unavoidable conflict—and this is more or less where standard neo-Darwinism stops, never-ending conflict waiting for a miracle to tip the balance in favor of one of the combatants.

Our story of evolutionary time and space leaves us with an essential incompleteness. Coping with conflict is possible due to ecological fitting in sloppy fitness space, which allows members of living systems to explore and then persist in marginal portions of fitness space. Over-use of preferred fitness space creates the conflicts and drives the change from exploitation-biased to exploration-biased behavior which is diagnostic of ecological fitting. But what would happen if the marginal fitness space were filled? Ecological fitting in sloppy fitness space by itself would lead to evolutionary stasis (Fig. 7.6). We would have a world of conflicts with no guarantee of resolution, and those in conflict that could afford the conflict would continue to exist in conflict (Maynard Smith 1974). This does not lead easily to an expectation of diversification, to the Tree of Life metaphor that Darwin deemed so important that it was the only illustration ever to appear in any edition of *The Origin of Species*.

The existence of a Tree of Life suggests that the history of life is more than never-ending conflict, more even than conflict and replacement. With an appreciation of Evolvable Space–Time in hand, the stage is set for our final metaphorical dualism, *conflict resolution*.

References

Agosta S (2006) On ecological fitting, plant-insect associations, herbivore host shifts, and host plant selection. Oikos 114:556–565

Agosta S, Klemens JA (2008) Ecological fitting by phenotypically flexible genotypes: implications for species associations, community assembly and evolution. Ecol Lett 11:1123–1134

Agosta S, Janz N, Brooks DR (2010) How generalists can be specialists: resolving the "parasite paradox" and implications for emerging disease. Fortschr Zool 27:151–162

Andrade E (1999) Maxwell demon's and natural selection: semiotic approach to evolutionary biology. Semiotica (special issue Biosemiotica) 127:133–149

Andrade E (2000) From external to internal measurement a form theory approach to evolution. Biosystems 57:49–62

Armbruster WS (1996) Exaptation, adaptation and homoplasy: evolution of ecological traits in *Dalechampia* vines. In: Sanderson MJ, Hufford L (eds) Homoplasy: the recurrence of similarity in evolution. Academic, New York, pp 227–243

Armbruster WS (1997) Exaptations link evolution of plant-herbivore and plant-pollinator interactions: a phylogenetic inquiry. Ecology 78:1661–1672

Arnold EN (1994) Investigating the origins of performance advantage: adaptation, exaptation and lineage effects. In: Eggleton P, Vane-Wright RI (eds) Phylogenetics and ecology. Academic, London, pp 123–168

Badiali JP (2005) Entropy, time-irreversibility and the Schrödinger equation in a primarily discrete spacetime. J Phys A Math Gen 38:2835–2847

Boucot AJ (1982) Paleobiologic evidence of behavioral evolution and coevolution. By the author, Corvallis

Boucot AJ (1983) Does evolution take place in an ecological vacuum? J Paleontol 57:1–30

Boucot AJ (1990) Community evolution: its evolutionary and biostratigraphic significance. In: Miller W III (ed) Paleocommunity temporal dynamics: the long-term development of multi-species assemblies. The Paleontological Soc. Special Publ. No 5, pp 48–70

Bradshaw AD (1965) Evolutionary significance of phenotypic plasticity in plants. Adv Genet 13:115–155

Brooks DR (1979) Testing the context and extent of host-parasite coevolution. Syst Zool 28:299–307

Brooks DR (1985) Historical ecology: a new approach to studying the evolution of ecological associations. Ann Mo Bot Garden 72:660–680

Brooks DR, Hoberg EP (2007) Darwin's necessary misfit and the sloshing bucket: the evolutionary biology of emerging infectious diseases. Evol Edu Outreach 1:2–9

Brooks DR, McLennan DA (2002) The nature of diversity: an evolutionary voyage of discovery. University of Chicago Press, Chicago

Brooks DR, Wiley EO (1986) Evolution as entropy: toward a unified theory of biology, 1st edn. University of Chicago Press, Chicago

Brooks DR, Wiley EO (1988) Evolution as entropy: toward a unified theory of biology, 2nd edn. University of Chicago Press, Chicago

Brooks DR, Hoberg EP, Boeger WA (2019) The Stockholm paradigm: climate change and emerging disease. University of Chicago Press, Chicago
Collier J (1986) Entropy in evolution. Biol Philos 1:5–24
Collier J (1988a) The dynamics of biological order. In: Weber BH, Depew DJ, Smith JD (eds) Information, entropy and evolution: new perspectives on physical and biological evolution. MIT, Cambridge, pp 227–242
Collier J (1988b) Non-equilibrium evolution. In: Proceedings of the 13th international Wittgenstein symposium, pp 290–292
Collier J (1990a) Two faces of Maxwell's demon reveal the nature of irreversibility. Stud Hist Philos Sci 21:257–268
Collier J (1990b) Intrinsic information. In: Hanson PP (ed) Information, language and cognition. Oxford University Press, Oxford, pp 390–409
Collier J (1992) Incorporating adaption into the unified theory. http://www.newcastle.edu.au/department/pl/Staff/JohnCollier/papers/iaut.pdf
Collier J (1993) Out of equilibrium: new approaches to biological and social change. Biol Philos 8:445–455
Collier J (1996a) Information originates in symmetry breaking. Symmetry: Sci Cult 7:247–256
Collier J (1996b) Looking beyond the veil of natural selection. Biol Philos 12:89–99
Collier J (1998) Information increase in biological systems: how does adaptation fit? In: van de Vijver G, Salthe SN, Delpos M (eds) Evolutionary systems: biological and epistemological perspectives on selection and self-organization. Kluwer Academic, Dordrecht, pp 129–140
Collier J (1999a) Causation is the transfer of information. In: Sankey H (ed) Causation, natural Laws and explanation. Kluwer, Dordrecht, pp 215–245
Collier J (1999b) Autonomy in anticipatory systems: significance for functionality, intentionality and meaning. In: Dubois DM (ed) Proceedings of the American Institute of Physics from CASYS '98 – second international conference on computing anticipatory systems. Woodbury, New York, pp 75–81
Collier J (2000a) Is there any virtue in modern science? Biol Philos 15:773–784. https://doi.org/10.1023/A:1006647719572
Collier J (2000b) The dynamical basis of information and the origins of semiosis. In: Taborsky E (ed) Semiotics, evolution, energy. Shaker, Aachen, pp 111–138
Collier J (2000c) Autonomy and process closure as the basis for functionality. In: Chandler JLR, van de Vijver G (eds) Closure: emergent organizations and their dynamics (Ann NY Acad Sci 901:280–291)
Collier J (2002) What is autonomy? Partial proceedings of CASYS '01: fifth international conference on computing anticipatory systems, vol 12, pp 212–221
Collier J (2003) Hierarchical dynamical information systems with a focus on biology. Entropy 5:100–124
Collier J (2004) Interactively open autonomy unifies two approaches to function. In: Dubois DM (ed) Proceedings of the American Institute of Physics from CASYS '03 – sixth international conference on computing anticipatory systems. Melville, New York, pp 228–235
Collier J (2008) Information in biological systems. In: Adriaans P, van Benthem J (eds) Handbook of philosophy of science, Philosophy of information, vol 8. North-Holland, Dordrecht, pp 763–787
Collier J (2010) A dynamical approach to identity and diversity in complex systems. In: Cilliers P, Preiser R (eds) Complexity, difference and identity: an ethical perspective. Springer, Berlin, pp 83–97
Collier J (2011) Information, causation and computation. In: Crnkovic GD, Burgin M (eds) Information and computation: essays on scientific and philosophical understanding of foundations of information and computation (World Scientific Series in Information Studies). World Scientific, Singapore, pp 89–105
Collier J (2013) Supervenience and reduction in biological hierarchies. Can J Philos 18(Suppl 1):209–234

Collier J (2015) What must the world be like to have information about it? In T9.2006. MDPI, Basel. https://doi.org/10.3390/isis-summit-vienna-2015-T9.2006
Crother BI, Murray CM (2018) Linking a biological mechanism to evolvability. J Phylogenet Evol Biol 6:192. https://doi.org/10.4172/2329-9002.1000192
Crother BI, Murray CM (2019) Early usage and meaning of evolvability. Ecol Evol 9:3784. https://doi.org/10.1002/ece3.5002
Daniels BC, Chen Y-J, Sethna JP, Gutenkunst RN, Myers CR (2008) Sloppiness, robustness, and evolvability in systems biology. arXiv:0805.2628v1 [qbio.QM]
Darwin C (1872) The origin of species, 6th edn. Murray, London
de Vladar HP, Santos M, Szathmáry E (2017) Grand views of evolution. Trends Ecol Evol 32:324–334
Dunham AE (1993) Population responses to global change: physiological structured models, operative environments, and population dynamics. In: Karieva P, Kingsolver J, Huey R (eds) Biotic interactions and global change. Sinauer Associates, Sunderland, pp 95–119
Dunham AE, Grant BW, Overall KL (1989) Interfaces between biophysical and physiological ecology and the population ecology of terrestrial vertebrate ectotherms. Physiol Zool 62 (2):335–355
Dunham AE, Overall KL (1994) Population responses to environmental change: life history variation, individual-based models, and the population dynamics of short-lived organisms. Am Zool 34:382–396
Eggleton P, Belshaw R (1993) Comparisons of dipteran, hymenopteran and coleopteran parasitoids: provisional phylogenetic explanations. Biol J Linn Soc 48:213–226
Frautschi S (1982) Entropy in an expanding universe. Science 217:593–599
Frautschi S (1988) Entropy in an expanding universe. In: Weber B, Depew DJ, Smith JD (eds) Entropy, information and evolution: new perspectives on physical and biological evolution. MIT, Cambridge, pp 11–22
Garcia-Robledo C, Horvitz CC, Kress WJ, Carvajal-Acosta AN, Erwin T, Staines CL (2017) Experimental assemblage of novel plant-herbivore interactions; ecological host shifts after 40 million years of isolation. Biotropica 49:803–810
Hansen TF, Solvin TM, Pavlicev M (2019) Predicting evolutionary potential: a numerical test of evolvability measures. Evolution 73:689–703
Hilgers L, Hartmann S, Hofreiter M, von Rintelen T (2018) Novel genes, ancient genes and gene co-option contributed to the genetic basis of the radula, a molluscan innovation. Mol Biol Evol 35:1638. https://doi.org/10.1093/molbev/msy052
Jablonka E, Lamb MJ (1995) Epigenetic inheritance and evolution. Oxford University Press, Oxford
Jablonka E, Lamb MJ (2014) Evolution in four dimensions: genetic, epigenetic, behavioral and symbolic variation in the history of life. MIT, Cambridge
Janiak A, Kwasniewski M, Sowa M, Gajek K, Zmuda K, Koscielniak J, Szarejko I (2018) No time to waste: transcriptome study reveals that brought tolerance in barley may be attributed to stressed-like expression patterns that existed before the occurrence of stress. Front Plant Sci 8:2212. https://doi.org/10.3389/fpls.2017.02212
Janzen DH (1968) Host plants as islands in evolutionary and contemporary time. Am Nat 102:592–595
Janzen DH (1973a) Host plants as islands. Competition in evolutionary and contemporary time. Am Nat 107:786–790
Janzen DH (1973b) Comments on host-specificity of tropical herbivores and its relevance to species richness. In: Heywood VH (ed) Taxonomy and ecology. Academic, New York, pp 201–211
Janzen DH (1980) When is it coevolution. Evolution 34:611–612
Janzen DH (1981) Patterns of herbivory in a tropical deciduous forest. Biotropica 13:271–282
Janzen DH (1983) Dispersal of seeds by vertebrate guts. In: Futuyma DJ, Slatkin M (eds) Coevolution. Sinauer Associates, Sunderland, pp 232–262
Janzen DH (1985) On ecological fitting. Oikos 45:308–310

Janzen DH, Martin PS (1982) Neotropical anachronisms: the fruits the Gomphotheres ate. Science 215:19–27
Kardon G (1998) Evidence from the fossil record of an antipredatory exaptation: conchiolin layers in corbulid bivalves. Evolution 52:68–79
Kauffman SA (1993) The origins of order: self-organization and selection in evolution. Oxford University Press, New York
Kauffman SA (1995) At home in the universe: the search for laws of self-organization and complexity. Oxford University Press, New York
Kauffman SA (2000) Investigations. Oxford University Press, New York
Kolmogorov A (1933) Grunbegriffe der Wahrsheinlichkeitsrechnung. Julius Springer, Berlin
Landsberg PT (1984a) Is equilibrium always an entropy maximum? J Stat Phys 35:159–169
Landsberg PT (1984b) Can entropy and "order" increase together? Phys Lett 102A:171–173
Lebedeva MA, Tvorogova VE, Tikhodeyev ON (2017) Epigenetic mechanisms and their role in plant development. Russ J Genet 53:1057–1071
Mason PA (2015) On the role of phenotypic plasticity in host shifting by parasites. Ecol Lett 19:121–132. https://doi.org/10.1111/ele.12555
Mayall J (1970) Room to move
Maynard Smith J (1974) The theory of games and the evolution of animal conflicts. J Theor Biol 47:209–221
Maynard Smith J (1976) What determines the rate of evolution? Am Nat 110:331–338
Maynard Smith J (1978) The evolution of sex. Cambridge University Press, Cambridge
Maynard Smith J, Szathmáry E (1995) The major transitions in evolution. Oxford University Press, Oxford
Maynard Smith J, Szathmáry E (1999) The origins of life. Oxford University Press, Oxford
McLennan DA (2008) The concept of co-option: why evolution often looks miraculous. Evol Edu Outreach 1:247–258
O'Dea RE, Noble DW, Johnson SL, Hesselson D, Nakagawa S (2016) The role of non-genetic inheritance in evolutionary rescue: epigenetic buffering, heritable bet hedging and epigenetic traps. Environ Epigenet 2:dvv014. https://doi.org/10.1093/eep/dvv014
Ross HH (1972a) The origin of species diversity in ecological communities. Taxon 21:253–259
Ross HH (1972b) An uncertainty principle in ecological evolution. Univ Arkansas Mus Occ Papers 4:133–157
Schlichting CD, Pigliucci M (1998) Phenotypic evolution: a reaction norm perspective. Sinauer Associates, Sunderland
Sidney A (1698) Discourses concerning government. By the author, London
Smith JDH (1988) A class of mathematical models for evolution and hierarchical information theory. Inst Math Appl Preprint Series 396:1–13
Smith JDH (1998) Canonical ensembles, competing species, and the arrow of time. In: Van de Vijver G, Salthe SN, Delpos M (eds) Evolutionary systems: biological and epistemological perspectives on selection and self-organization. Kluwer Academic, Dordrecht, pp 141–154
Smith JDH (2000) On the evolution of semiotic capacity. In: Taborsky E (ed) Semiotics, evolution, energy. Shaker, Aachen, pp 283–309
Smith JDH (2001) Some observations on the concepts of information-theoretic entropy and randomness. Entropy 3:1–11. https://doi.org/10.3390/e3010001
Sniegowski PD, Murphy HA (2006) Evolvability. Curr Biol 16:831–834
Soberón J, Arroyo-Peña B (2017) Are fundamental niches larger than the realized? Testing a 50-year-old prediction by Hutchinson. PLoS One 12:e0175138
Solà E, Álvarez-Presas M, Frías-López C, Littlewood DTJ, Rozas J, Riutort M (2015) Evolutionary analysis of mitogenomes from parasitic and free-living flatworms. PLoS One 10:e0120081
Spomer GP (1973) The concepts of "interaction" and "operational environment" in environmental analysis. Ecology 54:608–615
Stearns SC (1989) The evolutionary significance of phenotypic plasticity. Bioscience 39:436–445

Stearns SC (2015) The concept of phenotypic plasticity and the evolution of phenotypic plasticity in life history traits. In: Love AC (ed) Conceptual change in biology. Springer, Dordrecht, pp 131–146

Stout R (1935) League of frightened men. Farrar and Rhinehart, New York

Szathmary E (2000) The evolution of replicators. Philos Trans R Soc Lond B 355:1669–1676

Szathmary E (2015) Toward major evolutionary transitions theory 2.0. Proc Nat Acad Sci USA 112:10104–10111

Thomas GW, Dohmen E, Hughes DS, Murali SC, Poelchau M, Glastad K et al (2020) Gene content evolution in the arthropods. Genome Biol 21:1–14

Tikhodeyev ON (2018) The mechanisms of epigenetic inheritance: how diverse are they? Biol Rev 93:1987–2005

Trouvé S, Sasal P, Jourdane J, Renaud F, Morand S (1998) The evolution of life-history traits in parasitic and free-living platyhelminthes: a new perspective. Oecologia 115:370–378

Ulanowicz RE (1997) Ecology, the ascendent perspective. Columbia University Press, New York

Vasas V, Fernando C, Szilagyi A, Zachar I, Santos M, Szathmary E (2015) Primordial evolvability: impasses and challenges. J Theor Biol 381:29–38

Wanntorp H-E (1983) Historical constraints in adaptation theory: traits and non-traits. Oikos 41:157–160

Wanntorp H-E, Brooks DR, Nilsson T, Nylin S, Ronquist F, Stearns SC, Wedell N (1990) Phylogenetic approaches in ecology. Oikos 57:119–132

Weinstein DJ, Allen SE, Lau MC, Erasmus M, Asalone KC, Walters-Conte K, Deikus G, Sebra R, Borgonie G, van Heerden E, Onstott TC, Bracht JR (2019) The genome of a subterrestrial nematode reveals adaptations to heat. Nat Commun 10:5268. https://doi.org/10.1038/s41467-019-13245-8

West-Eberhard MJ (1989) Phenotypic plasticity and the origins of diversity. Ann Rev Ecol Syst 20:249–278

West-Eberhard MJ (2003) Developmental plasticity and evolution. Oxford University Press, New York

Wolf PG, Sessa EB, Marchant DB, Li FW, Rothfels CJ, Sigel EM, Gitzendanner MA, Visger CJ, Banks JA, Soltis DE, Soltis PS, Pryer KM, Der JP (2015) An exploration into fern genome space. Genome Biol Evol 7:2533–2544

Yeh PJ, Price TD (2004) Adaptive phenotypic plasticity and the successful colonization of a novel environment. Am Nat 164:531–542

Chapter 8
Conflict Resolution

Abstract There is an inevitable conflict between organisms and their surroundings: the nature of the organism is to overexploit by means of inheritance systems that overproduce highly similar offspring, creating a struggle for survival. Living systems cope with this conflict by exploring new options in fitness space, even when this leads to reduced fitness. But simply coping with conflict is not enough to produce evolutionary diversification. That requires conflict resolution resulting in permanent changes signified by the emergence of two or more novel inheritance systems from a single ancestral system. Evolution is thus less about the cost of conflict (organisms pay no more than they can afford) and more about whether or not conflict is resolved. When conflict resolution occurs, it is a two-step process involving compensatory changes—density-dependent responses that break inheritance systems apart—followed by selection for novel forms of cohesion—functional integration that brings inheritance systems together. Compensatory changes decrease correlations in the inheritance system leading it to become more generalized (diffused) in fitness space. Cohesion increases correlations in the inheritance system leading it to become more specialized (concentrated) in fitness space. When a recently generalized inheritance system becomes isolated in a new part of fitness space due to a compensatory change, selection favoring cohesion can produce newly emergent inheritance systems. Yet, because the nature of the organism always predominates, conflict resolution always leads to new conflict. This ensures that evolution continues. Natural selection is a pervasive emergent property providing positive feedback for metabolic and inheritance phenomena, facilitating exploration when it is favored and exploitation when it is favored.

As metabolic systems, every living organism on this planet has special needs. As inheritance systems, organisms always are capable of producing more offspring than can be supported by the resources necessary to satisfy those metabolic needs, no matter how abundant they are. One of Darwin's insights was that all organisms overexploit their surroundings through the overproduction of offspring, and so they all struggle for survival. This is the source of the inevitable conflict between the nature of the organism and the nature of the conditions. And because the nature of

S. J. Agosta, D. R. Brooks, *The Major Metaphors of Evolution*, Evolutionary Biology – New Perspectives on Its Development 2,
https://doi.org/10.1007/978-3-030-52086-1_8

the organism supersedes the nature of the conditions, not all offspring survive. Those that survive do so because they are better than others at exploiting the conditions, or because they have the capacity to explore new options in fitness space. Those that die lack the capacity to cope or explore adequately, or the luck to get to the dinner table on time.

Key to coping with these self-generated conflicts are capacities to explore new options in fitness space. This includes the ability to persist even if exploration leads to reduced fitness; a marginal existence is still existence (Olenina et al. 2015; Holland et al. 2017). At the same time, coping with conflict is not solving the problem that gave rise to it. Coping behavior, by itself, gives rise to a kind of dynamic stasis, in which overproduction is balanced by death and accident. Such a system cannot give rise to the abundance of life stemming from a process of diversification whose existence is evident in the Tree of Life. That requires conflict resolution. Conflict resolution results in a permanent change in the status quo, signified by the emergence of two or more independent inheritance systems from a single ancestral inheritance system. The efficacy of the evolutionary process depends less on the cost of the conflict than on whether or not the conflict is resolved. Whatever eliminates a given conflict will be favored in the evolutionary game of indefinite survival with benefits rippling outward in space and time. This does not happen on demand, and there is no guarantee that any particular conflict will ever be resolved evolutionarily. But the record of phylogeny tells us that conflict resolution occurs often enough to produce abundant biodiversity.

When conflict resolution occurs, it is a complex two-step process in which ecological fitting in sloppy fitness space sets the stage.

8.1 Compensatory Changes: Diversifying Your Portfolio

Physical scientists like a stable and predictable world, one in which for every action there is an equal and opposite reaction. Everything is conserved, there are no untidy bits left lying around. A world like that allows them to predict the outcome of any action before it occurs. The biological world is maddeningly unruly in this regard. While there is always a reaction to any stimulus, internal or external, the reaction is rarely equal or opposite in any reliable sense. The reactions are always contextualized by historically conservative inheritance and the nature of the conditions, so they can be easily explained after the fact but never fully predicted.

Evolvable life is full of such phenomena. Darwin ascribed this kind of phenomenon to what he called the mysterious laws of the correlation of parts, what many now call evolutionary developmental dynamics (evo-devo: Carroll 2005). Brooks and Wiley 1986, 1988) called them compensatory changes, suggesting that they existed throughout living systems, but produced no general principle to link them. Brooks and Agosta (2012) realized that all such compensatory changes were, in some form, density-dependent responses. For example, whenever there is gene duplication, chromosomal duplication, or even genome duplication, the total amount of genetic activity often remains relatively constant, a phenomenon known as dosage compensation (Muller 1914; Sturtevant 1919; for a review see Gartler 2014). Another example is cell division, which is initiated in large part by an increase in cell volume, creating a density-dependent compensatory change. More generally, the deformation of tubular structures into hexagons when densely packed occurs widely in living systems, from microfiber bundles in muscle tissues to the cells in beehives.

The outcomes of density-dependent phenomena can be largely predicted for developmental processes because they occur repetitively. Effects stemming from conflicts within fitness space cannot be anticipated to the same extent. Despite historical conservatism, capacity space is variable, and organisms are capable of substantial plasticity, and opportunity space is likewise complex and variable. Some things persist, but nothing repeats. Density-dependent interactions in fitness space are thus always highly contingent and context dependent (Santos et al. 1997; Juarrero 1999; Pie et al. 2006; Hiermes et al. 2015; Kelemen et al. 2019).

Spea bombifrons is a species of spadefoot toad. Fossorial creatures, males and females emerge from underground burrows to mate and lay eggs in temporary pools created by massive rainstorms. The larvae are herbivorous initially. If the temporary pool evaporates too quickly, increasing larval density, they will forsake their vegetarian lifestyle for aggressive cannibalism. With each sibling consumed, the successful cannibals ingest a thyroid gland, which increases the concentration of thyroxin in its body. The effect of the increased thyroxin concentration is accelerated rates of metamorphosis that allow the larvae to escape the Scylla of vanishing water and the Charybdis of cannibalism in an evaporating pool. The compensatory changes are aspects of ecological fitting: (1) staying in the same spatial place but occupying a new part of fitness space (cannibalism) that would not be occupied except for density-dependent compensatory changes; and (2) rapid metamorphosis allowing spadefoot toads to escape the cannibalism trap by dispersing to a spatially-distinct portion of fitness space.

Density-dependent compensatory changes catalyzed by conflicts within fitness space give us our starting point for understanding conflict resolution. We begin where we ended the last chapter, with ecological fitting in sloppy fitness space. Reproductive overrun produces unsustainably high densities of organisms all wanting the same resources. This leads to conflicts in fitness space, which catalyze exploration-biased activities allowed by ecological fitting, which we understand as a form of compensatory change (Mueller et al. 1991). Exploration by ecological fitting has two outcomes. Geographic dispersal, simply moving away from the area in which conflict is most intense, is the most general (Bitume et al. 2013). The second is the exploration of novel, though marginal, fitness space, determined by the suite of capacities for ecological fitting available to a given system. But there is a cost to coping with conflicts in fitness space in this way. In both classes of outcomes, realized fitness space increases. And while that eases the initial conflict by reducing the density of organisms wanting the same things, it also reduces connections within the inheritance system.

Changes from exploitation-biased to exploration-biased behavior occur when ecological fitting allows some members of a living system to move away (spatially and/or functionally) from a densely populated portion of fitness space to one more marginal in terms of resources but more congenial in terms of density of those wanting the same resources. Such compensatory changes do not guarantee conflict resolution. They do, however, lay the groundwork for conflict resolution by making the realized fitness space larger and more complex; they are the means by which inheritance systems maintain a diversified portfolio of evolutionary options (Schindler et al. 2015). Compensatory changes in fitness space thus weaken correlations between the system and its surroundings (connectivity) as well as among members of the system itself. Once some members of a living system have relocated to a portion of fitness space removed from the original conflict, therefore, we expect it to switch from exploration-biased back to the default exploitation-biased behavior but in a new context. In that new fitness space, exploitation may be based on capacities inherent in the bag of tricks of ecological fitting that were not used in the original portion of fitness space. And that is the basis for the next stage in conflict resolution.

8.2 Cohesion: Making Distinctions

Homologous parts tend to vary in the same manner, and homologous parts tend to cohere.—Darwin (1872)

Cohesive properties are deeply embedded throughout living systems, comprising capacities that produce boundaries between system and surroundings. Cohesive properties allow aspects of the nature of the organism to enter into tighter functional correlations with each other than with elements of the external environment. Cohesive properties embodied in inheritance systems thus allow the nature of the organism to be relatively autonomous from, or at least insensitive to, the nature of the conditions. They are what allow organisms to be cohesive wholes, from single cells (membranes) to genes organized into chromosomes, to chromosomes sequestered within a membrane within a cell (eukaryotes), to multicellular organisms (cell–cell adhesion, recognition, and specialization), to sexual reproduction and specific mate recognition systems. They are the built-in evolutionary capacities that increase the efficiency of storing and transmitting inherited information about exploiting environmental resources necessary for survival (Maynard Smith and Szathmary 1995). Cohesive properties include the evolutionary succession of functional traits that enhance the ability of the living system to survive by exploiting available resources (Caianiello 2015). Collier (1998, 2000) used the metaphor of a kite to explain his perception of cohesion—the more cohesive the various components of a kite, the greater the lift capacity and the better it will fly.

Collier suggested that cohesive properties could be generalized as those which increase the mutual information between the system needs and environmental resources, in which case the biological equivalent of "lift" for the kite is survival. At the same time, the cohesive properties embodied in inheritance create the conflicts in fitness space from which natural selection emerges. Paradoxically, cohesive properties are also an essential complement to compensatory changes in resolving conflicts. Exploitation-biased behavior, permitted by cohesive properties, inevitably leads to conflict which triggers exploration-biased behavior, permitted by compensatory changes, in the form of ecological fitting. In a complementary fashion, when compensatory changes establish a living system in a low-conflict region of fitness space, cohesive properties allow the system to shift from exploration to exploitation. Exploitation always trumps exploration and exploitation tends to increase mutual information between organisms and their surroundings. Once that behavior has allowed a system to move into a part of fitness space where exploration-biased behavior is no longer essential, we expect the system to return to the default exploitation-biased behavior, switching from reliance on compensatory changes to reliance on cohesion.

Exploitation-biased activities include forms of cohesion that lead to (a) decreased autonomy of lower levels and increased opportunities at emergent higher levels or (b) division of labor. These cohesive properties thus resolve conflicts by creating distinctions from which new diversity emerges and they are solidified by making it easier to exploit new opportunities. In this way, cohesion also creates new fitness space. The new living system becomes more cohesive, thus more isolated from its relatives. This produces conflict resolution by creating a portion of the fitness space in which conflict is weak enough for innovations to emerge and become integrated into the newly isolated and cohesive inheritance system. Innovations that increase cohesion (by increasing the efficiency of storing and transmitting information or by increasing the functional fit between system and surroundings—increasing mutual information) reinforce the conflict resolution and become integrated into the inheritance system. Darwin recognized a sense of the importance of cohesion in exploitation-biased behavior when accounting for the geographic range restrictions of many inheritance systems

> From these considerations I think we need not greatly marvel at the endemic species which inhabit the several islands of the Galapagos Archipelago, not having all spread from island to island. On the same continent, also, preoccupation has probably played an important role in checking the commingling of the species which inhabit different districts with nearly the same physical conditions.—Darwin (1872)

This is minimal conflict resolution. The newly distinct living system becomes more cohesive, and thus more distinct from its relatives in the original portions of fitness space. At the same time, new fitness space is generated. Changes in capacity are not necessary for compensatory changes and cohesion to combine to resolve conflicts.

8.3 Visualizing Conflict Resolution

We now attempt to visualize the process of conflict resolution described above. As we tell this part of the story using simple diagrams as visual metaphors, it is important to remember three things. First, the diagrams represent fitness space, which is an abstract space including functional as well as spatial distinctions. Second, low-density marginal fitness space can be more survivable than high-density preferred fitness space ("he that fights and runs away..."). Third, the localization and behavior of living systems in fitness space can be envisioned in a number of ways. We can think of this in terms of matter and energy, of diffusion along resource gradients as suggested by the mass flow and resource–consumer models of Lotka and Volterra, more recently called "entropy gradients" by some. This appeals to more traditional thermodynamic views, thinking in terms of fluxes and flows of energy and materials, emphasizing that there is a cost to every action by a living system. We can also think of this in terms of correlations, in which conflicting interactions among organisms are analogous to particle collisions, more of a statistical mechanical perspective, in which correlations can be thought of in terms of frequency or probability. This view emphasizes that there is an effect of every action by a living system. Finally, we can think of this in terms of information, in which the interactions producing correlations are exchanges of inherited information. This emphasizes that every action by a living system involves the actualization of some information paid for by the dissipation of potential. We believe the differences between these three views of the nature of the organism in fitness space are only nomenclatural, different ways of representing the same fundamental concept.

We begin our hypothetical visualization with an inheritance system maximally localized and tightly correlated in fitness space (Fig. 8.1). At this point, the system's

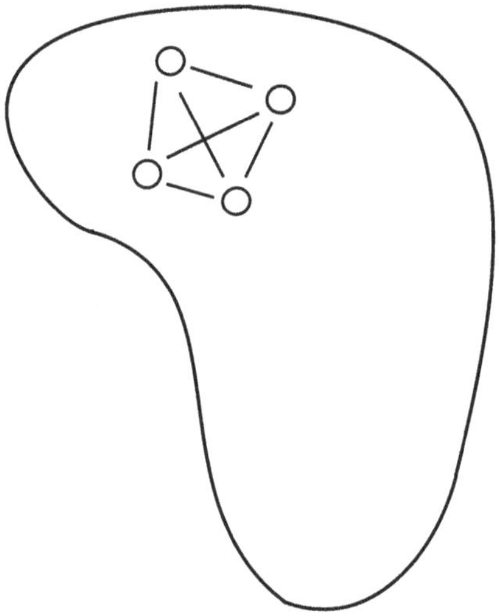

Fig. 8.1 Visualizing conflict resolution in fitness space: Step 1. The inheritance system (circles connected by lines) is maximally cohesive and therefore specialized in its fundamental fitness space. As a result, fitness space is maximally sloppy

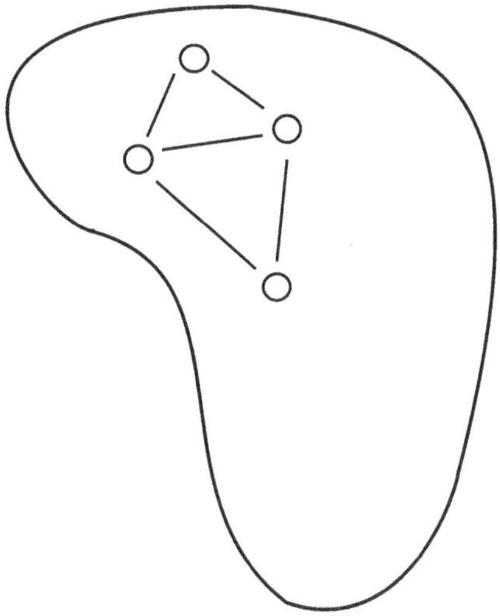

Fig. 8.2 Visualizing conflict resolution in fitness space: Step 2. In response to conflict within fitness space, the inheritance system begins to expand by some members exploring new parts of fitness space. This causes the system to become less cohesive and therefore more generalized in its fundamental fitness space. As a result, fitness space becomes less sloppy

realized fitness space is maximally sloppy because it occupies the smallest possible subset of its fundamental fitness space. The system is maximally cohesive (specialized) in fitness space, and we expect it to exhibit high fitness and low genetic variance as a result of the maximal exchange of inherited information among the localized members of the system.

The inevitable overproduction of offspring relative to available resources leads to density-dependent triggering of compensatory changes. Some members of the inheritance system move via ecological fitting to marginal fitness space (Fig. 8.2). This expands the system in fundamental fitness space, so the system is more generalized (diffuse) and fitness space is less sloppy. At the same time, correlations with the original system decrease in frequency, so correlation density decreases overall as the system becomes more spread out in fitness space (Fannjiang et al. 2004).

The process continues, and as the expanding portion of the system finds survivable fitness space, cohesive properties lead to the emergence of a new focus of exploitation, indicated jointly by an increase in correlations among members of the new subsystem and continuing decreases in correlations with members of the original system (Figs. 8.3 and 8.4). These effects can be extended in time through inherited connectivity (Bitume et al. 2013).

Conflict resolution occurs when connectivity between the two subsystems is lost (Fig. 8.5).

Two systems have now emerged from the original, indicating that conflict has been resolved (Fig. 8.6). There are now two separate systems, each maximally restricted in its new fundamental fitness space, as cohesion produces localized (specialized) emergent systems. Each system has its own fundamental fitness space, which is maximally sloppy, thus having the maximum capacity to produce conflict, but also to cope with and potentially resolve, those new conflicts. Loss of

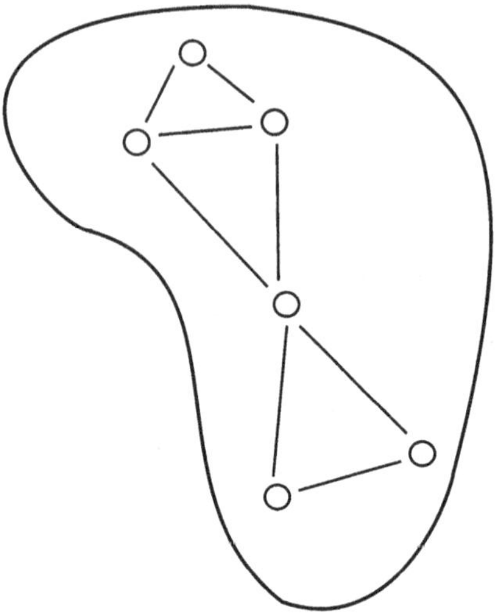

Fig. 8.3 Visualizing conflict resolution in fitness space: Step 3. Continued exploration and expansion of the inheritance system into new parts of fundamental fitness space leads to the emergence of a new focus of exploitation, increasing cohesion among members of the emerging subsystem while simultaneously decreasing cohesion with the original system

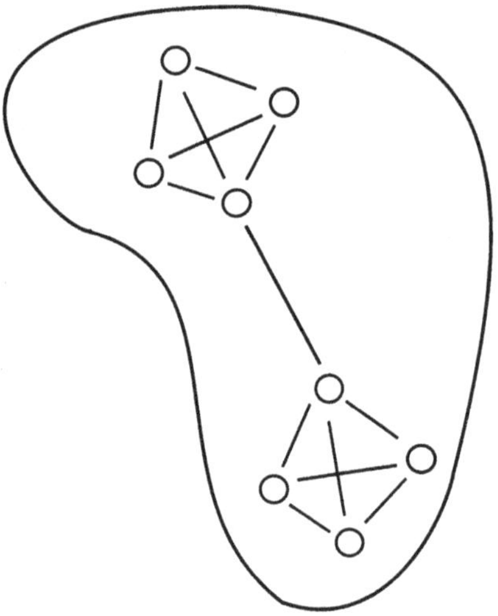

Fig. 8.4 Visualizing conflict resolution in fitness space: Step 4. The new focus of exploitation eventually produces two distinct subsystems where within-system cohesion is much higher than between-system cohesion

ancestral connectivity need not lead to a loss of genetic diversity as some fear (Jangjoo et al. 2016), and in fact is essential for setting the stage for the emergence of new genetic diversity.

Most significantly, evolutionary innovation emerges in what was originally marginal, not the most preferable, fitness space.

These visualizations can be described by algorithmic entropy formulations we discussed in Chaps. 6 and 7. That allows us to understand that all the nomenclatural variants describing what is depicted in Figs. 8.1, 8.2, 8.3, 8.4, 8.5, and 8.6 refer to the same phenomenon. Think of Fundamental Fitness Space as the boundary of the system (the realm of maximum possible entropy, H_{max}) and Realized Fitness Space

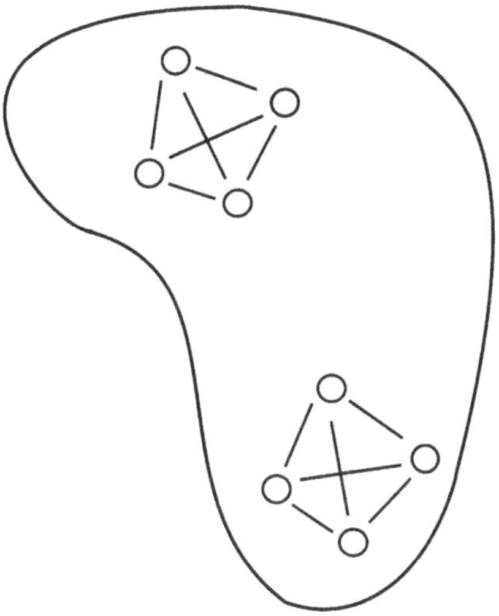

Fig. 8.5 Visualizing conflict resolution in fitness space: Step 5. Conflict resolution occurs when connectivity between the new system and original system in fitness space is lost

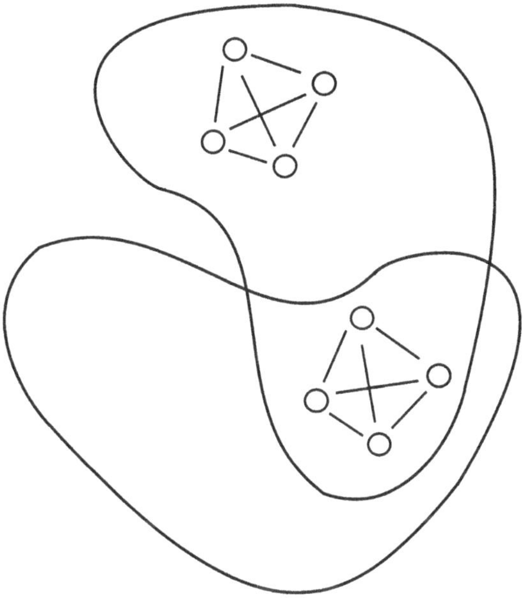

Fig. 8.6 Visualizing conflict resolution in fitness space: Step 6. Conflict is resolved when two new inheritance systems emerge from the original system. Each system now has its own separate fundamental fitness space in which it is maximally cohesive and specialized. As a result, fitness space is again maximally sloppy, returning each inheritance system back to Step 1 where new conflict arises

as the entropy of the system as it exists within the boundaries (H_{obs}). In exploration mode, entropy increases as the system becomes extended in fitness space. When exploitation mode emerges in a new portion of fitness space, entropy still increases, now manifested as a new focus of locally increasing density. At this point, there is still a single system (though more complex than initially) and a single Fundamental Fitness Space. Realized Fitness Space has grown, so it seems that the cost of coping with conflict by moving into marginal fitness space is decreasing the sloppiness of fitness space. Entropy increases are thus inversely proportional to the sloppiness of fitness space.

The difference between H_{max} and H_{obs} (sometimes called the relative entropy) fluctuates, however, because capacity and opportunity are relatively autonomous, leading to dynamic phenotypic variance gradients (Pertoldi et al. 2014). New Fundamental Fitness Space does not emerge until conflict resolution is achieved. At that point, conflict resolution generates a meta-system of two independently operating systems rather than one. The boundary conditions (H_{max}) of the combination of the two new systems (Fig. 8.6) is higher than the ultimate configuration of the ancestral system (Fig. 8.5), but each system is in a low-entropy configuration relative to that previous state. In addition, because each new system now generates its own Fundamental Fitness Space (Fig. 8.6), each exists in a relatively specialized (sloppy fitness) state, which embodies the potential for new episodes of conflict, due to reproductive overrun, as well as new episodes of coping with conflict and the potential for conflict resolution.

8.4 The Meaning of Conflict: Beauty Is in the Eye of the Beholder

Organisms talk to themselves constantly. This is not literally true, of course, but organisms engage in sophisticated communication with and about themselves, with and about their surroundings. They cope with a variety of conditions through self-referential features, physicochemical mechanisms including feedback loops and homeostatic mechanisms that signal their internal state. Most of these internal-state signaling systems enhance the cohesion and functional integration of the organism. Cell, molecular, and developmental biologists routinely describe molecules, cells, and tissues as signaling each other during ontogeny and reproduction. In addition, life would not exist, and evolution could not occur in the absence of communication between the organism and its surroundings. Signaling mechanisms are thus of two types, those in which the organism signals itself about itself and those in which the organism signals itself about its surroundings (Brooks and McLennan 1997; Maynard Smith 2000; Witzany 2018). Organisms are, as Darwin put it, preoccupied with the conditions of life. All of that signaling involves a single objective—survival.

Organisms are simultaneously in the environment and part of the environment; indeed, most of each living system's fitness space comprises other living systems (Maynard Smith 1976). As a result, some internally originating signaling systems affect the organism in which they originate in ways that are apparent to other organisms, and those organisms may respond in ways that affect the system in which the signals originated. Just as the effects of internal signaling in the sender may have external manifestations that are irrelevant to the organism that produced them, the ability to receive such a signal and respond to it are properties of the receiving organism's internal signaling system.

8.4.1 *Intention in Biological Signals: The Sender*

Much biological signaling is talking to yourself, with unintentional broadcasting into the surroundings. Some signaling, however, seems to be intentional (Juarrero 1999, 2000). Intentionality in signaling requires an impulse to send a particular signal and the use of attributes of the organism to send it. In a Darwinian world, intentionality emerges when the impulse to send a particular signal originates, and the signal is sent using existing attributes of the organism. New signals are not constructed in order to express the impulse. For example, regardless of our perceptions of how significant mimicry is, mimic species did not evolve because ancestral species desired to become mimics. A male stickleback fish is ready to mate and signals his readiness. Did the intention to mate emerge and somehow cause the signal to appear? Or did signal and intention emerge as a result of the same non-intentional changes in photoperiod and water temperature that produced physiological and hormonal changes in the fish, resulting in the emergence of the intention to mate and the signaling of those intentions as a by-product (emergent property)? Death is surely unintentional in most cases, and yet it is a highly unambiguous signal to myriads of organisms whose lives depend on decomposing dead organic matter.

8.4.2 *Meaning in Biological Signals: The Receiver*

Regardless of intentionality, a sender cannot control the way in which a signal is interpreted. A male stickleback's signal to conspecific females that he is ready to mate is also a signal to himself of his having ingested a substantial quantity of high energy food and is also a signal to predators in the environment that he is edible. He has eaten high energy food, and that fact is manifested in an unintentional change in his body color. He intends to mate, and his intention is signaled in the same change in body color. He does not intend to be eaten, but he signals to some receivers that he is edible. So long as mating occurs more often than, or prior to, being eaten, sticklebacks can cope with the conflict in fitness space produced by predators (just because you look like lunch does not mean you are on the menu today).

Most signals can, and do, mean different things to different observers. The sense of meaning, that of the information conveyed by the signal, requires us to examine the receiver rather than the sender. If a signal is perceived but the receiver does not respond to it, the signal has no meaning. If a tree falls in the woods and there are no humans present, there are still signal-produced soundwaves. But without humans, the soundwaves will have no meaning—there will be no sound, if by sound we mean something perceived by the human ear–brain complex. This is because meaning can emerge only from the actions of the receiver after receiving a signal. When there is a response to a signal, it will have meaning for the sender only when the sender is the receiver, i.e., in cases of self-signaling. Otherwise, meaning will rest entirely outside of the sender.

In a world in which receivers evolve independently of the desires and intentions of senders, intention may not be evolutionarily relevant. Only the outcome of signal–receiver interactions matter. And those interactions comprise a major portion of fitness space.

8.5 Fitness Space: A Complex Mix of Signals and Messages

The more closely aligned signal and response are, the more mutual information is shared between signaler and receiver. It might seem at first glance that any evolutionary process increasing mutual information would be highly beneficial. But this is only conditionally true.

Compensatory changes require some mutual information between the organism and its surroundings, but too much mutual information risks becoming an evolutionary cul-de-sac; the closer the "fit" between biological systems and their environments, the less space there is for coping with, or resolving conflicts (Collier's Paradox: Collier 1986; Brooks 2001; see also Ellers et al. 2012). How do inheritance systems avoid this trap? Ecological fitting in sloppy fitness space.

Mutual information will tend to increase due to cohesive properties and exploitation of the conditions, which takes precedence over the exploration of fitness space. But the autonomous overproduction of offspring means that the more mutual information the more rapidly density-dependent effects will emerge, triggering compensatory changes and movement into marginal fitness space, where there is reduced density of organisms and lowered mutual information. Clonal organisms living in preferred fitness space exhibit maximum mutual information, yet quickly overproduce, again triggering compensatory changes in some offspring that move them into marginal fitness space, where mutual information is reduced (Darwin's Necessary Misfit).

Compensatory changes move members of a given system away from conflict without knowing the consequences, and where they end up in fitness space and what they end up doing are the result of compatibilities and incompatibilities between the sender (intention) and receiver (meaning). There cannot have been conflict resolution without compensatory changes, so conflict resolution always begins with relatively low mutual information, and evolutionary innovations that increase mutual information arise in marginal fitness space.

Fitness space can be envisioned as the portions of opportunity space in which a given system has sufficient mutual information about its surroundings to survive and reproduce—this is the domain in which members of the system have fitness. The region of fitness space in which the highest level of mutual information is achieved is the region of maximum fitness. That state is always temporary and contingent. This means that being maximally fit—having maximum mutual information—cannot be more meaningful evolutionarily than simply being fit—having an adequate amount of mutual information.

Although signals do not have to be intentional on the part of the sender, they must have meaning to a receiver. Meaning, however, need not be singular or unambiguous. Organisms always mean something to themselves (the sender is the receiver), though that meaning changes over time and according to the "conditions of life" (Darwin 1872). Organisms also mean something to their surroundings (the sender and the receiver are different organisms), but that meaning is as varied as the number of different organisms in the environment and the ways in which their internal signaling systems perceive the nature of the conditions. In addition, organisms are both signals and signal-bearers, because some of their attributes may be perceived by other organisms in their surroundings. If external signals are by-products of internal signaling, then the effects of such "internal conversations" may have more than one meaning to observers in the environment. The evolutionary fate of signal bearers will be an outcome of the ways in which other organisms perceive the signal bearer. And those perceptions may well conflict with each other (Brooks and McLennan 1997).

Consider an organism, like the male stickleback above, that changes size or color or behavior when it eats a particular type of high-energy food. The change in color is an unintentional by-product of an internal signal of well-being (more energy coming into the organism's metabolic system) that may also make the organism highly visible to predators. Mechanisms that enhance the efficiency of obtaining high-energy food and minimize the external manifestations of that success will be cohesive. Signals are indications of internal state and may or may not indicate some intention. As a result, what creates conflict in one part of fitness space may resolve it in another, because intention and meaning in biology are decoupled. The possible outcomes of all interactions create an array of possible meanings for any signal projected into the environment. All meaning in this sense is contextual and contingent. The range of meanings, and the richness of meanings, will be proportional to the diversity and sophistication of the sensory systems and complexity of the surroundings.

Fitness space is thus a complex and contingent mix of signals produced by organisms preoccupied with their own existence, in which there are multiple possible outcomes associated with the same effect (Brooks and McLennan 1997; Kaiser et al. 2018). Different observers, each with a different construction and constitution, monitor their surroundings in different ways. They may perceive the signaler in different ways because they perceive different attributes (no organism is capable of perceiving all the signals in its surroundings), they may perceive the same attribute of the signaler in different ways (e.g., different photoreceptors have different spectral sensitivities), and they may respond to the same signal in different ways. It is in the nature of organisms to assess themselves ("How am I? I am hungry.") and their surroundings ("There is a red stickleback, that means food to me. I will eat him.") through elaborate self-signaling. The hunter did not intend to seek a red signal in order to eat (indeed this is one reason predators can make mistakes and try to eat something distasteful). The stickleback did not intend to send a signal meaning "I am edible, come eat me." But the temporal and spatial conjunction of two different self-absorbed and self-preoccupied systems, one acting as a "sender" and the other as a "receiver" in any particular interaction, routinely leads to predictable outcomes in the biosphere. These responses are not dictated by the senders, so the question of meaning is a question of the properties of the receivers ("beauty is in the eye of the beholder:" Brooks and McLennan 1997; Hutton et al. 2015). A lack of appreciation for the Darwinian relationship between the sender and receiver could lead one to believe that ambiguities in meaning on the part of receivers were the result of intentional dishonesty on the part of the sender (Popat et al. 2015).

8.6 The Nature of Selection

If intention and meaning are decoupled, natural selection becomes a metaphor for the process of coping with conflict and conflict resolution. Selection is not an imposition on the nature of the organism by the nature of the conditions. It is an emergent property of organisms imposing themselves on the surroundings, including other organisms. There is a duality in the manifestations of selection, which we will call metabolic and inheritance selection. This duality is more commonly known by other names such as environmental and sexual selection (Fisher 1930, 1958); interactors and replicators (Hull 1988); and the ecological and genealogical hierarchies (Eldredge and Salthe 1984; Eldredge 1985, 1986; Salthe 1993). Both manifestations of selection are associated with increased cohesion (mutual information) within the inheritance system. This can occur indirectly by increasing cohesion between the inheritance system and the surroundings (external cohesion of Collier 1998, 2000; Collier and Hooker 1999), or directly by increasing the efficiency of storing and transmitting inherited information (Maynard Smith and Szathmary 1995) (for a review of genomic evidence of the ubiquity of selection, see Haasl and Payseur 2015).

Metabolic selection (think of activities involving getting food and avoiding becoming food) improves the efficiency of the inheritance system indirectly by increasing mutual information with the surroundings. We expect macroscopic information (I), or organization, to increase over time in evolving systems; therefore, we expect to see an increase in the macroscopic order (Landsberg 1984), or redundancy (Gatlin 1972) of the system (Q):

$$Q = 1 - (H_{\text{obs}}/H_{\text{max}}) = I/H_{\text{max}}$$

If the environment is the primary source of macroscopic ordering for biological systems, then the mutual information between the system and environment should be high. This leads to Collier's Paradox (Collier 1986): if mutual information between metabolic systems and their surroundings is high, evolutionary change will be negligible or stochastic with respect to the environment. Darwin's Necessary Misfit resolves Collier's Paradox. The power of metabolic selection will then be inversely proportional to the amount of mutual information between the inheritance system and its surroundings.

Inheritance selection (think of activities involved in searching for sex and having sex) increases the cohesion (mutual information) between potential mates, thereby increasing the efficiency of the inheritance system without regard for the surroundings; these are the selective increases in efficiency of storing and transmitting information of Maynard Smith and Szathmary (1995). Inheritance selection emerges from increased mutual information between members of the genealogical hierarchy; thus, it is quite different from natural selection, as Darwin (1871) thought. In order for inheritance selection to be effective, potential mates must interact with each other in ways that influence the flow of inherited information from one generation to another. The degree to which sexual selection is effective, therefore is directly proportional to the degree of mutual information, and it is genealogical processes that are responsible for shared information, so sexual selection is an evolutionarily cohesive process.

Metabolic selection emerges as a result of a lack of sensitivity (low mutual information) between the genetic system and the environment. Inheritance selection, by contrast, emerges as a result of increased sensitivity (high mutual information) among members of the inheritance system. Metabolic and inheritance selection arise in different ways and affect different parts of the information flow system, so it is possible for them to be in conflict (e.g., Fisherian runaway sexual selection theory), or to complement each other (e.g., "truth in advertising" views of sexual selection) (Brooks 2001). In a study of male and female courtship behavior in a species of stickleback fish, McLennan (1996, 2000) concluded that "...the origin and the intensity of female nuptial coloration is tightly coupled with ovulation and the origin and the intensity of female courtship behavior, and is as distinct from the male signal as possible. Both the redundancy between female colour and female behaviour, and the disparate natures of the male and female signal, interact to reinforce the female's message of courtship readiness. This, in turn, increases the probability that the encoded information will be received..." This high efficiency of mate assessment, choice, and spawning (inheritance selection) leads to the fish being exposed for a shorter time to potential predators (metabolic selection).

Similar lines of investigation and explanation invoking selection as a positive feedback mechanism have been developed for coping with and resolving conflicts within cells (intracellular fitness space) and between cells (developmental fitness space) (Maynard Smith and Szathmary 1995 have been the most comprehensive proponents of this perspective). The combination of these different kinds of fitness spaces with similar emergent forms of selection is sometimes referred to as multilevel selection theory (e.g., Tuomi et al. 1988; Wilson et al. 2008).

8.7 Summary

Darwinism is summarized by the Jagger principle: You can't always get what you want, but if you try sometimes, you just might find you get what you need.—Daniel R. Brooks and Deborah A. McLennan (2002)

Changes in the nature of the organism do not occur in response to, and occur more slowly than, changes in the nature of the conditions, so fitness space will always be sloppy. Coupled with the capacity for ecological fitting, there will always be the potential for inheritance systems to cope with and resolve conflicts in fitness space. Historically conservative inheritance is the agency, and survival creates a feedback loop perpetuating the lessons of the living, and never repeating the lessons of the dead. Natural selection is any positive feedback in that loop. Ecological fitting in sloppy fitness space allows us to see natural selection as an emergent property producing positive feedback in which conflict resolution sets the stage not only for new conflict but for new conflict resolution. Natural selection thus co-emerged with genetically autonomous living systems and has always been part of the process (Maynard Smith and Szathmary 1995, 1999).

Compensatory changes in the form of ecological fitting leading to coping with conflict is natural selection. Compensatory changes in the form of ecological fitting leading to conflict resolution sets the stage for evolution by natural selection. Compensatory changes initiate conflict resolution, and cohesion completes it. Cohesion will always override compensatory changes because exploitation overrides exploration, so no matter how much preexisting correlation structure is disrupted by compensatory changes, cohesion will lead to a new correlation structure. New cohesion thus creates the potential for new conflicts. Natural selection thus plays a synergistic role in all episodes of conflict resolution (Maynard Smith 1982; Maynard Smith and Szathmary 1995; Corning 2003, 2005; Corning and Szathmary 2015).

Conflict resolution changes the conflict arena but does not eliminate it—so we expect all conflict resolutions to set the stage for new conflicts. Intention and meaning will always be decoupled, so there will always be conflict. And because intention is in the sender and meaning is in the receiver, all conflict resolutions will set the stage for new conflicts: you cannot escape selection because you cannot escape fitness space, you can only alter it. Conflict may not bring resolution, but conflict resolution will bring new conflict. That new conflict will focus on the form of cohesion that characterized the previous resolution. A full cycle of compensatory changes breaking old correlations leading to new correlations in the form of novel cohesion producing conflict resolution and a new form of selection is an evolutionary transition (Darwin 1872; Maynard Smith and Szathmary 1995; Szathmary 2015). Oscillations between compensatory changes and cohesion may occur without the permanent resolution of conflicts. But when conflict resolution does occur, permanent increases in diversity result.

Compensatory changes decrease old correlations due to inheritance, allowing an inheritance system to become generalized in fitness space—loss of correlations alone, however, does not produce new inheritance systems. Cohesion leads to specialization and then isolation, emergent autonomous systems. Conflict resolution is characterized by the emergence of a new form of cohesion, from which a new element of selection emerges (i.e., evolutionary transitions). Evolvable life is constant conflict, widespread coping with conflict, occasional conflict resolution, and rare transitions leading to Darwin's "endless forms most beautiful and most wonderful."

The astute reader will realize that our metaphorical framework for understanding conflict in fitness space, coping with those conflicts, and conflict resolution is consistent with phenomena well known to neo-Darwinism up to the Extended Hardened Synthesis of today. We do not think the similarities between our perspective and Sewall Wright's shifting balance theory (Wright 1932, 1956, 1978, 1982) are coincidental (Hodge 2011; Ishida 2017). Compensatory changes result in decreased fitness and increased variance, while cohesion results in increased fitness and decreased variance. We also think it might be useful to wonder if Fisher had used additive genetic variance rather than fitness as his token of entropy, would he have produced the statistical mechanics of evolution he sought? Would that have allowed him to see that there was a biological arrow of time and that it was the same as the thermodynamic arrow of time, as suggested by Boltzmann in 1905 (see Broda 1983)?

Our perspective flows from a Darwinian base, driven by the nature of the organism, rather than a neo-Darwinian base, driven by the nature of the conditions. As a result, selection can be characterized as an emergent property, an effect, an outcome, a constraint, or a feature depending on your nomenclatural preferences. From this perspective, everything about natural selection can be explained in the manner that Darwin himself discussed it. Natural selection is an emergent property of preoccupied organisms imposing themselves on an indifferent environment. As a result, there will always be conflict. This does not sound very optimistic. And yet, just as we see reproductive overrun not as wasteful but the cost of doing business, never costing more than the inheritance system can afford, we see selection not as a brutal culling and vicious fight for ascendancy but rather a means of positive feedback enhancing the abilities to cope with the self-generated conflicts and reinforcing episodes of conflict resolution. Natural selection is a "perfecting" mechanism in the sense that it reinforces life's ability to persist and evolve.

The most fundamental form of compensatory change is running away, and the most fundamental form of cohesion is inbreeding. So, the most basic form of conflict resolution is localized inbreeding in marginal fitness space. Historical ecological assessments suggest that the vast majority of conflict resolution is of this form, the emergence of divergent inheritance systems living in different places but mostly engaged in the same kind of activities inherited from a common ancestor (Ross 1972a, b; Boucot 1975a, b, 1978, 1981, 1982, 1983, 1990; Wanntorp 1983; Brooks 1985; Brooks and McLennan 1991, 2002; Brooks and Boeger 2019). But phylogeny also highlights much more interesting and complex episodes in the history of life on this planet. These evolutionary transitions are the sagas of evolution, embodied in Darwin's first major visual metaphor, the Tree of Life. Next, we begin telling those sagas.

References

Bitume EV, Bonte D, Ronce O, Bach F, Flaven E, Olivieri I, Nieberding CM (2013) Density and genetic relatedness increase dispersal distance in a subsocial organism. Ecol Lett 16:430–437

Boucot AJ (1975a) Standing diversity of fossil groups in successive intervals of geologic time viewed in the light of changing levels of provincialism. J Paleontol 49:1105–1111

Boucot AJ (1975b) Evolution and extinction rate controls. Elsevier, New York

Boucot AJ (1978) Community evolution and rates of cladogenesis. In: Hecht MK, Steere WC, Wallace B (eds) Evolutionary biology, vol 11. Plenum, New York, pp 545–655

Boucot AJ (1981) Principles of benthic marine paleoecology. Academic Press, New York

Boucot AJ (1982) Paleobiologic evidence of behavioral evolution and coevolution. Author, Corvallis, OR

Boucot AJ (1983) Does evolution take place in an ecological vacuum? II. J Paleontol 57:1–30

Boucot AJ (1990) Community evolution: its evolutionary and biostratigraphic significance. In: Miller W III (ed) Paleocommunity temporal dynamics: the long-term development of multispecies assemblies. The Paleontological Society Special Publications 5: 48–70

Broda E (1983) Darwin and Boltzmann. In: Geissler E, Scheler W (eds) Darwin today: the 8th Kühlungsborn colloquium on philosophical and ethical problems of biosciences. Abhandlungen der Akademien der Wissenschaften der DDR. Akademie, Berlin, pp 61–70

Brooks DR (1985) Historical ecology: a new approach to studying the evolution of ecological associations. Ann Mo Bot Gard 72:660–680

Brooks DR (2001) Evolution in the information age: rediscovering the nature of the organism. SEED 1: 37 pp. http://www.library.utoronto.ca/see

Brooks DR, Agosta AJ (2012) Children of time: the extended synthesis and major metaphors of evolution. Fortschr Zool 29:497–514

Brooks DR, Boeger WA (2019) Climate change and emerging infectious diseases: evolutionary complexity in action. Curr Opin Syst Biol 13:75–81

Brooks DR, McLennan DA (1991) Phylogeny, ecology and behavior: a research program in comparative biology. University of Chicago Press, Chicago

Brooks DR, McLennan DA (1997) Biological signals as material phenomena. Rev pensee d'aujord d'hui 25:118–127. [in Japanese]

Brooks DR, McLennan DA (2002) The nature of diversity: an evolutionary voyage of discovery. University of Chicago Press, Chicago

Brooks DR, Wiley EO (1986) Evolution as entropy: toward a unified theory of biology, 1st edn. University of Chicago Press, Chicago

Brooks DR, Wiley EO (1988) Evolution as entropy: toward a unified theory of biology, 2nd edn. University of Chicago Press, Chicago

Caianiello S (2015) Succession of functions, from Darwin to Dohrn. Hist Philos Life Sci 36:335–345

Carroll SP (2005) Endless forms most beautiful: the new science of evo devo and the making of the animal kingdom. W.W. Norton, New York

Collier J (1986) Entropy in evolution. Biol Philos 1:5–24

Collier J (1998) Information increase in biological systems: how does adaptation fit? In: van de Vijver G, Salthe SN, Delpos M (eds) Evolutionary systems: biological and epistemological perspectives on selection and self-organization. Kluwer Academic, Dordrecht, pp 129–140

Collier J (2000) The dynamical basis of information and the origins of semiosis. In: Taborsky E (ed) Semiotics, evolution, energy. Shaker Verlag, Aachen, pp 111–138

Collier J, Hooker C (1999) Complexly organised dynamical systems. Open Syst Inf Dyn 6:241–302

Corning PA (2003) Nature's magic: synergy in evolution and the fate of humankind. Cambridge University Press, New York

Corning PA (2005) Holistic Darwinism: synergy, cybernetics and the bioeconomics of evolution. University of Chicago Press, Chicago

Corning PA, Szathmary E (2015) "Synergistic selection": a Darwinian frame for the evolution of complexity. J Theor Biol 371:45–58
Darwin C (1871) The descent of man, and selection in relation to sex. John Murray, London
Darwin C (1872) The origin of species, 6th edn. John Murray, London
Eldredge N (1985) Unfinished synthesis. Columbia University Press, New York
Eldredge N (1986) Information, economics and evolution. Ann Rev Ecol Syst 17:351–369
Eldredge N, Salthe SN (1984) Hierarchy and evolution. In: Dawkins R, Ridley M (eds) Oxford surveys in evolutionary biology, vol 1. Oxford University Press, Oxford, pp 182–206
Ellers J, Kiers ET, Currie CR, Bradon R, McDonald BR, Visser B (2012) Ecological interactions drive evolutionary loss of traits. Ecol Lett 15:1071. https://doi.org/10.1111/j.1461-0248.2012.01830.x
Fannjiang A, Nonnenmacher S, Wołowski L (2004) Dissipation time and decay of correlations. Nonlinearity 17:1481–1508
Fisher RA (1930) The genetical theory of natural selection. Clarendon Press, Oxford
Fisher RA (1958) The Genetical theory of natural selection, 2nd edn. Dover, New York
Gartler SM (2014) A brief history of dosage compensation. J Genet 93:591–595
Gatlin LL (1972) Information theory and the living system. Columbia University Press, New York
Haasl RJ, Payseur BA (2015) Fifteen years of genomewide scans for selection: trends, lessons and unaddressed genetic sources of complication. Mol Ecol 25:5. https://doi.org/10.1111/mec.13339
Hiermes M, Bakker TCM, Mehlis M, Rick IP (2015) Context-dependent dynamic UV signaling in female three spine sticklebacks. Sci Rep 5:17474. https://doi.org/10.1038/srep17474
Hodge J (2011) Darwinism after Mendelism: the case of Sewall Wright's intellectual synthesis in his shifting balance theory of evolution (1931). Stud Hist Phil Biol Biomed Sci 42:30–39
Holland BS, Chiaverano LM, Howard CK (2017) Diminished fitness in an endemic Hawaiian snail in nonnative host plants. Ethol Ecol Evol 29:229–240
Hull DL (1988) Science as a process. University of Chicago Press, Chicago
Hutton P, Seymoure BM, McGraw KJ, Ligon RA, Simpson RK (2015) Dynamic color communication. Curr Opin Behav Sci 6:41–49
Ishida Y (2017) Sewall Wright, shifting balance theory, and the hardening of the modern synthesis. Stud Hist Phil Biol Biomed Sci 61:1–10
Jangjoo M, Matter SF, Roland J, Keyghobadia N (2016) Connectivity rescues genetic diversity after a demographic bottleneck in a butterfly population network. Proc Natl Acad Sci U S A 113:10914–10919
Juarrero A (1999) Dynamics in action. MIT Press, Boston, MA
Juarrero A (2000) Dynamics in action: intentional behavior as a complex system. Emergence 2:24–57
Kaiser K, Boehlke C, Navarro-Pérez E, Vega A, Dudgeon S, Robertson JM (2018) Local preference encoded by complex signaling: mechanisms of mate preference in the red-eyed treefrog (*Agalychnis callidryas*). Behav Ecol Sociobiol 72:182. https://doi.org/10.1007/s00265-018-2597-0
Kelemen A, Tölgyesi C, Valkó O, Deák B, Miglécz T, Fekete R, Török P, Balogh N, Tóthmérész B (2019) Density-dependent plant–plant interactions triggered by grazing. Front Plant Sci 10:876. https://doi.org/10.3389/fpls.2019.00876
Landsberg PT (1984) Can entropy and "order" increase together? Phys Lett 102A:171–173
Maynard Smith J (1976) What determines the rate of evolution? Am Nat 110:331–338
Maynard Smith J (1982) Evolution and the theory of games. Cambridge University Press, Cambridge
Maynard Smith J (2000) The concept of information in biology. Philos Sci 67:177–194
Maynard Smith J, Szathmáry E (1995) The major transitions in evolution. Oxford University Press, Oxford
Maynard Smith J, Szathmáry E (1999) The origins of life. Oxford University Press, Oxford

McLennan DA (1996) Integrating phylogenetic and experimental analyses: the evolution of male and female nuptial coloration in the Gasterosteidae. Syst Biol 45:261–277

McLennan DA (2000) The macroevolutionary diversification of female and male components of the stickleback breeding system. Behaviour 137:1029–1045

Mueller LD, Guo P, Ayala FJ (1991) Density-dependent natural selection and trade-offs in life history traits. Science 253:433–435

Muller HJ (1914) A gene for the fourth chromosome of *Drosophila*. J Exp Zool 17:325–336

Olenina I, Vaičiukynas E, Šulčius S, Paškauskas R, Verikas A, Gelžinis A, Bačauskienė M, Bertašiūtė V, Olenin S (2015) The dinoflagellate *Prorocentrum cordatum* at the edge of the salinity tolerance: the growth is slower but cells are larger. Estuar Coast Shelf Sci 168:71. https://doi.org/10.1016/j.ecss.2015.11.013

Pertoldi C, Bundgaard J, Loeschcke V, Barker JSF (2014) The phenotypic variance gradient—a novel concept. Ecol Evol 4(22):4230–4236. https://doi.org/10.1002/ece3.1298

Pie MR, Engers KB, Boeger WA (2006) Density-dependent topographical specialization in *Gyrodactylus anisopharynx* (Monogenoidea, Gyrodactylidae): boosting transmission or evading competition? J Parasitol 92:459–463

Popat R, Pollitt EJG, Harrison F, Naghra H, Hong K-W, Chan K-G, Griffin AS, Williams P, Brown SP, West SA, Diggle SP (2015) Conflict of interest and signal interference lead to the breakdown of honest signaling. Evolution 69:2371–2383

Ross HH (1972a) The origin of species diversity in ecological communities. Taxon 21:253–259

Ross HH (1972b) An uncertainty principle in ecological evolution. In: Allen RT, James FC (eds) A symposium on ecosystematics. Occasional paper 4. University of Arkansas, Little Rock, pp 133–157

Salthe SN (1993) Development and evolution: complexity and change in biology. MIT Press, Boston, MA

Santos M, Borash DJ, Joshi A, Bounlutay N, Mueller LD (1997) Density-dependent natural selection in *Drosophila*: evolution of growth rate and body size. Evolution 51:420–432

Schindler DE, Armstrong JB, Reed TE (2015) The portfolio concept in ecology and evolution. Front Ecol Environ 13:257–263

Sturtevant AH (1919) Contributions to the genetics of *Drosophila melanogaster*. III. Inherited linkage variations in the second chromosome. Carnegie Inst Wash Publ 278:305–341

Szathmáry E (2015) Toward major transition theory 2.0. Proc Natl Acad Sci U S A 112:10104–10111

Tuomi J, Vuorisalo T, Laihonen P (1988) Components of selection: an expanded theory of natural selection. In: de Jong G (ed) Population genetics and evolution. Springer, Berlin, pp 109–118

Wanntorp H-E (1983) Historical constraints in adaptation theory: traits and non-traits. Oikos 41:157–160

Wilson DS, Van Vugt M, O'Gorman R (2008) Multi-level selection theory and major evolutionary transitions. Curr Dir Psychol Sci 17:6–9

Witzany G (2018) 2.4 Communication as the main characteristic of life. In: Handbook of astrobiology. CRC Press, Boca Raton, FL, p 91

Wright S (1932) The roles of mutation, inbreeding, crossbreeding and selection in evolution. In: Proceedings of the 6th International Congress of Genetics, pp 356–366

Wright S (1956) Modes of selection. Am Nat 90:5–24

Wright S (1978) Modes of speciation. Paleobiology 4:373–379

Wright S (1982) The shifting balance theory and macroevolution. Annu Rev Genet 16:1–19

Chapter 9
Evolutionary Transitions

Abstract Evolutionary transitions are episodes of conflict resolution resulting in irreversible changes in inheritance systems. The Tree of Life is a record of these transitions, a story of the ongoing saga of evolvable life on this planet since its inception some 3.5 billion years ago. Each episode in the saga highlights the open-endedness of evolution and the indefinite capacity of evolvable life to cope with changes in the conditions by using inherited information to buy time for continued survival. It is during these periods that evolutionary innovations emerge from the indefinite variation that is a property of inheritance systems, resolving conflicts and resulting in selective diversification. The highly conservative nature of the inheritance systems allows us to retrace this history using phylogenetic analysis. Phylogeny tells life's story "from the beginning," giving primacy to the nature of the organism and providing the essential context and polarity needed to understand the significance of evolutionary transitions. Some transitions are more significant than others, reverberating throughout the Tree of Life. These are the milestones of evolution that emerge, spread, and become widely embellished through time. The truly major transitions such as the origin of eukaryotic cells, heterotrophy, photoautotrophy, herbivory, and terrestriality change "the game of life" itself. Such transitions happen rarely, being limited both by the nature of the organism (the information is difficult to achieve, requiring the emergence of novel capacities) and the nature of the conditions (opportunities for selection favoring the transition are rare). But the evolutionary payoff for such transitions can be very large, leaving a lasting impact on the Tree of Life and strong signals for phylogenetic analysis to detect.

> The present tense. . .takes the story out of time. . .Physics is normally written in the present tense, in part because it generalizes. . .but also because it deals so much with non-directional time. Time for a physicist is quite likely to be reversible. . .No beginning, no end. All middle. The past and present tenses become useful to science when it gets involved with irreversible events, when beginning, middle and end move only in that order. "Once upon a time" tells you things are going to change.—Ursula Le Guin (1989)

S. J. Agosta, D. R. Brooks, *The Major Metaphors of Evolution*, Evolutionary Biology – New Perspectives on Its Development 2,
https://doi.org/10.1007/978-3-030-52086-1_9

Phylogeny is the once upon a time of evolvable life's indefinite capacity for survival. The sagas of phylogeny are episodes in which coping with conflict in fitness space leads to conflict resolution. Conflicts in fitness space that cannot be resolved or are not resolved in the context in which they occur pass away, leaving behind those which are fit to survive. If a particular conflict resolution is costly but the system can afford the cost, it will be fit enough to persist. If it is costly but results in an evolutionary transition, it will be fit enough to persist. If it is costly but compensated by even greater benefits, it will be fit enough to persist.

Our ability to detect each conflict resolution episode and especially the ones that produced evolutionary transitions depends mostly on how much of the history of those events has been retained since their origin, and little on where members of the system live today and what they currently are doing. Because retained history is an evolutionary agent, we cannot understand the stories from "top-down," i.e., from the present backward. The sagas of evolvable life on this planet must, therefore, be told "from the beginning" as much as possible (Brooks 2010).

When we approach the story of evolution from the standpoint of the conditions that exist in the here and now, we find that the organisms we study are either capable of doing what they are doing (e.g., Zhang et al. 2019; Lin et al. 2019)—which is not very enlightening—or that they are capable of doing more than they are doing—which begs an important question. How did they come to be able to do more than they need to do? Do they know something we do not know about what is ahead in time and are they taking proactive measures to cope with what is coming? Not likely—if living systems had that capacity, nothing would ever go extinct. Abandoning that bit of wishful and magical thinking about looking into the future, we are forced to look back in time and ask about origins and history. And when we do that, we find two things about those "extra capacities." First, they are old, ancient, and persistent (Woese 1987). And second, they are often very important traits that allow coping across a wide range of conditions (Jordan et al. 2002; Galperin et al. 2018; Weinstein et al. 2019). Sometimes knowing the origins tells us about conditions when those capacities first emerged (Xue et al. 2015; Kwong et al. 2019); sometimes, it allows us to uncover hidden episodes of co-option during the evolutionary journey to the current state of affairs (Liebeskind et al. 2011; Nutman et al. 2016; Ruszczycky and Lui 2017; Wilburn et al. 2017; Chevalier et al. 2019); and sometimes, it shows us how long ago evolution set the stage for what emerged later (Mukherjee et al. 2018; Martin and Servais 2019). And as we follow these traces deeper in time, adding the panoramic view of the evolutionary saga that Darwin represented with his metaphorical Tree of Life, something amazing begins to appear. It seems that evolution has been episodic, at times wildly creative and exuberant. Some of those episodes are as ancient as life itself, leading researchers to see in the episodes of the early evolution of life echoes of the Big Bang that marked the origin of the universe in which it lives (Koonin 2007; David and Alm 2011; Droser and Gehling 2015; Slater 2015; Nutman et al. 2016; Morris et al. 2018; Betts et al. 2018; Garcia-Pichel et al. 2019; Staps et al. 2019).

Each saga begins when conflict resolution leads to the emergence of innovations in an inheritance system irreversibly split from its antecedent. The fundamental basis of irreversibility in the universe has been hotly debated for more than 200 years, mostly by physicists and chemists. Biologists have tended to accept the reality of irreversibility as a matter of course and not worry about it. Maynard Smith (1974), however, recognized that there was an issue related to irreversibility in the neo-Darwinian conceptual framework, specifically the focus on maximum fitness as the driving force of evolution (*Fisher's Fundamental Theorem*). Maynard Smith realized that because fitness is relative, rather than absolute, the view of selection embodied in Fisher's Fundamental Theorem implies no temporal directionality in evolution. And yet, there is a phylogeny.

Evolutionary transitions are temporally irreversible, so Maynard Smith and Szathmary (1995) proposed a minimal criterion for the evolutionary irreversibility pertaining to the episodes of conflict resolution. They suggested that the integration of two or more phenomena, each of which is improbable a priori, into an inheritance system is sufficient to make a conflict resolution event irreversible. Our criterion for irreversibility is consistent with theirs. All biological processes that generate time are irreversible, whether the circular time of metabolism or the linear time of inheritance. Many things in biology repeat but nothing goes backward. Conflict resolution produces two or more surviving systems each with its own cohesive inheritance system, ensuring that the conflict resolution event will be at least conditionally irreversible. Conflict resolution viewed thusly produces irreversible behavior in two ways. The first is changes solely in the nature of the organism that become integrated into the emergent inheritance system, and the second is changes in the nature of the organism that become integrated into the emergent inheritance system in the context of particular conditions. Lagload increases the odds that such conjunctions will happen—the conservative nature of inherited history allows a priori improbable traits to be carried along until they can become integrated into the same inheritance system. It is less important that innovations are coincidences or rare than it is that they are concatenated by being integrated into a common historical flow.

Which evolves first, a new form of compensatory change or a new form of cohesion, does not matter, nor does it matter if one of them evolves prior to a particular conflict resolution; both can be integrated into an evolutionary transition. And if such traits are widespread persistent ancestral traits in multiple inheritance systems, there are many chances for an evolutionary transition to occur when the complementary innovation arises. History does not simply make time an agency of evolution, it makes it an agency for turning the improbable into the probable and from that into the actual. This is how concatenations of coincidences become concatenations of historical persistence.

Some conflict resolutions turn out to be more significant than others—these have been dubbed *evolutionary transitions* (Maynard Smith and Szathmary 1995; Queller 1997; Brooks and McLennan 2002; Koonin 2007; Wilson et al. 2008; Carmel and Shavit 2020). Maynard Smith and Szathmary's examples of major transitions in evolution were meant to be so self-evidently significant that they avoided issues of what actually counted as a "transition" and as a "major transition." They all involved the evolution of mechanisms that increase the efficiency of storing and transmitting inherited information, including a combination of division of labor and cooperation.

The *stochastic corrector*—catalytic templates occurring in two cellular compartments—replication and metabolism—providing protection against destructive template mutants by means of the template with the higher genetic information constraining the template with the lower genetic information, is perhaps the most fundamental example (Szathmary and Demeter 1987; Zintzaras et al. 2002). *Division of labor* is a manifestation of what we earlier called *compensatory changes*, which enhance exploration, while *cooperation* is a manifestation of *cohesion*, which enhances exploitation (see also Gorelick et al. 2004). If a conflict resolution needs one new form of each to count as a transition, then transitions are those conflict resolutions from which emerge both enhanced exploitation and exploration capacities, thereby increasing the odds that the system will successfully cope with new conflicts. And because the only biological criterion for success is persistence in time and space, evolutionary transitions are cases in which conflict resolution leads to unusually successful inheritance systems. Carmel and Shavit (2020) have suggested that such conflict resolutions are the source of emergent levels of individuated systems that many refer to as hierarchical structure (Allen and Starr 1982, 2017; Eldredge and Salthe 1984; Salthe 1985, 1993, 1998; Eldredge 1986, 1995).

The issue of irreversibility underlies the recognition that while everything evolves, nothing evolves *into something else*; new inheritance systems emerge from conflict resolution leading to selective diversification. The origin and influence and fate of those evolutionary transitions are recorded in phylogeny, and the visual metaphor for that is the Tree of Life. Phylogeny is the ghost of conflict past, a never-ending story of conflict resolution that leaves indelible tracks in the sands of time. The cohesion and conservatism of inheritance makes those tracks legible.

9.1 Phylogenetic Analysis as a Reflection of the Dynamics of Conflict Resolution

> As it is difficult to show the blood relationship between the numerous kindred of any ancient and noble family even by the aid of genealogical trees, and almost impossible to do so without this aid, we can understand the extraordinary difficulty which naturalists have experienced in describing, without the aid of a diagram, the various affinities which they perceive between the living and extinct members of the same great natural class.—Charles Darwin (1872)

We made a big deal about the significance of the Tree of Life in Chaps. 2–4 and of the phylogenetics revolution in Chap. 5. Then we went largely silent on those topics while we discussed three major metaphorical dualities of evolution. Telling the stories of evolutionary diversification from the perspective of those dualities requires that we have a good map of the routes various inheritance systems have taken through time and space. That map is phylogeny and we next show how phylogenetic analysis is linked to our metaphorical perspective. Once again, we find that the common abstraction for making the connection is information theory.

Darwin likened the Tree of Life to a genealogy of noble families. Our perspective is that the status of the entities existing along the way in evolutionary history is not the point. All of them are temporary eddies of semi-stability that eventually disappear through modification or extinction. What is important is the flow of the process of evolutionary diversification, and phylogenies use particular events and entities as guideposts. As DeLisle (2019) has pointed out, modern perceptions of the Tree of Life differ in some ways from Darwin's; we hope our telling of this part of the story is at least consistent with his.

We begin with Lila Gatlin (1972), whose primary interest was the use of information theory to help explain the evolutionary significance of diversity and organization in nucleotide sequences. She began by introducing two forms of informational redundancy. *R*-redundancy results from the repeated occurrence of the same symbol in order to get a message across. This is one way to ensure proper communication of a message, but since each symbol must be repeated, *R*-redundancy is also associated with reduced message variety. *D*-redundancy, or Shannon redundancy, results when a single symbol arises once but subsequently always stands for the same thing. *D*-redundancy is associated with increased message variety, since no symbol need be repeated and thus a larger number of possible messages could be transmitted using the same number of symbols than for situations in which some symbols were repeated. This state may also be associated with reduced message fidelity, however, since missing the initial symbol results in a loss of information subsequently, without hope of a recurrence of the symbol. Gatlin associated *D*-redundancy with the search for optimal coding in communication systems, such as inheritance systems based in nucleotides. Finally, *R*-redundancy is associated with low information density per symbol (each symbol represents only itself) and *D*-redundancy with high information density per symbol (each symbol represents many observations).

Brooks (1981) provided the first bridge between Gatlin's ideas and methods for recovering evolutionary history, imagining that recovered pieces of the Tree of Life were messages about evolutionary history. He reasoned that *R*-redundancy would be maximized if every trait in every species arose independently, whereas *D*-redundancy would be maximized if a single coding event produced all the versions of a given trait. And if common inheritance is the most fundamental form of cohesion, maximizing D would amount to maximizing correlation structure based on phylogenetic cohesion, which ought to be the best estimate of phylogenetic history. This corresponds to the state of the greatest difference between H_{max} and H_{obs} (see Figs. 6.5 and 7.2), given the data. Brooks (1985) later formalized this framework with what they called the D measure as an homage to Gatlin. Smith (1988) provided a model and mathematical proof showing that if phylogenetic diversification followed this kind of dynamic generally, the process would be one in which both realized and potential diversity of inherited information would evolve together. This provided one connection with phylogenetic systematic analysis that developed following the protocol for phylogenetic inference first formalized by Hennig (1950). But there is an additional connection between our metaphorical framework and contemporary phylogenetic analysis as well.

The Reverend Thomas Bayes was an eighteenth-century English mathematician who was interested in the concept of using what we know today to predict future events. His paper, "An Essay Towards Solving a Problem in the Doctrine of Chances," published 2 years after his death in 1761, introduced what would become known as Bayes' theorem (Barnard 1958), in which the probability of the hypothesis (H) given the observations, or data (D), [$P(H|D)$], is called the posterior probability. The likelihood of a hypothesis is $P(D|H)$, a parameter in the calculation of the posterior probability. $P(H)$ is the prior probability of the hypothesis before the observation, data, or analysis, and reflects the original beliefs regarding the problem. $P(D)$ is the probability of the data, equal to the sum of the nominator for all considered hypotheses, and acts as a normalizing factor to ensure the sum of all posterior probabilities equals 1 or 100%. Bayes' Theorem describes the relationship between the prior and posterior probabilities. The prior probability of the hypothesis is updated to take into account the observations, producing a new estimate of the hypothesis that may form the prior for subsequent calculations if more observations are then considered. Bayes' Theorem thus acts in an iterative way, altering the posterior probability to reflect the effects of all available data or the likelihood they will change the hypothesis.

Tribus and McIrvine (1971) suggested that van der Waals first linked Bayesian reasoning to statistical concepts of entropy in 1911, proposing that the highest entropy state for a system was its a priori most probable state. Lewis (1930) subsequently suggested that increases in entropy were associated with losses of information, and Shannon (1948) used the statistical formulation of entropy as a synonym for expected uncertainty as a foundation of information theory. Kullbach (1951) reinforced the link between information theory and Bayesian reasoning by using the term *surprisal* indicating a departure from the most probable/most expected state. Brillouin (1951, 1953a, b, 1962) showed that there was consistency between Shannon's use of entropy in information theory and the use of entropy in statistical mechanics and probability theory. Jaynes (1957a, b) derived the formal relationship underlying this consistency and proposed the first entropy maximization principle, in which the maximum entropy state of a system could be formally construed as the most probable state. Parenthetically, Jaynes reasoned that adopting the most complex model among all those that explained a system would expose our ignorance of possibilities while adopting the simplest would give us a false sense of security about how much knowledge we actually had.

Although not presented in those terms, the D measure is Bayesian in nature (as perhaps are all efforts to apply statistical mechanical formalisms to historical reconstructions: Miyzaki et al. 1996). Bayesian approaches in information theory are thus those for which the a priori subjective hypothesis is determined by the entropy maximum principle—the a priori most probable result is H_{max}, in direct analogy with the maximum entropy state being the most probable for a closed system. This becomes Bayesian if we stipulate that the set of observations we are using in any analysis is a closed subset of all possibilities (i.e., our estimate of H_{max} is based on a subjective subsample of an imperfectly known universe of characters) and that we will not introduce additional observations during the testing procedure. The entropy maximum principle is therefore not only analogous to the a priori expected most probable state, it is also the state of lowest information density of the observations, hence least informative, hence least surprising (in a Bayesian sense). For any set of observations (subjective in the sense that they are a subjectively selected subset of all observations, drawn from a universe for which we do not have any sense of the actual size or distribution of variables—even if the observations are obtained objectively), we can objectively compute the most probable state (H_{max}). We can also objectively compute the least probable state (H_{min}), which is the state of greatest information density for the observations, and thus the state of greatest surprise. The most powerful analysis of such data is one that seeks to find the most improbable/highest information density configuration permitted by all the data at hand. For phylogenetic analysis, H_{max} and H_{min} can be calculated from the basic data matrix (hence H_{max} is a priori), whereas H_{obs} is calculated over a set of trees (hence, it is a posteriori). The preferred result is the one in which H_{obs} approaches H_{min} as closely as possible.

Joseph Felsenstein briefly discussed Bayesian likelihood ideas as they could apply to phylogeny reconstruction in his Ph.D. thesis in 1968, but the statistical and computational framework with which to derive reliable approximations of posterior probabilities was not available at the time (see Huelsenbeck et al. 2002). Harper (1979) had considerable foresight in devising a Bayesian framework for choosing between competing phylogenetic hypotheses. His version of Bayes' Theorem sought to determine the probability that some taxa were monophyletic given the observation of a synapomorphy between them. Harper's calculation was unique in including estimates of error due to misinterpreting plesiomorphies or homoplasies as synapomorphies. In 1996, three independent groups introduced working Bayesian methods for phylogenetics that are similar to those currently in use (Li 1996; Mau 1996; Rannala and Yang 1996). All three evaluate phylogenetic hypotheses using the posterior probabilities of different phylogenetic trees.

Bayesian Likelihood approaches bear a strong similarity to the *D* measure, and informational measures are now being advocated for use in choosing preferred models for Bayesian likelihood analyses. A key operation in the computer program Mr. Bayes (Ronquist and Huelsenbeck 2003; Ronquist 2004) is "data compression," which is equivalent to finding the most information-dense configuration of the observations in a data set (see also McCulloch et al. 2018). Huelsenbeck and Rannala (2004), for example, adopted Jaynes' perspective in suggesting that the best Bayesian likelihood results would be obtained if one chose the most complex model, much in the same sense as the proposals by Lockhart et al. (1994) for maximum likelihood. These views also complement the use of the Akaike Information Criterion or the Bayesian Information Criterion when choosing the best model for maximum likelihood analysis (Posada and Buckley 2004).

Most importantly, historical correlations due to inheritance (homology) and to the cohesion associated with evolutionary transitions are so pronounced that phylogenetic analysis using any of the above variations on Bayesian philosophy produce significant agreement regardless of the sources of data used (Brooks et al. 2007; Wiley and Lieberman 2011; Wright and Hillis 2014). This reinforces the Darwinian notion that when formulating evolutionary explanations, we should first maximize those based on common inheritance. And it also connects the technical process of providing the most defensible hypothesis of phylogeny with the way that inheritance systems work (see Campbell 2016 for a discussion of Darwinian evolution as a Bayesian process).

Evolutionary history is the embodiment of the Nature of the Organism through time. Phylogeny is the biological information highway, whose landmarks are evolutionary transitions.

9.2 An Initial Taxonomy of Transitions

Like all useful biological taxonomies, this one is meant to be dualistic and nested rather than dichotomous and exclusionary.

9.2.1 Maynard Smith and Szathmary: What Is the Limiting Factor?

Maynard Smith and Szathmary (1995; Szathmary 2015) distinguished two classes of evolutionary transitions, based on the kinds of limiting factors overcome in accomplishing them. *Selection-limited transitions* are those for which it is relatively easy to access part of capacity space for two improbable traits, but the evolutionary success of the innovations requires special conditions. *Variation (information, or diversity)-limited* transitions are those for which it is difficult to access a particular part of capacity space for those improbable traits, but once they have been achieved, special conditions are not needed for them to become entrenched evolutionarily. Variation-limited transitions are not immune from selection, as all conflict resolution events involve selection. They do not, however, rely on selection to explain the initial success of the transition that accompanies the conflict resolution event.

9.2.2 Queller: How Are the Participants Related?

Queller (1997) observed that Maynard Smith and Szathmary's transitions could be divided into two different kinds of classes based on the source of the building blocks used to produce evolutionary innovations. *Fraternal transitions* arise when the building blocks have a common origin. A good example would be cell differentiation in the ontogeny of multicellular organisms. *Egalitarian transitions*, by contrast, involve elements having independent origins.

Perhaps the most significant egalitarian transitions were the cooperation and division of labor emerging from the initial parasitism of one group of prokaryotes by others. The parasites that were heterotrophs became mitochondria and those that were photoautotrophs became chloroplasts. And when the replicating portion of the host cell became sequestered inside its own membrane, the eukaryotic condition emerged, setting the stage for, among other things, us.

Queller's recognition of egalitarian transitions focuses attention on the fact that Maynard Smith and Szathmary's framework, with its emphasis on cooperation, casts doubt on the notion that cooperation among non-kin must always lead to conflicts that are too costly to persist. We can now acknowledge that when the costs of cooperation are lower than the benefits, cooperation will be favored, regardless of the genealogical origins of the cooperating elements. Also, because intention is not meaning (Brooks and McLennan 1997), the concept of maximal selfishness, initiated by Spencer and refined by Dawkins (1976), can now be understood as something that is overcome in evolution rather than something that drives the process. Evolutionary novelties all originate within kin groups and if they are co-opted for positive interaction between groups, the lines between kin and group selection become blurry indeed (Powers et al. 2011; Szathmary 2011). Human language may have emerged initially by facilitating cooperation and transfer of information among members within kin groups (Maynard Smith and Szathmary 1995), but it became a major transition when it facilitated communication and cooperation and division of labor across groups (Bickerton and Szathmary 2011). This shifts our focus to questions of how cooperation occurs and what are its outcome—who are the cooperators, who are the non-cooperators, and why?

9.2.3 *Brooks and McLennan: What Is the Degree of Difficulty?*

Brooks and McLennan (2002) proposed four classes of evolutionary transitions based on their phylogenetic effects. The most common are those that are not difficult to achieve and do not require special conditions. Brooks and McLennan referred to these as "*taking advantage of the rules of the game*" *(new capacities in preexisting conditions)*. These transitions are the easiest to achieve because they occur within a preexisting set of conditions. They are transitions that occur when conflict resolution takes the form of isolation in different spatial portions of relatively homogeneous fitness space. They require only a change in capacities that enhance the means of coping with conflicts in fitness space comprising preexisting conditions. This could mean simply an augmentation of traits already used in coping with those conditions, or it could involve the evolution of novel traits that could replace the ones previously used in those conflicts.

The second two classes of transitions are those for which the traits associated with the transition are easy to achieve but require special conditions (one version of

Maynard Smith and Szathmary's *selection-limited transitions*) and those that are difficult to achieve but do not require special conditions (one version of Maynard Smith and Szathmary's *variation-limited transitions*). Brooks and McLennan called these transitions "*changing the rules of the game*." This kind of transition is more difficult to achieve, because it requires changes in the nature of the organism, in the nature of the conditions, or both. Inheritance systems making this kind of transition may evolve capacities in preexisting conditions that allow them to alter how they impose themselves on the surroundings, thereby changing the fitness space that determines the rules of the game. Or, they may evolve capacities that allow them to impose themselves on new conditions in ways that allow them to persist in the new conditions that determine the rules of the game.

These are tactical evolutionary transitions and characterize inheritance systems that persist and diversify moderately in space and time. The truly major transitions are strategic, so Brooks and McLennan referred to them as "*changing the game*." Such transitions are the rarest of all because they involve the emergence of novel capacities to cope with novel conditions. They are both *variation-limited* and *selection-limited*. The evolutionary payoff for such transitions, rare as they are, can be very large, which may be why they diagnose highly successful evolutionary events. The major transitions associated with the origin of life discussed by Maynard Smith and Szathmary (1995) are of this type.

9.3 Some Sagas

So there is nothing but the present. . .the rich, real, stable present, the moment now. And you think it is something that can be possessed. . . But it is not real, you know. . . Things change, change. You cannot have anything. . . And least of all can you have the present, unless you accept with it the past and the future. . . Because they are real: only their reality makes the present real.—Ursula Le Guin (1974)

Humans celebrate their history because they recognize that it has had a great impact on their present existence. These histories include heroes and heroic events, and it is the job of the historical ecologist to find and to explain the heroic episodes in the history of life on this planet.—Daniel R. Brooks and Deborah A. McLennan (2002)

The Tree of Life provides the narrative line for telling the sagas of the evolutionary epic. Making certain our sagas are fully consistent with phylogeny helps us keep them attuned to the evidence and avoid swerving into romantic or even magical invocations. They also set the stage for a variety of experimental and field investigations that enrich our understanding of the biosphere and its dynamics (Brooks and McLennan 2002). The sagas are episodes about processes and events, using places and events as signposts. We will next tell some sagas incorporating our metaphorical framework. We chose them because they intrigue us, and because they are not the major transitions discussed with great authority by Maynard Smith and Szathmary (1995, 1999; Szathmary 2015). Each reader should try telling the saga of their focus of attention in this way at least once, to see how it turns out, what it sounds like, and if it gives new insights.

9.3.1 Making a Living

We now do cruel injustice to an enormous number of researchers engaged in fascinating and compelling research on what is likely to become a research field of its own that we think of as the *Cosmology of Life*—the Big Bang explosion from a few critical steps producing the system of evolvable life on this planet (for excellent summaries, see Maynard Smith and Szathmary 1995; Szathmary 2015). We encourage readers to become familiar with as much of this literature as possible, not only for the basic information it provides but also because insights from research efforts in this realm inform many areas of molecular biology and biotechnological developments. The findings of this group of researchers will be invaluable if our species manages to survive long enough to effectively explore planets within and beyond our solar system.

From the beginning, what distinguished living systems was a sense of self, the Gantian sense of information plus metabolism inside a membrane. This created a distinction between the nature of the organism and the nature of the conditions, and established Darwin's Necessary Misfit, because informational autonomy made the nature of the organism more important than the nature of the conditions. Life emerged from chemical chaos, took control of its destiny and thrives on conflict resolution.

Let us begin with how organisms make a living. The first living systems were chemoautotrophs, synthesizing what they needed from the primordial chemical broth in which they emerged. One of the next major transitions was *heterotrophy*, living systems that use other genomes for metabolism rather than for information storage and transmission. Those that were capable of letting others do much of the synthesizing work, and then absorbing their products by ingesting them, would have been at a tremendous advantage, assuming that they spent less energy attacking and absorbing chemoautotrophs than they would have synthesizing their own nutrients. Finding, subduing, and ingesting another living system is no small undertaking and might seem to require a number of innovations. But it seems that it was accomplished by one of the major examples of Darwinian co-option. Darwin believed that this type of dynamic was "an extremely important means of transition" in evolution.

The emergence of heterotrophy is an example of evolutionary transitions (we use the plural because heterotrophy may have evolved more than once) that change the rules of the game. And in this case, the change was significant. Heterotrophy was not necessarily "better" than chemoautotrophy, and we would have expected heterotrophs to evolve and flourish along with chemoautotrophs, who began to perform a new function as food for the heterotrophs. Also, along with heterotrophy came the division of labor and additional autonomy within the information system. Even today, many heterotrophic microbes are capable of digesting part of their own information system in times of energy deprivation. Importantly, the portions of the information system digested are those dealing with metabolism, not with storing and transmitting information to subsequent generations. Those same microbes are capable of acquiring new sets of metabolic genes through heterotrophy.

The next, and arguably most impactful of the transitions in modes of making a living occurred when some chemoautotrophs became *photoautotrophs*, tapping into an enormous, ubiquitous energy source—photons from the sun. From their standpoint, all that changed was the power source for autotrophy. From the standpoint of the chemoautotrophs and heterotrophs, the emergence of photoautotrophs meant the world would never be the same.

This is because photoautotrophs produce oxygen as a metabolic by-product—organic waste—but they are highly tolerant of it. The release of oxygen into the primordial soup and from there into the primordial atmosphere, changed a chemically reducing environment suitable for chemoautotrophic life into a chemically oxidative environment they could not tolerate. The majority of chemoautotrophs and their dependent heterotrophs went extinct—today, there are only remnants of that life gone by. The entire structure of the biosphere changed; the only heterotrophs to survive were those already capable of coping with an oxygen-rich environment (Ruszczycky and Lui 2017). Chemoautotrophs did not disappear completely; they exist today as the microorganisms responsible for recycling organic matter. They remain critical components of ecosystem structure, but they are no longer in charge.

Photosynthesis arose as a result of a series of complex evolutionary changes in the nature of the organism. The organisms in which these changes occurred took advantage of opportunities offered by the nature of the conditions, but the conditions did not create photosynthesis. Moreover, the conditions under which photosynthesis originally evolved were at best indifferent to photosynthetic life. Photoautotrophs survived originally because they had the capacity to cope with one set of surroundings and also evolved the capacity to change the game. Equally important, the capacity to change the game flourished because a by-product of the new mode of making a living changed the nature of the conditions. Today's environment is more supportive of photoautotrophs as a result of their own activities. But that created the existential conflict within the fitness space of most chemoautotrophs, resulting in the first mass extinction event. Just as all episodes of conflict resolution set the stage for new conflict, major transitions in evolution have unanticipated consequences.

9.3.2 Origins of Herbivory

We often marvel at how powerful ecosystems can be when they are powered by photonic energy. This is because the emergence of photoautotrophs created a new potential food source, driven by a virtually unlimited energy source, for the heterotrophs that survived the oxygen catastrophe. But as evolution continued, feeding on photoautotrophs became a mixed blessing. The cell walls of photoautotrophs are made of complexes of glycoproteins and polysaccharides. Cellulose is one of those and it makes the cell walls of many eukaryotic photoautotrophs—from algae to multicellular plants—sturdier. Having cell walls made inflexible by cellulose provides many benefits, including the ability to be multicellular, to live on land, and to point unerringly at the source of their energy. Having cellulose-laced cell walls, however, does not make them immune to being eaten. A number of heterotrophic systems have persisted in eating plants for a very long time. Some microbial chemoautotrophs are capable of digesting cellulose and do so regularly in their roles as saprophages and decomposers. But this is not the case for all heterotrophs. No metazoan system has ever evolved the ability to digest cellulose. This prevents most heterotrophs that feed on plants from gaining maximum energy resources from the experience. On rare occasions in the course of time, some animals have wandered into complex but workable ways to obtain nutrition from cellulose.

The evolution of herbivory in metazoans is a good example of a variation-limited transition—difficult to achieve but with a large payoff if it happens (e.g., Dearing et al. 2000; Bellwood 2003; Labandeira 2007; Wiens et al. 2015). Five obstacles must be overcome in order to produce a true metazoan herbivore. First, a cellulose-digesting microbe that could live in the intestine of the metazoan must be accessible in the environment of the metazoan. Second, the metazoan must encounter and ingest (but not digest) the microbe. Third, the metazoan intestine must be a suitable living place for the microbe. Fourth, at this point, the microbes must experience a conflict resolution episode that results in the emergence of a cohesive inheritance system comprising only the microbes living in the metazoan intestine and not connected with their free-living relatives. Fifth, and finally, the metazoan must have a mechanism for transmitting the now-symbiotic microbes to subsequent generations. Only if all of these elements are combined can a metazoan become a true herbivore. A conjunction of all five of these phenomena is surely sufficiently improbable to explain why herbivory does not emerge more often and to ensure that the transition to true herbivory is irreversible when it does occur. Not everything that feeds on plants does so exclusively. Some routinely eat a mixed diet of plant and animal material (we call them omnivores) while others, such as giant pandas, eat plant material almost exclusively but retain the ancestral capacity to eat animal material (Xue et al. 2015).

Despite the fact that most insects feed on plants, phytophagy has originated only a few times in the history of the Insecta (Mitter et al. 1988; Mitter and Farrell 1991; Wiegmann et al. 1993; Farrell and Mitter 1994; Farrell 1998). Contemporary phytophagous insects that are not herbivores show us how the stage is set for the transition to true herbivory. They damage plant tissues when feeding but derive fewer nutrients than they would if they could digest cellulose. This creates opportunities for microbes on the plants that feed on damaged plant material and digest cellulose but cannot damage the plant material themselves. The microbes tend to show up around places where the insects damage the plants, so the insects will encounter them regularly. If the insects ingest any of those microbes, and if some of them end up living in the insect gut, the microbes will continue to get damaged plant material delivered to them but in a sheltered environment in which the by-products of their cellulose digestion can be absorbed by the insect gut. If a mechanism for transmitting the now-symbiotic microbes to offspring emerges, the transition from mechanical phytophagy to true herbivory has been accomplished, the cost of living has been lowered for both the microbes and the insect. Interestingly, the insects and microbes were never in conflict, i.e., they could have continued to live without the symbiotic association ever forming. And yet with the transition to true herbivory, the cost of living for both was reduced through an egalitarian transition (group selection) and division of labor.

Plants did not "evolve cellulose to protect them from being eaten" and herbivores never "evolved the ability to digest cellulose." And despite the enormous potential energetic payoff, few animals ever evolved the microbial symbiont system that allowed them to tap into that potential. Thus, the existence of an enormous unexploited resource did not "cause herbivory to evolve" in any form, but it did provide the opportunity. Also, in each instance the microbes differ, the intestinal environment differs, and the mode of transferring the microbes to the next generation differs. However, each time those rare transitions occurred, the herbivores benefitted in a manner we would expect, by persisting a long time and becoming highly diverse and widespread along the way. This is a great example of the Jagger Principle in action.

9.3.3 The "Conquest of Land"

> ... tempt one to imagine them [amphibians] as having slowly evolved, in the midst of the Carboniferous marshes, from true fishes; first wriggling helplessly among the slime, and afterwards generally acquiring lungs for breathing air and limbs for locomotive purposes, and lastly, their strong and peculiar teeth for masticating the vegetation on which they may be presumed to have principally lived.—B. Webster Smith (1926)

There is substantial fossil evidence that modern terrestrial ecosystems owe much to the marine ecosystems from which their constituent inheritance groups arose (see Behrensmeyer et al. 1992). The first heterotrophs and photoautotrophs that moved onto land had little to do with each other for about 10 million years. It is not so much that marine organisms conquered land as it is that they brought their marine ecosystems onto land with them, became established, and then produced the innovations that are particular to terrestrial ecosystems today.

McLennan (2008) discussed at length numerous cases of evolutionary trait co-option, including the transition from aquatic water-breathing to terrestrial air-breathing vertebrates. In the transition from water to land, it is clear from phylogenetic and fossil evidence that tetrapod limbs and lungs existed *before* vertebrates moved onto land (Clack 2006) and so did their ability to lay eggs in terrestrial soils (Skulan 2000). What's more, even before the transition to land, each character was already an example of long-term evolutionary tinkering in shallow water environments. Extraordinary paleontological studies have demonstrated that the limbs tetrapods use for walking on land were modified from limbs used for swimming and crawling in shallow waters that evolved from fins originally used for swimming (Fig. 9.1).

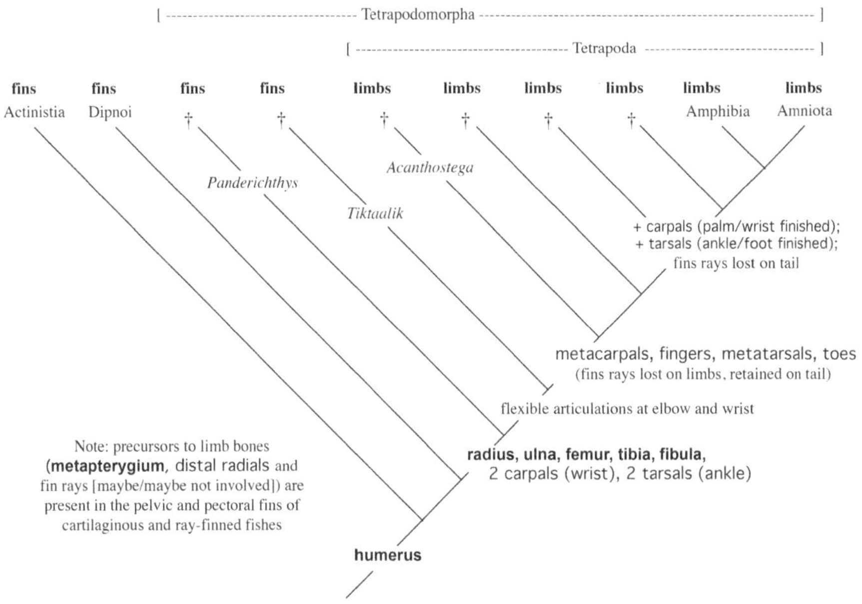

Fig. 9.1 Phylogeny showing that traits involved in the evolution of tetrapod limbs from fish fins arose before the transition onto land. Redrawn and modified from McLennan (2008)

Similarly, lungs used for terrestrial air-breathing were modified from lungs used for living in shallow waters which evolved from gas bladders used for buoyancy, while aerial scent detection emerged from tinkering with a breathing tube (Fig. 9.2).

Each step of the way, these innovations arose from—possibly even were part of the cause of—some kind of conflict resolution. We have no real evidence about what the conflicts might have been, but we can be certain that none of this was the result of some inevitability that because land existed, terrestrial vertebrates would evolve. In fact, most of these innovations would have been relatively easy to accomplish because they involved mostly recycling and modifying existing parts. A "lung" is what you get when you tinker with a "gas bladder," for example (Tatsumi et al. 2016).

So, now we know that the suite of traits related to "terrestriality" was built up over long periods of time by evolutionary tinkering and co-option, and *everything needed to survive on land was in place before tetrapods made the move* (see also Triques and Christoffersen 2009). What happened then? There is, of course, a neo-Darwinian perspective:

> Until recently it was widely assumed that the origin of tetrapods was synonymous with the invasion of land, a view epitomized by countless popular illustrations of lobe-finned fishes hauling themselves laboriously onto Devonian pond margins and river banks... the new discoveries ... suggest that tetrapods evolved initially to exploit a shallow-water environment.—Per E. Ahlberg and Andrew R. Milner (1994)

This account is better than Smith (1926) but still appeals from effect to cause, from current utility to reason for the origin. All is not lost, however. Implicit in the

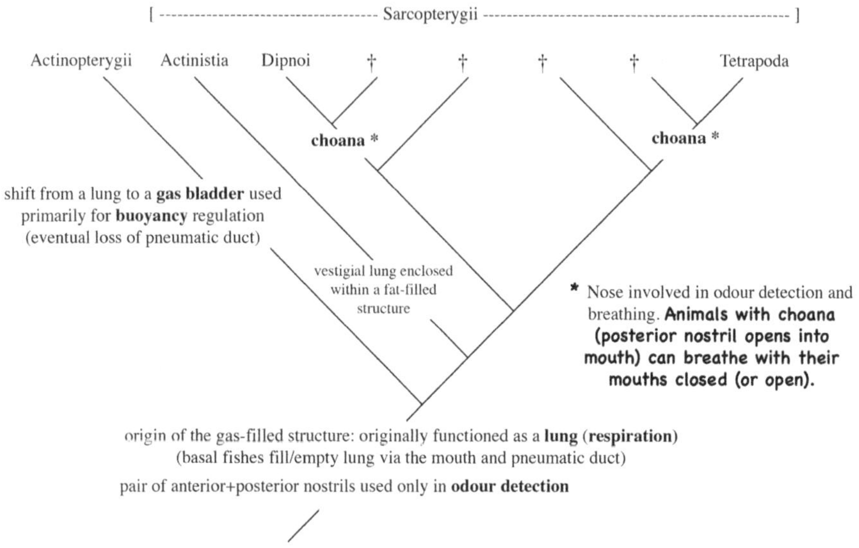

Fig. 9.2 Phylogeny showing that traits involved in the evolution of the tetrapod "lung" used for air-breathing from the fish "gas bladder" used for respiration/buoyancy arose before the transition onto land

statement above is the notion that being able to survive in shallow-water environments set the stage for emergence onto land, but that final bit is still left dangling. We still have no idea how it happened, what the trigger was. Look at it from the standpoint of ecological fitting—what would they have been capable of besides living in shallow-water systems? Living on land. How do you go from shallow-water systems to land? Drought. Some fish that had lungs, necks, hands, and feet were able to move from drying areas to wetter places, where they survived a severe drying spell through ecological fitting by co-option. It is not the warm shallow seas, it is their periodic drying that was the key. Think of lungfish today.

forewarned, forearmed.—Robert Greene (1591)

This saga of tetrapods colonizing land is an amazing story full of beauty and wonder, but it is not magical or heroic. It is a great example of the power of evolutionary co-option to make things *survivable*. Not magical, just functional and effective. Just Darwinian.

9.3.4 Filling Niches or the Nature of the Organism?

The Compositae plant genus *Montanoa* comprises approximately 30 taxa living throughout Central America, extending as far north as central Mexico and as far south as central Colombia. Twenty-one of these species are shrubs, five are "daisy

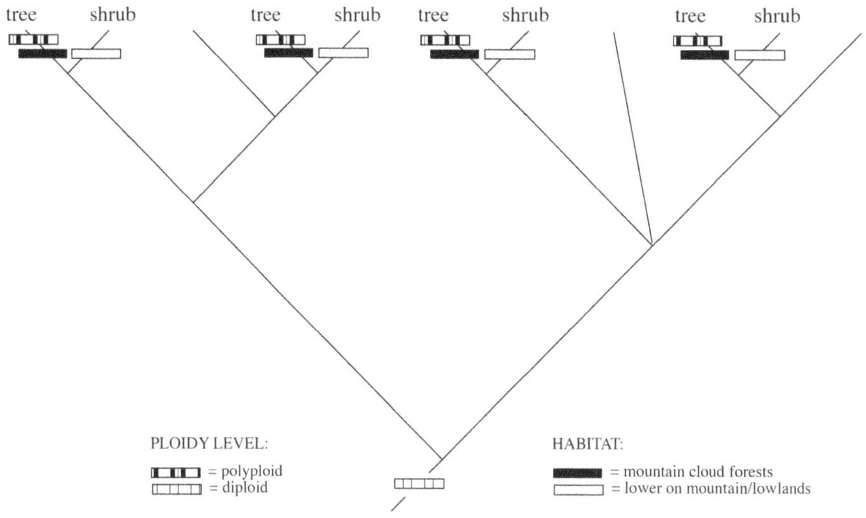

Fig. 9.3 Phylogeny of the plant genus *Montanoa* showing a recurring pattern of association between the emergence of polyploidy and the evolution of the tree growth form. Redrawn and modified from Brooks and McLennan (1991)

trees" reaching 20 m in height, and four are vines. Vicki Funk, an intrepid field biologist and pioneer of plant phylogenetics, performed the first phylogenetic analysis of the genus (Funk 1982), discovering that the shrub-like habit is plesiomorphic (ancestral) in *Montanoa*. Seven of 47 branches on her phylogenetic tree are associated with a change in habit (tree forms arose 4 times and vines arose 3 times). All five tree species have a number of similar morphological and physiological characters permitting them to survive in cloud forests and none has ever been found outside a cloud forest. They are members of four different clades, their sister-species being shrubs living at adjacent lower elevations in each case. Examination of the distribution of the *Montanoa* habit and ploidy level on the phylogenetic tree reveals a recurring pattern of association between the appearance of tree forms and increased polyploidy (Fig. 9.3).

Diploidy is the ancestral condition in *Montanoa*, but all the tree species, and only the tree species, are high-level polyploids (Funk and Raven 1980; Funk 1982; Funk and Brooks 1990). It is common for diploids to produce polyploid seeds in the Compositae under a variety of environmental conditions, and for polyploids to be viable, so the repeated appearance of polyploids is not a surprising phylogenetic pattern. It is also common for polyploids to be larger than diploids, so the tree habit is likely a by-product of polyploidization. The restriction of *Montanoa* tree forms to cloud forests suggests that special conditions are required for the survival of those forms when they evolve. Presumably, the polyploids require significantly more moisture than their shrubby diploid relatives. Additionally, *Montanoa* seeds have limited ability to be dispersed except by water. It would seem that only those polyploid offspring produced by shrubby diploids on the margins of cloud forests would have a chance of getting into survivable fitness space and also that polyploid seeds washed downstream out of the cloud forests would not be able to survive. This is an example of a transition that is easy to achieve but requires special conditions, and those special conditions circumscribe extremely limited fitness space.

9.3.5 Transitions in Context

Modern heterotrophy has two major manifestations—overpower the prey, kill and eat it, or underpower the prey and live on or within it while eating some of it. These options arose early (Cong et al. 2017) and often (Weinstein and Kuris 2016) in many lineages, and today's diversity represents a long history of elaborations on those ancient themes. In addition to the indefinite variation arising from the nature of the organism, evolution is fundamentally open-ended because each new inheritance system that emerges from an episode of conflict resolution is a potential fitness space for other inheritance systems. Products of the nature of the organism comprise much of fitness space, so the evolution of diverse inheritance systems increases the magnitude and complexity of fitness space. It is not surprising, therefore, that heterotrophic diversity is heavily skewed toward the second option. In fact, more than half the inheritance systems on this planet are symbionts of some form, including an extraordinary diversity of pathogens.

The full story of the significance of the evolutionary transitions that produced the overwhelming array of underpowering heterotrophs lies in knowing not only how the initial transitions occurred but also how successful they were in diversification and persistence in space and time. For an illustration, we will once again belabor the world of interactions between plants and animals. There is no question that becoming phytophagous, and especially herbivorous, are rare and major evolutionary transitions. The evolutionary diversification of a group of leaf beetles (*Ophraella*) in the context of their host plants (Funk et al. 1995a, b; Knowles et al. 1999) provides a nice illustration of our perspective (Fig. 9.4). *Ophraella* are herbivores, a trait they share due to inheritance from a common ancestor in which that major transition occurred initially, so that is background context for our saga. Many of the events associated with the phylogenetic diversification within *Ophraella* evidently occurred in North America during the Plio-Pleistocene. That period was a time of substantial environmental change, initially involving a decrease and then a re-expansion in the physical dimensions of fitness space. As the spatial dimensions of fitness space decreased, the density of organisms of varying degrees of relatedness all with similar resource requirements increased. The result would have been increased conflict in shrinking fitness space. At the same time, however, functional diversity within that smaller fitness space increased as the number of potential alternative hosts living in one place increased. This would have increased the chances of host colonization through ecological fitting if many of the newly sympatric plant hosts shared the same or sufficiently similar specific and conservative resources preferred by the insects. Secondary spatial expansion of fitness space would have provided opportunities for

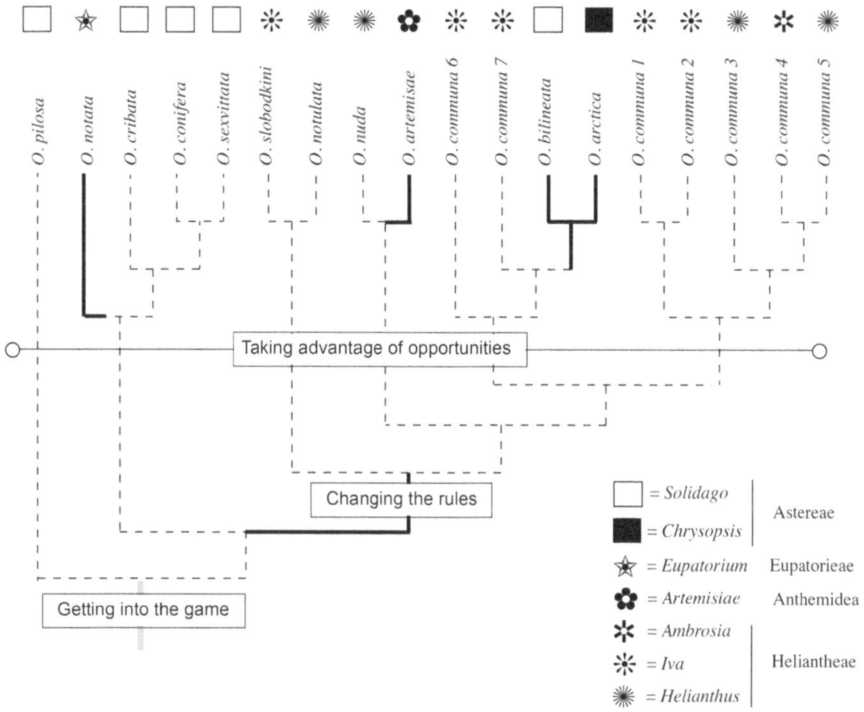

Fig. 9.4 Phylogeny of the beetle genus *Ophraella* showing the evolution of associations with their food plant hosts (symbols). Note the two types of evolutionary transitions where beetles either "take advantage of opportunities" by forming new associations with closely related hosts (dotted lines) or "change the rules of the game" by forming new associations with more distantly related hosts (solid lines). Redrawn and modified from Brooks and McLennan (2002) and Brooks et al. (2019)

newly established insect–plant systems to diffuse in newly expanded fitness space, potentially to the point at which low density would have decreased conflict, favoring cohesive properties, and leading to the emergence of novel independent inheritance systems.

Ophraella communa 1 + 2 and *O. conifera* + *O. sexvittata* are evidently the result of the minimalist form of conflict resolution. They are each other's closest relatives and they feed on plants that are each other's closest relatives. Their common ancestors managed to survive the decrease in fitness space while maintaining their original host associations. The re-expansion of fitness space led insects and plants to explore the new fitness space, decreasing the density of conflict and ultimately diverging into pairs of inheritance systems.

Conflict resolution events in the remaining members of *Ophraella* involved colonization of a new species of host plant. The species whose relationships are indicated by dotted lines in Fig. 9.4—other than *Ophraella communa* 1 + 2 and *O. conifera* + *O. sexvittata*—represent cases in which ecological fitting in the form of conservative resource capacities allowed colonization of new host plants that were closely related to the original host lant

The final category—those indicated by bold solid lines in Fig. 9.4—encompasses fewer events. Each of these transitions involved, minimally, colonization of a new species of host plant that is distantly related to the original host plants, likely representing a new kind of resource in fitness space, thereby requiring some innovation in the inherited capacities of the insects. There is no indication that these events were correlated with each other or with any particular episode of environmental change, so they seem to have been independent innovations in each ancestral inheritance system.

In the context of our metaphorical construct, we would say that half a century ago, Ross (1972a, b) suggested that only 1 in 30 conflict resolutions produced an evolutionary transition. The other 29 cases represented the emergence of two descendant inheritance systems each separated in relatively homogeneous fitness space and cohesive enough internally to preclude communication with the other. The most general means by which this kind of situation can happen is separation in the physical space component of fitness space. This encompasses all forms of "divergence by distance," including various forms of allopatric speciation (see Brooks and McLennan 2002 and references therein). In this case, 14/18 contemporaneous systems indicated in the phylogeny of *Ophraella* (78%) and 28/34 total events (82%) represented conflict resolutions without transitions. While this is not the 97% suggested by Ross, it nonetheless suggests that a transition early in phylogeny can facilitate later persistence and diversification.

Our saga of *Ophraella* covers only those that successfully weathered the storm of the Plio-Pleistocene. Environmental changes of that magnitude expose the inheritance systems that lack the capacity to cope with or to take advantage of the altered nature of the conditions. Here the conservative nature of inheritance systems can be a disadvantage. The inability to explore new fitness space may be the result of adjacent fitness space being occupied by a relative already in control of the resources required by a potential colonizer. In this example, we would anticipate that *Ophraella communa* 1 + 2 and *O. conifera* + *O. sexvittata* would be most capable of excluding each other if the conditions arose.

9.4 Summary

All the news that's fit to print.—*New York Times* masthead

The Tree of Life is the ongoing saga of evolvable life on this planet. Each episode in that saga highlights the indefinite—but not infinite—capacity of evolvable life to cope with changes in the surroundings, making use of preexisting traits carried in inheritance to buy time for continued survival. It is in those intervals of time that conflict resolution is achieved by evolutionary innovations emerging from the indefinite variation that is a property of inheritance systems. The result is selective diversification of life. The conservative nature of inheritance systems allows us to retrace the history of evolutionary sagas using phylogenetic analysis, and tell them from the beginning, sometimes in great detail. This allows us to see that some evolutionary innovations involved in conflict resolutions have been more significant than others. They are the great themes of life, laid down and then embellished through time. These *evolutionary transitions* are the milestones of evolution, the heroic episodes in the history of life. Phylogenies help us put those transitions in context, allowing us to determine not only "what happened" but also "where," "when," "in what sequence," and "how often" they happened. We must tell these from beginning to end—the origins tell us about how things survive today but the reverse does not tell us how they arose.

There will always be evolutionary conservatism, so while we might never have been able to predict a given transition, we can certainly understand it in hindsight; and in some cases, the persistent historical elements may give us some degree of "retrodiction"; the paradox of "in evolution nothing is conserved yet conservatism plays an integral role." At the same time, we ought always to assume there will be unanticipated outcomes.

> History makes space between accident and necessity.—Susan Neiman (2002)

Phylogeny may be the biological information highway, but life on this planet is not just diversification of inheritance systems. It is diversified inheritance systems that live together in the complex communities and ecosystems making up the biosphere. All living systems are intimately tied together in the structure of that biosphere. Our story of evolution is therefore still incomplete—we need an understanding of how the various members of the Tree of Life form the Entangled Banks that characterize the biosphere. In order to tell the full evolutionary epic, we must first localize living systems in time, and then in space. In the next chapter, we continue our proposal for a metaphorical platform for discussion exactly where Darwin ended *The Origin of Species*, with the metaphor of the *Entangled Bank.*

References

Ahlberg PE, Milner AR (1994) The origin and early diversification of tetrapods. Nature 368:507–513

Allen TFH, Starr TB (1982) Perspectives for ecological complexity, 1st edn. University of Chicago Press, Chicago, IL

Allen TFH, Starr TB (2017) Perspectives for ecological complexity, 2nd edn. University of Chicago Press, Chicago, IL

Barnard GA (1958) Studies in the history of probability and statistics: IX. Thomas Bayes's essay towards solving a problem in the doctrine of chances. Biometrica 45:293–315

Behrensmeyer AK, Damuth JD, DiMichele WA, Potts R, Sues H-D, Wing SL (eds) (1992) Terrestrial ecosystems through time. University of Chicago Press, Chicago

Bellwood DR (2003) Origins and escalation of herbivory in fishes: a functional perspective. Paleobiology 29:71–83

Betts HC, Puttick MN, Clark JW, Willliams TA, Donoghue PCJ, Pisani D (2018) Integrated genomic and fossil evidence illuminates life's early evolution and eukaryote origin. Nat Ecol Evol 2:1556–1562
Bickerton D, Szathmary E (2011) Confrontational scavenging as a possible source for language and cooperation. BMC Evol Biol 11(1):261. http://www.biomedcentral.com/1471-2148/11/261
Brillouin L (1951) Physical entropy and information. J. Appl Phys 22:338–343
Brillouin L (1953a) Negentropy principle of information. J Appl Phys 24:1152–1163
Brillouin L (1953b) Science and information theory, 2nd edn. Academic Press, New York
Brillouin L (1962) Science and information theory, 2nd edn. Academic Press, New York
Brooks DR (1981) Classifications as languages of empirical comparative biology. In: Funk VA, Brooks DR (eds) Advances in cladistics: proceedings of the first meeting of the Willi Hennig society. New York Botanical Garden, New York, pp 61–70
Brooks DR (1985) Historical ecology: a new approach to studying the evolution of ecological associations. Ann Mo Bot Garden 72:660–680
Brooks DR (2010) Sagas of the children of time: the importance of phylogenetic teaching in biology. Evo Edu Outreach 3:495–498
Brooks DR, McLennan DA (1991) Phylogeny, ecology and behavior: a research program in comparative biology. University of Chicago Press, Chicago
Brooks DR, McLennan DA (1997) Biological signals as material phenomena. Rev pensee d'aujord d'hui 25:118–127. [in Japanese]
Brooks DR, McLennan DA (2002) The nature of diversity. University of Chicago Press, Chicago
Brooks DR, Bilewitch J, Condy C, Evans DC, Folinsbee KE, Fröbisch J, Halas D, Hill S, McLennan DA, Mattern M, Tsuji LA, Wahlberg N, Ward J, Zamparo D, Zanatta D (2007) Quantitative phylogenetic analysis in the 21st century. Rev Mex Biodiv 78:225–252
Brooks DR, Hoberg EP, Boeger WA (2019) The Stockholm paradigm: climate change and emerging disease. University of Chicago Press, Chicago
Campbell JO (2016) Universal Darwinism as a process of Bayesian inference. Front Syst Neurosci 10:49. https://doi.org/10.3389/fnsys.2016.00049
Carmel Y, Shavit A (2020) Operationalizing evolutionary transitions in individuality. Proc R Soc B 287:20192805. https://doi.org/10.1098/rspb.2019.2805
Chevalier FD, Le Clec'h W, McDew-White M, Menon V, Guzman MA, Holloway SP, Cao X, Taylor AB, Kinung'hi S, Gouvras AN, Webster BL, Webster JP, Emery AM, Rollinson D, Djirmay AG, Khalid M, Mashikhi KM, Al Yafae S, Idris MA, Mone H, Mouahid G, Hart PJ, LoVerde PT, Anderson TJC (2019) Oxamniquine resistance alleles are widespread in Old World Schistosoma mansoni and predate drug deployment. PLoS Pathog 15(10):e1007881. https://doi.org/10.1371/journal.ppat.1007881
Clack JA (2006) The emergence of early tetrapods. Palaeogeogr Palaeoclimatol Palaeoecol 232:167–189
Cong P, Ma X, Williams M, Siveter DJ, Siveter DJ, Gabbott SE, Zhai D, Goral T, Edgecombe D, Hou X (2017) Host-specific infestation in early Cambrian worms. Nat Ecol Evol 1:1465–1469. https://doi.org/10.1038/s41559-017-0278-4
Darwin C (1872) On the origin of species, 6th edn. John Murray, London
David LA, Alm EJ (2011) Rapid evolutionary innovation during and Archaean genetic expansion. Nature 469:93–96
Dawkins R (1976) The selfish gene. Oxford University Press, Oxford
Dearing MD, Mangione AM, Karasov WH (2000) Diet breadth of mammalian herbivores: nutrient versus detoxification constraints. Oecologia 123:397–405
DeLisle R (2019) Charles Darwin's incomplete revolution: the origin of species and the static worldview. Springer, New York
Drosera ML, Gehling JG (2015) The advent of animals: the view from the Ediacaran. Proc Natl Acad Sci U S A 112:4865–4870
Eldredge N (1986) Information, economics and evolution. Annu Rev Ecol Syst 17:351–369

Eldredge N (1995) Reinventing Darwin: the great debate at the high table of evolutionary theory. Wiley, New York

Eldredge N, Salthe SN (1984) Hierarchy and evolution. In: Dawkins R, Ridley M (eds) Oxford surveys in evolutionary biology, vol 1. Oxford University Press, Oxford, pp 182–206

Farrell BD (1998) "Inordinate fondness" explained: why are there so many beetles? Science 281:555–559

Farrell BD, Mitter C (1994) Adaptive radiation in insects and plants: time and opportunity. Am Zool 34:57–69

Funk VA (1982) Systematics of *Montanoa* (Asteraceae: Heliantheae). Mem N Y Bot Gard 36:1–135

Funk VA, Brooks DR (1990) Phylogenetic systematics as the basis of comparative biology. Smithson Contrib Bot 73:1–45

Funk VA, Raven PH (1980) Polyploidy in *Montanoa* Cerv. (Compositae, Heliantheae). Taxon 29:417–419

Funk DJ, Futuyma DJ, Orti G, Meyer A (1995a) A history of host associations and evolutionary diversification for *Ophraella* (Coleoptera: Chrysomelidae): new evidence from mitochondrial DNA. Evolution 49:1017–1022

Funk DJ, Futuyma DJ, Orti G, Meyer A (1995b) Mitochondrial DNA sequences and multiple data sets: a phylogenetic study of phytophagous insects (Chrysomelidae: *Ophraella*). Mol Biol Evol 12:627–640

Galperin MY, Makarova KS, Wolf YI, Koonin EV (2018) Phyletic distribution and lineage-specific domain architectures of archaeal two-component signal transduction systems. J Bacteriol 200: e00681-17. https://doi.org/10.1128/JB.00681-17

Garcia-Pichel F, Lombard J, Soule T, Dunaj S, Wu SH, Wojciechowski MF (2019) Timing the evolutionary advent of cyanobacteria and the later great oxidation event using gene phylogenies of a sunscreen. mBio 10:e00561-19. https://doi.org/10.1128/mBio.00561-19

Gatlin L (1972) Information theory and the living system. Columbia University Press, New York

Gorelick R, Bertram SM, Killeen PR, Fewell JH (2004) Normalized mutual entropy in biology: quantifying division of labor. Am Nat 164:677–682

Greene R (1591–2) A notable discovery of coosnage, 1591; The second part of conny-catching, 1592. 1923 reprint. Harrison GB (ed). John Lane, London

Harper CWJ (1979) A Bayesian probability view of phylogenetic systematics. Syst Zool 28:547–533

Hennig W (1950) Grundzüge einer theory der phylogenetischen Systematik. Deutscher Zentralverlag, Berlin

Huelsenbeck JP, Rannala B (2004) Frequentist properties of Bayesian posterior probabilities of phylogenetic trees under simple and complex substitution models. Syst Biol 53:905–913

Huelsenbeck JP, Larget B, Miller RE, Ronquist F (2002) Potential applications and pitfalls of Bayesian inference in phylogeny. Syst Biol 51:673–688

Jaynes ET (1957a) Information theory and statistical mechanics I. Phys Rev 106:620

Jaynes ET (1957b) Information theory and statistical mechanics II. Phys Rev 108:171

Jordan IK, Rogozin IB, Wolf YI, Koonin EV (2002) Essential genes are more evolutionarily conserved than are nonessential genes in bacteria. Genome Res 12:962–968

Knowles LL, Futuyma DJ, Eanes WF, Rannala B (1999) Insight into speciation from historical demography in the phytophagous beetle genus *Ophraella*. Evolution 53:1846–1856

Koonin EV (2007) The biological big bang model for the major transitions in evolution. Biol Direct 2:21. https://doi.org/10.1186/1745-6150-2-21

Kullbach S (1951) Information theory and statistics. Wiley, New York

Kwong WK, del Campo J, Mathur V, Vermeij MJA, Keeling PJ (2019) A widespread coral-infecting apicomplexan with chlorophyll biosynthesis genes. Nature 568:103–110

Labandeira C (2007) The origin of herbivory on land: initial patterns of plant tissue consumption by arthropods. Insect Sci 14:259–275

Le Guin U (1974) The dispossessed. Harper and Row, New York

Le Guin U (1989) Dancing at the edge of the world: thoughts on words, women, places. Grove Press, New York

Lewis GN (1930) The symmetry of time in physics. Science 71:569–577

Li S (1996) Phylogenetic tree construction using Markov Chain Monte Carlo. Ph.D. dissertation, Ohio State University, Columbus, OH

Liebeskind BJ, Hillis DM, Zakon HH (2011) Evolution of sodium channels predates the origin of nervous systems in animals. Proc Natl Acad Sci U S A 108:9154–9159

Lin Z, Chen L, Chen X, Zhong Y, Yang Y, Xia W, Liu C, Zhu W, Wang H, Yan B, Yang Y, Liu X, Kvie KS, Røed KH, Wang K, Xiao W, Wei H, Li G, Heller R, Gilbert MTP, Qiu Q, Wang W, Li Z (2019) Biological adaptations in the Arctic cervid, the reindeer (*Rangifer tarandus*). Science 362:eaav6312

Lockhart PJ, Steel MA, Hendy MD, Penny D (1994) Recovering evolutionary trees under a more realistic model of sequence evolution. Mol Biol Evol 11:605–612

Martin RE, Servais T (2019) Did the evolution of the phytoplankton fuel the diversification of the marine biosphere? Lethaia. https://doi.org/10.1111/let/12343

Mau B (1996) Bayesian phylogenetic inference via Markov Monte Carlo methods. Ph.D. dissertation. University of Wisconsin, Madison, WI

Maynard Smith J (1974) The theory of games and the evolution of animal conflicts. J Theor Biol 47:209–221

Maynard Smith J, Szathmáry E (1995) The major transitions in evolution. W.H. Freeman, Oxford

Maynard Smith J, Szathmáry E (1999) The origins of life. W.H. Freeman, Oxford

McCulloch AF, Jauregui R, Maclean PH, Ashby RL, Moraga RA, Laugraud A, Brauning R, Dodds KG, McEwan JC (2018) An entropy-reducing data representation approach for bioinformatic data. Database. https://doi.org/10.1093/database/bay029/4962528

McLennan DA (2008) The concept of co-option: why evolution often looks miraculous. Evol Educ Outreach 1:247–258

Mitter C, Farrell B (1991) Macroevolutionary aspects of insect-plant relationships. In: Bernays E (ed) Insect-plant interactions, vol 3. CRC Press, Boca Raton, FL, pp 35–78

Mitter C, Farrell B, Weigmann B (1988) The phylogenetic study of adaptive zones: has phytophagy promoted insect diversification? Am Nat 132:107–128

Miyazaki S, Sugawara H, Ohya M (1996) The efficiency of entropy evolution rate for construction of phylogenetic trees. Genes Genet Syst 71:323–327

Morris JL, Puttick MN, Clark JW, Edwards D, Kenrick P, Pressel S, Wellmane CH, Yang Z, Harald Schneider H, Donoghue PCJ (2018) The timescale of early land plant evolution. Proc Natl Acad Sci U S A 115:E2274. https://doi.org/10.1073/pnas.1719588115

Mukherjee I, Large RR, Corkrey R, Danyushevsky LV (2018) The boring billion, a slingshot for complex life on earth. Sci Rep 8:4432. https://doi.org/10.1038/s41598-018-22695-x

Neiman S (2002) Evil in modern thought: an alternative hstory of philosophy. Princeton University Press, Princeton, NJ

Nutman A, Bennett V, Friend C, Martin J, Van Kranendonk MJ, Chivas AR (2016) Rapid emergence of life shown by discovery of 3,700-million-year-old microbial structures. Nature 537:535–538

Posada D, Buckley TR (2004) Model selection and model averaging in phylogenetics: advantages of Akaike information criterion and Bayesian approaches over likelihood ratio tests. Syst Biol 53:793–808

Powers ST, Penn AS, Watson RA (2011) The concurrent evolution of cooperation and the population structures that support it. Evolution 65(6):1527–1543

Queller DC (1997) Cooperators since life began. Q Rev Biol 72:184–188

Rannala B, Yang Z (1996) Probability distribution of molecular evolutionary trees: a new method of phylogenetic inference. J Mol Evol 43:304–311

Ronquist F (2004) Bayesian inference of character evolution. Trends Ecol Evol 19:475–481

Ronquist F, Huelsenbeck JP (2003) MrBayes 3: Bayesian phylogenetic inference under mixed models. Bioinformatics 19:1572–1574

Ross HH (1972a) The origin of species diversity in ecological communities. Taxon 21:253–259

Ross HH (1972b) An uncertainty principle in ecological evolution. Univ Arkansas Mus Occ Pap 4:133–157

Ruszczycky MW, Lui H-W (2017) The surprising history of an antioxidant. Nature 551:37–38

Salthe SN (1985) Evolving hierarchical systems: their structure and representation. Columbia University Press, New York

Salthe SN (1993) Development and evolution: complexity and change in biology. MIT Press, Boston

Salthe SN (1998) The role of natural selection theory in understanding evolutionary systems. In: Van de Vijver G, Salthe SN, Delpos M (eds) Evolutionary systems: biological and epistemological perspectives on selection and self-organization. Kluwer Academic, Dordrecht, pp 13–20

Shannon C (1948) A mathematical theory of communication. Bell Syst Tech J 27:379–423

Skulan J (2000) Has the importance of the amniotic egg been overstated? Zool J Linnean Soc 130:235–261

Slater GJ (2015) Not-so-early bursts and the dynamic nature of morphological diversification. Proc Natl Acad Sci U S A 112:3595–3596

Smith BW (1926) The world in the past. Warne, London

Smith JDH (1988) A class of mathematical models for evolution and hierarchical information theory. Institute for Mathematics and its applications preprint series, vol 396, pp 1–13

Staps M, van Gestel J, Tarnita CE (2019) Emergence of diverse life cycles and life histories at the origin of multicellularity. Nat Ecol Evol 3:1197–1208. https://doi.org/10.1038/s41559-019-0940-0

Szathmáry E (2011) To group or not to group. Science 334:1648–1649

Szathmáry E (2015) Toward major evolutionary transitions theory 2.0. Proc Natl Acad Sci U S A 112:10104–10111

Szathmáry E, Demeter L (1987) Group selection of early replicators and the origin of life. J Theor Biol 128:463–486

Tatsumi N, Kobayashi R, Yano T, Noda M, Fujimura K, Okada N, Okabe M (2016) Molecular developmental mechanism in polypterid fish provides insight into the origin of vertebrate lungs. Sci Rep 6:30580. https://doi.org/10.1038/srep30580

Tribus M, McIrvine MC (1971) Energy and information. Sci Am 225:179–188

Triques ML, Christoffersen ML (2009) Exaptations in the conquest of land by Tetrapoda. Gaia Sci 3:69–74

Weinstein SB, Kuris AM (2016) Independent origins of parasitism in Animalia. Biol Lett 12:20160324. https://doi.org/10.1098/rsbl.2016.0324

Weinstein DJ, Allen SE, Lau MCY, Erasmus M, Asalone KC, Walters-Conte K, Deikus G, Sebra R, Borgonie G, van Heerden E, Onstott TC, Bracht JR (2019) The genome of a subterrestrial nematode reveals adaptations to heat. Nat Commun 10:5268. https://doi.org/10.1038/s42467-019-13245-8

Wiegmann BM, Mitter C, Farrell B (1993) Diversification of carnivorous insects: extraordinary radiation or specialized dead end? Am Nat 142:737–754

Wiens J, Lapoint R, Whiteman N (2015) Herbivory increases diversification across insect clades. Nat Commun 6:8370. https://doi.org/10.1038/ncomms9370

Wilburn DB, Arnold SJ, Houck LD, Feldhoff PW, Feldhoff RC (2017) Gene duplication, co-option, structural evolution, and phenotypic tango in the courtship pheromones of plethodontid salamanders. Herpetologica 73:206–219

Wiley EO, Lieberman BS (2011) Phylogenetics: theory and practice of phylogenetic systematics, 2nd edn. Wiley-Blackwell, New York

Wilson DS, Van Vugt M, O'Gorman R (2008) Multi-level selection theory and major evolutionary transitions. Curr Dir Psychol Sci 17:6–9

Woese CR (1987) Bacterial evolution. Microbiol Rev 51:221–271

Wright AM, Hillis DM (2014) Bayesian analysis using a simple likelihood model outperforms parsimony for estimation of phylogeny from discrete morphological data. PLoS One 9(10): e109210. https://doi.org/10.1371/journal.pone.0109210

Xue Z, Zhang W, Wang L, Hou R, Zhang M, Fei L, Zhang X, Huang H, Bridgewater LC, Jiang Y, Jiang C, Zhao L, Pang X, Zhang Z (2015) The bamboo-eating giant panda harbors a carnivore-like gut microbiota, with excessive seasonal variations. mBio 6(3):e00022-15. https://doi.org/10.1128/mBio.00022-15

Zhang T, Wiao Q, Novikova PY, Wang Q, Yue J, Guan Y, Ming S, Liu T, De J, Al-Shehba IA, Sun H, van Montagu M, Hunag J, Van de Peer Y, Qiong L (2019) Genome of *Crucihimalaya himalaica*, a close relative of *Arabidopsis*, shows ecological adaptation to high altitude. Proc Natl Acad Sci U S A 116:7137–7146

Zintzaras E, Santos M, Szathmáry E (2002) "Living" under the challenge of information decay: the stochastic corrector vs. hypercycles. J Theor Biol 217:167–181

Chapter 10
The Stockholm Paradigm

Abstract The full richness of Darwinian evolution emerges when conflicts in fitness space driven by reproductive overrun are coupled with changes in the conditions. Perturbations in the conditions create changes in fitness space—old opportunities disappear and new ones arise, catalyzing a shift from exploitation-biased to exploration-biased behavior by inheritance systems, including entire biotas which may be affected by the same external event. This produces widespread *biotic mixing* as each inheritance system explores both the geographical and functional dimensions of its newly altered fitness space. Exploration takes the form of *generalizing* in fitness space, moving away from parts that have deteriorated to new parts previously inaccessible or nonexistent, using the capacity for ecological fitting inherent to the nature of the organism. After the perturbation, parts of the original inheritance system become isolated and begin *specializing* in the newly explored parts of fitness space when selection for exploitation-biased behavior takes over. Repeated cycles of expansion/exploration/generalization and isolation/exploitation/specialization driven by repeated external perturbations are what produce the complex web of interacting species—*Darwin's Entangled Bank*—that we observe. Changing the conditions thus catalyzes the dynamics that allow biological systems to *oscillate* between being exploiters and explorers of their surroundings, giving them their fundamental capacity to cope with change and routinely escape complete extinction. Even after mass extinction, so long as some life still exists, there is massive potential for its evolutionary renewal. The *Stockholm Paradigm* is the integration of all these factors, describing the dynamic behind Darwin's Entangled Bank.

We arrive at this chapter of our story having talked largely about the *text* of evolution—*emergent properties* initiated by features of the nature of the organism imposed on their surroundings, played out in fitness space. We have talked about how alternations between exploitation and exploration allow inheritance systems to cope with conflicts in fitness space, and sometimes to resolve those conflicts, leading to phylogenetic diversification. This focused the spotlight on capacity, which changes and grows with a high degree of autonomy from the conditions, allowing

S. J. Agosta, D. R. Brooks, *The Major Metaphors of Evolution*, Evolutionary Biology – New Perspectives on Its Development 2,
https://doi.org/10.1007/978-3-030-52086-1_10

biological systems to impose themselves on their surroundings rather than be driven by them. The fact that organisms and the inheritance systems that produce them are largely autonomous from their surroundings and can "fit" where they can "make a living" is enough to make the history of inheritance systems complex (Brooks and McLennan 2002). But, as Darwin well knew, this is not the whole story.

> Indefinite variability is a much more common result of changed conditions than definite variability. . .—Charles Darwin (1872)

> natural selection is the indirect action of changed conditions—Charles Darwin (1872)

The next critical phase in the evolutionary saga, therefore is, "What happens when the *nature of the conditions* changes?" To this point, we have talked primarily about how conflicts in relatively stable fitness space can lead to evolutionary innovation and diversification due solely to reproductive overrun. Now we discuss how external perturbations that alter the dimensions of fitness space catalyze the same kind of response except in this case, groups of inheritance systems and indeed entire biotas can be affected by the same perturbation. Focusing on changes in the conditions emphasizes the essential *context* of evolutionary diversification (Juarrero 1999), *systemic properties* that make fitness space neither simple (much of fitness space is made up of products of the nature of the organism) nor static (fitness space can be altered by external perturbations).

This part of the story will look very much like the basic dynamic of coping with and resolving conflicts in fitness space, with two important differences. The first is that external perturbations rather than demographics catalyze the shift between exploitation-biased and exploration-biased behavior. When the conditions change in ways that impose themselves on the nature of the organism fitness space changes. Old opportunities to achieve fitness disappear and new opportunities arise. Inheritance systems offered the opportunity to explore take advantage of the new opportunities by moving and exploring new parts of fitness space that were previously inaccessible or nonexistent. Just as for cases in which reproductive overrun leads to conflicts in fitness space, it is the capacity for ecological fitting that provides organisms with a high degree of responsiveness when external perturbations alter fitness space. The second is that external perturbations produce similar responses in multiple inheritance systems.

Demographically driven ecological fitting in static fitness space leads to self-limiting episodes of expansion and isolation that fill fitness space with maximally specialized inheritance systems. If this were the entire story of evolution, there would have been only a single and limited episode of phylogenetic diversification, not the indefinite diversification that has characterized the history of life. It would not produce the multitude of diversified inheritance systems living together that we see all around us. For that, we need repeated external perturbations to continually alter the "playing field" of fitness space. These perturbations catalyze the dynamics that allow biological systems to alternate—or *oscillate*—between exploiting and exploring two major dimensions of fitness space.

10.1 Altered Geographical Fitness Space: Taxon Pulses

> In considering the geographic distribution of organic beings over the face of the globe ... [w]e see . . .some deep organic bond, through space and time, over the same areas of land and water, independently of physical conditions. The naturalist must be dull who is not led to enquire what this bond is. The bond is simply inheritance, that cause which alone, as far as we positively know, produces organisms quite like each other, or, as we see in the case of varieties, nearly alike.—Charles Darwin (1872)

For Darwin, the most pervasive component of fitness space was the geographical location in which an inheritance system resided and of its closest relatives. Less than a century later, the geography of evolution was undergoing a decided paradigm shift. Philip Darlington was one of the prominent taxonomists who heeded Julian Huxley's (1940, 1942) call to view evolutionary diversification from an ecological perspective. His extensive observations of insect faunas on mountaintops and islands led him to conclude that new inheritance systems with new ecological characteristics arose in "centers of diversification" with environmental perturbations causing geographical ranges to fluctuate around an environmentally stable, continuously occupied core (Darlington 1943). The biotic fluctuations might be affected by the formation of barriers, producing episodes of isolation in which new species arose and ecological diversification might occur. Breakdown of those barriers produced new episodes of biotic expansion from the center, setting the stage for replacement of older species.

Darlington linked his ideas about the formation of inheritance systems to the *species–area relationship* (Cain 1938), the notion that any given area can support only a given number of species. That limitation, coupled with the production of new species in a *center of origin*, drove species out of the continuously occupied center. E.O. Wilson altered this perspective when he coined the term *taxon cycles* (Wilson 1959, 1961) for situations in which multiple species living in the same place colonized new areas when environmental changes enlarged the amount of suitable habitat, then contracted their ranges when environmental changes reduced the amount of suitable habitat. Taxon cycles occurred without producing new species and without new species arriving in the stable, continuously occupied center of the geographic distribution of the species involved. In this way, the issue of how species might be formed was de-coupled from assessments of how, and how many, species were coexisting and interacting ecologically in any given place. Wilson's collaboration with Robert MacArthur extended this line of thinking to produce the Equilibrium Theory of Island Biogeography (MacArthur and Wilson 1963, 1967), based on the view that the species–area relationship between species richness and the size of an "island" (i.e., any isolated area) is linear log-normal. Cain's notion and Darlington's generalizations finally had a quantitative basis, but at a price. MacArthur and Wilson envisioned a world of "sources" (Darlington's continuously occupied centers) and "islands" (areas colonized from the center) having a dynamic yet one-way history of colonization. The fitness space represented by various islands was static and only inheritance systems already in existence in the source area were involved in the colonization dynamics. The species–area relationship for any island resulted from a balance between colonization from a source and extinction. When the island had fewer than the equilibrium number of inheritance systems, it was open to colonization. When the equilibrium number of species was reached, no new inheritance system may colonize the area without displacement or extinction of one already in residence.

MacArthur and Wilson recognized that their theory worked best if one assumed that the geographical components of fitness space, and the inheritance systems available for colonization were fixed. They also realized that over long periods of time, neither of these assumptions could be true, because new species do evolve, but they assigned such phenomena to the "confounding effects of history," treating evolutionary diversification as akin to the error term in a statistical analysis of variance, something that needed to be partitioned out of the data to reveal the true patterns. This left their theory unable to adequately explain the potential contribution of the nature of the organism to the global change that preoccupied some of their contemporaries, such as Charles Elton (1958).

In the late 1970s, Terry Erwin proposed the *Taxon Pulse Hypothesis* (Erwin 1979), returning both a phylogenetic element and the notion of changeable fitness space to discussions about the dynamics of geographical fitness space. Taxon pulses are characterized by the expansion of multiple species into suitable habitat when perturbations in the conditions cause barriers to break down, increasing connectivity in geographic fitness space. During an expansion phase, different species within a biota encounter additional geographical heterogeneity. Further perturbations leading to the formation of new barriers lead to an isolation phase, during which inheritance systems diversify and innovations emerge. The next pulse cycle begins when a new perturbation breaks down those barriers, initiating a new phase of biotic expansion. Erwin's notion eliminated the need for a single center of origin, permitting the inheritance systems inhabiting a given area to have come from multiple sources at different times.

Erwin and colleagues developed his concept further in the early 1980s (Erwin 1981, 1985), including phylogenetically based criteria for distinguishing taxon pulses from taxon cycles (Liebherr 1988; Liebherr and Hajek 1990). An explicit methodology for examining the phylogenetic context of geographic distributions in a way that could detect taxon pulses if they occurred, however, was still not available. That changed in the first few years of the twenty-first century. Noted paleontologist Bruce Lieberman published a series of landmark studies that expanded our view of diversification on the deep-time scales that confront paleontology (Lieberman 2000, 2003a, b). He noted that large-scale biotic diversification would likely occur over large enough spatial scales that historical biogeographical relationships would comprise a combination of general episodes of isolation as, for example, pieces of continental landmasses under fragmentation, as well as general episodes of connection as continental landmasses collided and fused. This led to a unified phylogenetic methodology for documenting the historical signal of taxon pulses (Halas et al. 2005).

The taxon pulse provides a natural means by which inheritance systems diversify in isolation, then mix together during periods of biotic expansion. In studies of

Mexican terrestrial biotas (Halas et al. 2005) and Hawaiian and Caribbean island biotas (Eckstut et al. 2011), as much as two-third of those biotas comprise species that coexist with "related species." Evolutionary diversification by taxon pulses allows "island" inheritance systems to (re)colonize the "source," in the process changing the "equilibrium" number of inheritance systems that can live in an area in a manner that is not reducible to static parameters of space and distance. MacArthur and Wilson were indeed correct that historical effects would confound their model, but the taxon pulse showed that taking such historical effects into account would lead to a more comprehensive understanding of the context of geographic fitness space in evolutionary diversification. This is especially true when it comes to understanding that multiple inheritance systems co-occur not as an equilibrium number occupying an area of a given size, but as a nonequilibrium number occupying an area of any size based on how many taxon pulses affected the place (Heaney 2000; Lomolino 2000; Halas et al. 2005; Eckstut et al. 2011).

History doesn't repeat itself, but it often rhymes.—Attributed to Mark Twain

Taxon pulses are historically repetitive, meaning that biotas resulting from them are made up of species that have been associated with each other for varying lengths of time, and arrived under varying circumstances. Biotas assembled in this manner are complex mosaics resulting from the mixture of episodic expansion, isolation, and mixing during new expansion. Diversification is driven by waves of biotic expansion as well as isolation, so we expect to find general patterns associated with both phenomena. Episodes of biotic expansion, especially those involving large areas, will inevitably lead to complicated historical relationships among areas, characterized by biotas comprising species of different ages derived from different sources (e.g., Agosta and Janzen 2005). This has made modeling the evolutionary dynamics of taxon pulses difficult (e.g., Carvalho et al. 2015), although the problem is more tractable with respect to empirical data. In multigroup analyses (Spironello and Brooks 2003; Bouchard and Brooks 2004; Brooks and Ferrao 2005; Halas et al. 2005; Folinsbee and Brooks 2007; Hoberg and Brooks 2008; Lim 2008; Eckstut et al. 2011), not all groups participated equally in all the historical episodes. Nonetheless, the signature of the taxon pulse as an overarching process is always clearly legible. Most groups in each study agree on the episodic highlights but differ in the details. Those studies are only the tip of the iceberg; almost all recent studies in historical biogeography have identified alternating episodes of expansion and isolation, leading to the expected mosaic patterns of biotic assemblage within any given area (Hoberg and Brooks 2008; Brooks et al. 2019).

As an example, Folinsbee and Brooks (2007) discovered that the historical biogeography of hominoids, proboscideans (elephants and relatives), and hyaenids (hyaenas) since the Miocene epoch approximately 12 million years ago was explained by a series of taxon pulses (Fig. 10.1).

The results shown in Fig. 10.1 explain why the greatest diversity of fossil proboscideans occurs in Central Asia—that is the area affected by the largest number of expansion and isolation events. Another study of multiple groups of extant taxa in Mexico including insects, plants, and vertebrates showed that two relatively small

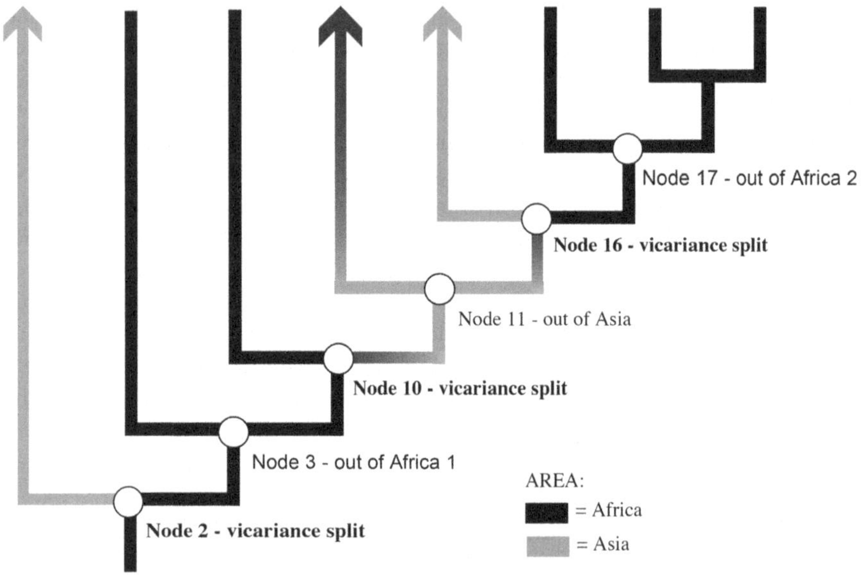

Fig. 10.1 Diagram of the historical biogeography of proboscideans, hominoids, and hyaenas since the Miocene. Note the series of taxon pulses between periods of geographic isolation between Africa and Asia (vicariance split nodes) and geographic expansion ("out of" nodes) exhibited by each group. Redrawn and modified from Folinsbee and Brooks (2007)

areas had more species for their size than predicted by the Equilibrium Theory of Island Biogeography (Halas et al. 2005). Like fossil elephants in Central Asia, this could be attributed to the fact that these areas had been involved in the largest number of expansion and isolation events.

10.2 Altered Functional Fitness Space: The Oscillation Hypothesis

Taxon pulses mix and match inheritance systems geographically. This in turn catalyzes episodes of intense exploration of newly connected fitness space during biotic expansion events followed by periods of intense exploitation of new habitat during isolation events. When multiple co-occurring inheritance systems are affected by the same external perturbation, much of their generalizing and specializing occurs in the context of functional interactions with each other. This is most easily visualized for groups of systems, such as insect–plant and parasite–host systems, whose functional interactions with each other are specific enough that we can arbitrarily call one member the *host* and the other the *parasite*. These terms are completely arbitrary, as noted by Price (1980) and Brooks and McLennan (1993).

Until the beginning of the twenty-first century, most orthogeneticists and neo-Darwinians assumed that parasites were so specialized with respect to their hosts as to be evolutionarily dependent on them. Among other things, this meant parasites should rarely have the capacity to add or move to a new host. By the end of the twentieth century, emerging results from parasite biodiversity inventories (e.g., Brooks and Hoberg 2000) and from public and agricultural health specialists were telling a different story. Parasites were appearing in novel hosts on a regular basis. This led to the recognition of what is called the *Parasite Paradox*: how can highly specialized parasites adopt new hosts with which they have had no prior evolutionary history? The answer was ecological fitting in sloppy fitness space (Agosta et al. 2010; Brooks et al. 2019). Parasites have highly specific capacities but, more importantly, they have highly specific resource requirements that are housed in their hosts. If those required resources are phylogenetically conservative traits of their hosts, there will always be other hosts with which a given parasite could exist quite happily, or at least adequately, given the opportunity. Host range is part of fitness space, so there should be a discrepancy between *fundamental host range* (all hosts that could be used) and a *realized host range* (hosts that are being used at a given place and time), i.e., host range will be sloppy.

> ... when a parasite arrives in a new habitat, it will feed on those species whose defense traits it can circumvent because of the abilities it carries at the time. Such a parasite cannot be distinguished from one that evolved the ability to circumvent a defense while in trophic contact with its host.—Daniel Janzen (1980)

Ecological fitting in sloppy fitness space for parasites would be manifested as host range changes (host switching) that do not track host phylogeny directly (cospeciation) but are constrained by phylogenetically restricted distributions of necessary resources (co-phylogeny) throughout evolution. Twenty-first-century studies have provided abundant evidence for this pattern (Brooks and McLennan 1993, 2002; Desdevises et al. 2002; Agosta and Janzen 2005; Brooks and Ferrao 2005; Agosta 2006; Banks et al. 2006; Janz et al. 2006; Brooks et al. 2006a, b, 2014, 2019; Wijova et al. 2006; Agosta and Klemens 2008, 2009; Hoberg and Brooks 2008, 2010, 2015; Janz and Nylin 2008; Nylin and Janz 2009; Agosta et al. 2010; Janz 2011; Mejia-Madrid 2012; de Vienne et al. 2013; Braga et al. 2014; Nylin et al. 2014; Ellis et al. 2015; Hoberg et al. 2015, 2017; Malcicka et al. 2015; Messenger et al. 2015; Estrada-Pena et al. 2016; Forbes et al. 2017; Hoberg et al. 2017; Patella et al. 2017; Jorge et al. 2018; Nylin et al. 2018; Chagnon et al. 2019; Kaczvinsky and Hardy 2019; Mellado and Zamora 2019; for additional studies see Brooks et al. 2019).

There is more to this pattern, however. Phylogenetic analyses of changes in host range over long time periods show that host range changes are not scattered throughout evolutionary history. Rather, they are clumped in time, so that the phylogenetic diversification of any parasite group exhibits alternating increases, then decreases in host range. The increase in host range allows the possibility of mismatches between parasite and host phylogenies when host ranges become restricted. This has been dubbed the *Oscillation Hypothesis* (Janz et al. 2006; Janz

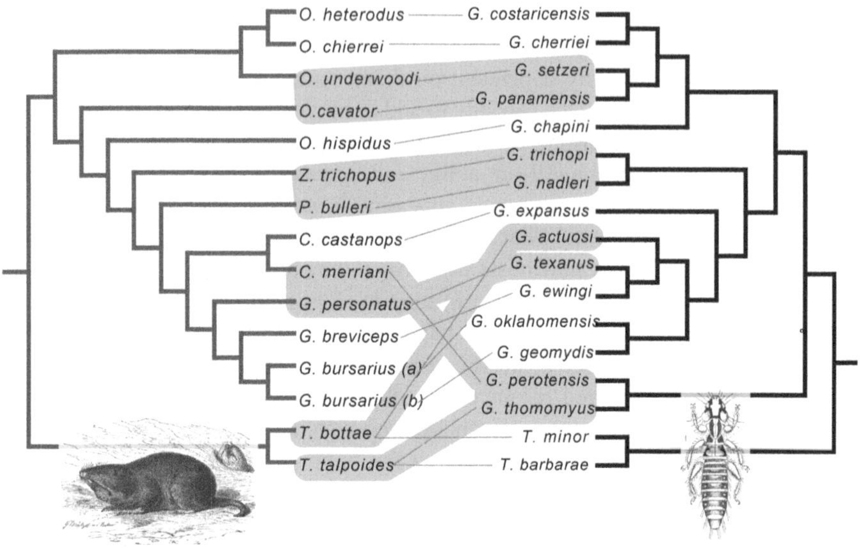

Fig. 10.2 Comparison between the phylogeny of pocket gophers and their parasitic lice showing evidence of oscillations in host range. Note the alternating episodes of host range expansion by lice (highlighted in gray) and isolation with cospeciation. Redrawn and modified from Brooks et al. (2019)

and Nylin 2008; Nylin and Janz 2009; Nylin et al. 2014), and though it was originally proposed to explain patterns of host range in insect–plant systems, which encompasses a lot of diversity, it seems to be far more general.

Pocket gophers are small rodents in the family Geomyidae, largely concentrated in temperate regions. Their current diversity arose between 4.2 and 1.8 million years ago (Spradling et al. 2004). In North America, pocket gophers are parasitized by lice of the genus *Geomydoechus*. Transmission of lice occurs when pocket gophers are in their nests, so the opportunities for the parasite to expand its host range are limited. This means that when lice are carried by their original host, the chances to transfer to new hosts are minimal even when perturbations such as climate change drive widespread geographic expansion (for a similar dynamic associated with bat parasites, see Dick and Patterson 2007). One way the opportunity to expand host range might arise is if a nest abandoned by one host species becomes occupied by another species—then there is a chance that some parasites that were left behind will come into contact with a novel but closely related host species. The burst of diversification in pocket gophers and their lice parasites coincided with a period of substantial climate variability and habitat change. Comparing lice and pocket gopher phylogenies (Fig. 10.2), half of the associations that emerged from that episode of diversification are due to switching to new hosts and subsequent differentiation in association with them. Moreover, the phylogenetic pattern shows clear evidence of an oscillation dynamic, with periods of host range expansion followed by periods of isolation and differentiation (for details, see Brooks et al. 2015).

If we use host range (the number of different host inheritance systems used) as a proxy for generalizing and specializing in fitness space, we should expect that during periods of biotic expansion, host range should increase (fitness space becomes less sloppy) and during periods of isolation, host range should decrease (fitness space becomes more sloppy). Computer simulations (Araujo et al. 2015), experimental studies (Pfennig et al. 2006; Horton et al. 2019; Peterson and Cipollini 2020), and field studies (Ando et al. 2013; Pariselle et al. 2014; Holland et al. 2017; Chan et al. 2018; Singer and Parmesan 2019) have shown the ease with which oscillations in fitness space can occur, revealing several key insights about the dynamics of parasite–host range. Colonization of new hosts can happen readily if those hosts have inherited the required resources for the parasite from a common ancestor. Parasite populations can persist for many generations in association with hosts that represent low-fitness habitat, and can even colonize and survive in association with new suboptimal hosts. This allows variation to accumulate that can later be co-opted to colonize new hosts. A parasite lineage can readily colonize a series of distantly related hosts through a "stepping-stone" process fueled by repeated ecological fitting events. The emergence of the coronavirus SARS-CoV-2 is an example of the stepping-stone dynamic. SARS-CoV-2 normally inhabits bats, and the genetic composition of the virus population in bats includes some low-fitness variants that cannot infect humans. When a founder population of the virus colonizes a nonhuman mammalian host, some of those low-fitness variants are able to thrive and produce new recombinant forms, and some of those can infect humans.

Araujo et al. (2015) showed that colonization of a new host involves first a contraction in a parasite species fitness space through a reduction in genotypic and phenotypic variance, followed by an expansion in fitness space through the accumulation of variation through recombination and innovations in the ability to use hosts. Braga et al. (2018) subsequently showed that with enough time, new parasite species may emerge from these colonizing populations. By expanding the basic model to allow repeated colonization events through continued propagule pressure, oscillations in parasite–host range emerge as an inherent property of the system. As long as new hosts continue to become available in the simulation, the cycle repeats itself. The biosphere comprises a diverse constellation of species exploiting existing resources while also exploring opportunities to use new resources by constantly probing the margins of fitness space. Alternations in host range are simply one manifestation of the general phenomenon of oscillating between exploiting current surroundings and exploring new opportunities in fitness space. Colonization of a new host is not a simple matter of propagule pressure, it also depends on the sloppiness of host fitness space (opportunity) when the parasite encounters a new host (D'Bastiani et al. 2021) and the phylogenetic distribution of resources required by the parasite among available hosts (Braga et al. 2020). This leads to complex networks of parasite–host associations (Hoberg and Brooks 2008; Nylin et al. 2018; Brooks et al. 2019; D'Bastiani et al. 2021).

10.3 Integrating Spatial and Functional Oscillations: The Stockholm Paradigm

Perturbations in the nature of the conditions affect geographic and functional fitness space for a multitude of inheritance systems. Perturbations that increase connectivity in fitness space lead to geographic expansion into novel areas as well as exploration of new opportunities in fitness space, based on preexisting capacities (ecological fitting). Affected inheritance systems *generalize* in fitness space (Kolbe et al. 2004; Carbonaro et al. 2018). Perturbations that decrease connectivity in fitness space lead to geographic isolation and exploitation-biased functional processes leading inheritance systems to *specialize* in fitness space (for an excellent example of this dynamic, see Moonjely and Bidochka 2019). Innovations in functional fitness space emerge when geographic space is not being explored, so the oscillations in geographic and functional fitness space are out of phase in complementary ways. Each new perturbation to the conditions leads to a new wave or *pulse* of geographic expansion, biotic mixing, and generalization, followed by episodes of isolation and specialization. Each new episode of expansion stems from a different source, leading to locally idiosyncratic and ecologically complex associations of inheritance systems.

It is during periods of geographic isolation and specialization in fitness space that the successful emergence of evolutionary innovations is favored, which sets the stage for more biotic expansion and exploration of fitness space the next time the conditions change. Such functional innovations are significant because they produce inheritance systems with reduced overlap in fitness space compared to that of their closest relatives living in adjacent areas, and that allows increased coexistence the next time biotic expansion is catalyzed by an external perturbation. Imagine an episode of expansion and isolation that gives rise to 30 isolated associations of inheritance systems and the splitting of all inheritance systems into 30 new inheritance systems. Now imagine that in only one of these places, a functional innovation emerges in a single inheritance system. That new inheritance system can now potentially coexist without conflict (or with reduced conflict) in the 29 remaining isolated places where their closest relatives exist functioning largely in the ancestral mode. Now, let us assume that in each of the 30 isolated places a single, but different, inheritance system evolves a functional innovation. The potential diversity of associations is immense, even without a high rate of functional innovation or diversification. So long as the conditions keep changing and life continues to have the capacity to respond, nature becomes very complex very easily.

To illustrate, we again lean on parasite–host systems as a convenient model (Fig. 10.3), patterned after a study by Patella et al. (2017) for some Brazilian freshwater fish and their monogenoidean parasites (for a compatible study based on butterflies and their host plants, see Audusseau et al. 2020). We begin with the expansion/generalization phase of two hypothetical host species and their associated parasites (upper left panel). Each species is widely distributed and has high connectedness among populations. The conditions change (e.g., sea levels rise) (upper right

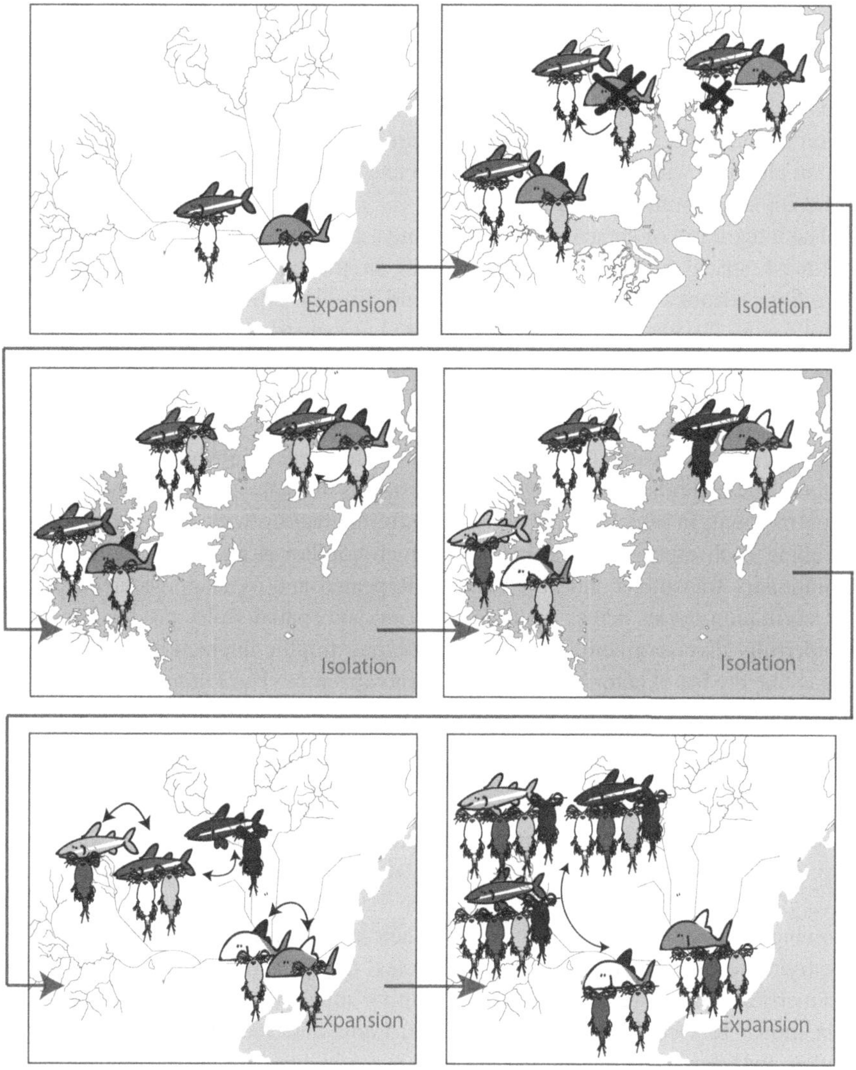

Fig. 10.3 Cartoon diagram of the combined taxon pulse and oscillation in functional fitness space dynamic in a hypothetical parasite–host system based on a study by Patella et al. (2017). Redrawn and modified from Brooks et al. (2019). See text for details

panel) so that populations become isolated. As a result, three things happen: a host goes extinct (X) in one of the isolated populations; its parasite persists by colonizing a new host (arrow) through ecological fitting; and a parasite goes extinct in another population leaving an "empty" host. In the middle left panel, a parasite expands its host range by ecological fitting to include the "empty" host, forming a new association. Isolation favors specialization and innovation, leading to diversification in both parasites and hosts (middle right panel). Finally, in the lower left and right panels, the conditions change again (e.g., sea levels fall). Barriers to dispersal break down. Host and parasite faunas become reconnected. Biotic expansion and mixing leads to increased opportunities to colonize new hosts, leading to further diversification and complexity in the system. Subsequent episodes of isolation and specialization alternate with episodes of expansion and generalization, increasing diversity and complexity in the system over time.

The integration of capacity and opportunity, and the complementary oscillations in fitness space arising from perturbations in the conditions is the *Stockholm Paradigm* (Brooks et al. 2014; Hoberg and Brooks 2015; Hoberg et al. 2015; Brooks et al. 2019). Although viewed as newly emerged, the Stockholm Paradigm has been in development since 1859. If the *Tree of Life* is a record of the history of conflict resolution, irreversible evolutionary transitions, and the selective accumulation of diversity, then the *Entangled Bank*—Darwin's final metaphor for the deployment of evolvable life on this planet—is the context in which the inheritance systems that comprise the *Tree of Life* coexist. It is the state of the biosphere at each given moment, in which organisms are busily interacting with their surroundings, including each other. It is the arena in which conflict is resolved, giving rise to evolutionary transitions and new conflict. Repeated and overlapping taxon pulses and oscillation cycles across multiple temporal and spatial scales produce both the wonderfully diverse inheritance systems and their trophic interactions. The *Tree of Life* is like the list of actors in a play and their capacities. The *Entangled Bank* is the play itself, the nature of the organism (capacity) meeting the nature of the conditions (opportunity) to produce complex associations of inheritance systems and the relationships among them. The dynamic interplay between the Tree of Life and the Entangled Bank produces the panoply of life on this planet in all its celebrated diversity. Neo-Darwinians admire the Entangled Bank far more than the Tree of Life. In doing so they ignore the fundamental role of history in explanations of the Entangled Bank. This, in effect, shifts the focus to explaining the play without knowing much about the actors beyond their current interactions on stage. It is like trying to explain the tragedy of Sophocles' *Oedipus Rex*—Oedipus kills Laius and marries Jocasta—without knowing anything about the characters' backstories—unbeknown to Oedipus until it is too late, Laius and Jocasta are his father and mother. In the end, Oedipus gouges his own eyes out, but without knowing history it is impossible to understand why.

The *Geographic Mosaic Theory of Coevolution* (Thompson 1988, 1994, 1997, 1999a, b, c, 2005) has many things in common with the Stockholm Paradigm. It emphasizes the role of geographic structure and functional connectedness among populations of host and parasite inheritance systems. Parasites may interact differently with the same or different hosts in different parts of each species' geographical range, giving rise to multiple opportunities for specialist parasites to evolve from other specialists forming a spatial mosaic of interactions.

From our perspective, the theory is hampered by assumptions that the world is comprised of inheritance systems that are inherently either generalists or specialists and that coevolutionary interactions produce increasingly specialized organisms with reduced evolutionary potential, indicated by narrow host range and limited capacity to switch to new hosts. Critically, there is no mechanism for the evolution of *new generalists from specialists*. Without this, there can be no oscillating between exploring and exploiting, no central coping mechanism for the Entangled Bank to deal with environmental change. Two things are missing. The first is the capacity for ecological fitting in sloppy fitness space which provides the essential fuel for "how specialists can be generalists" (Agosta et al. 2010) by allowing the colonization of new habitats and resources, such as novel hosts. The second is appreciation for external perturbations to catalyze taxon pulses, resetting the stage for the geographic mosaics of coevolution to play out. When viewed from this perspective, we see that what looks like associations comprised inheritance systems labeled as *specialists* and *generalists* is an illusion. Instead, at any given point, ecosystems comprise inheritance systems in various stages of *specializing* and *generalizing* in fitness space. There are no such things as "generalists" and "specialists," only inheritance systems that are more or less generalizing or specializing in fitness space at particular places and times. And the capacity for ecological fitting in sloppy fitness space allows inheritance systems to oscillate between generalizing and specializing according to the conditions at hand.

From the perspective of the Stockholm Paradigm, therefore, the very phenomenon of coevolution may be illusory. At the point when geographic isolation leads to newly emergent ecosystems, interactions between hosts and parasites are maximally specialized and the system will appear the most "coevolved." If we only studied very young, geographically localized associations, we might find patterns that fit the predictions of the Geographic Mosaic Theory of Coevolution very well. But this would only be a transient state, a thin slice in time that could easily give the illusion of a "delicately balanced" ecosystem with low potential to respond to change, when in fact the opposite was true (e.g., Hoberg and Zarlenga 2016). Fitness space is at its most sloppy when inheritance systems are at their most specialized, and the sloppier the fitness space, the greater the potential for an inheritance system to expand geographically and functionally in fitness space when the opportunities arise. Modeling the Stockholm Paradigm (Araujo et al. 2015; Braga et al. 2018; Braga et al. 2020; D'Bastiani et al. 2021) shows that coevolutionary arms races are only possible when capacities and conditions leave a single parasite and a single host no options other than to cope with each other; exploitation mode is maximized in isolation, leading to evolutionary stasis. But that is the most specialized situation possible, with the greatest potential for adding hosts whenever new opportunities present themselves. And once a parasite begins adding hosts to its repertoire, coevolutionary arms races devolve into Red Queen dynamics (Van Valen 1973; Araujo et al. 2015) in which the more hosts you have the less chance any one of them has to evolve an effective defense and greater the evolutionary advantage to the parasite. This is likely the reason more than 50% of all inheritance systems on the planet are parasites of some sort (Price 1980; Brooks et al. 2019) and why complex multi-trophic associations may be quite stable functionally and yet are assembled piecemeal evolutionarily (Brooks et al. 2006a, b; Malcicka et al. 2015; Burgess et al. 2018; Perez-Ramos et al. 2019).

Entangled banks are not assembled by coevolutionary mechanisms or by filling up some imaginary number of preexisting slots in a given area. They are assembled by oscillations of generalizing and specializing in fitness space catalyzed by repeated perturbations in the nature of the conditions (e.g., Dong et al. 2017). The deep-history signatures of this dynamic are not singular events, therefore, but dualities of expansion and isolation, generalizing and specializing, lagging slightly behind changes in the conditions (e.g., Brooks and McLennan 2002; Ackerly 2003; Stigall 2010, 2012a, b, 2014, 2017, 2019; Mejia-Madrid 2012; Brame and Stigall 2014; Nurnberg and Aberhan 2015; Economo et al. 2015; Estrada et al. 2015; Stigall et al. 2017, 2019; Bernardi et al. 2018; Kennedy et al. 2018; Lam et al. 2018; Moser et al. 2018; Rominger et al. 2019; Rull and Carnaval 2020).

10.4 Coping with Uncertainty

Gambler's Ruin: A gambler who raises his bet to a fixed fraction of his bankroll when he wins, but does not reduce it when he loses, will eventually go broke, even if he has a positive expected value on each bet. In evolutionary terms, the Gambler's Ruin means that no matter how well a species or population seems to be coping with a current environment, it can still go extinct if conditions change too rapidly or in ways which do not allow it to cope.

Darwinian evolution is amazing, but it is not perfect. It cannot anticipate the future, especially future external perturbations. And this can be costly. John Maynard Smith (Maynard Smith 1974; Maynard Smith and Price 1973) introduced the Gambler's Ruin metaphor to evolutionary biology to highlight the importance of understanding that evolution cannot anticipate future perturbations. Living systems appear anticipatory only in the sense that conservative inheritance means that they "anticipate" that tomorrow will be like today and yesterday. And on a day-to-day basis, this tends to be sufficient for survival. But periodically, it is disastrous. Extinction is part of the evolutionary saga.

During periods of environmental stasis, Darwinism is "survival of the adequate with the fittest dominating numerically." During periods of environmental perturbations, Darwinism is "survival of the adequate with the fittest becoming less numerous or going extinct, replaced by variants that had lower fitness in the previous environment." This is the Gambler's Ruin: no matter how successful you are today you can always be worse off tomorrow if the game changes. No matter how fit an organism, population, or species is at any given place and time, it can still go extinct if the conditions change. For a spectacular example of this general principle witness the demise of the non-avian dinosaurs.

The historical conservatism of inheritance systems, and the inherent mismatch between the nature of the organism and the conditions that this creates (*Darwin's Necessary Misfit*), is what underlies the production of sloppy fitness space and therefore the ability of species to cope with changing conditions. But the same conservatism also places severe constraints on the ways in which and extents to which organisms can respond to change. "Optimal" organisms do not exist because conservative inheritance ensures that the nature of the organism always lags behind the nature of the conditions. This fundamental constraint on the inheritance system is what makes life inherently vulnerable to the Gambler's Ruin, in spite of its propensity for indefinite persistence. Conservative inheritance arms organisms with a retained history of past success which may be useful in the present or future, but it also means that there can be environmental changes of such great magnitude or rate that the system cannot cope (mount a sufficient evolutionary response) and becomes extinct.

The Gambler's Ruin focuses our attention on the fact that all inheritance systems are capable of being overwhelmed by external perturbations, and yet life still persists. How is that possible? The answer is that life is evolvable and has the capacity to fluctuate between exploitation-biased behavior (making the best of the current situation) and exploration-biased behavior (making the best of new situations) as conditions, and therefore fitness space, changes. Although it never returns to the same state, this inherent oscillatory behavior allows life to persist in a state of perpetual novelty, never repeating itself but always rhyming with the past, continuously reaching new states that are simply "good enough" to survive the next change.

Novel inheritance systems form easily during periods of isolation, when conditions are stable. Every episode of evolutionary diversification thus begins with an external perturbation that catalyzes biotic expansion, decreasing fitness in the area of origin, but which also links previously unconnected pieces of fitness space, allowing each inheritance system a chance to flee for its life. The ones that go extinct lack the preexisting capacity to cope with the changes and take advantage of the opportunity to explore new connections in fitness space. The survivors are those that leave areas where their preferred conditions are deteriorating for areas where they can still find survivable conditions, in the process becoming generalized in fitness space. This diversifies their portfolios, increasing the odds of survival in an uncertain future (Schindler et al. 2015).

> New beginnings are often disguised as painful endings—Lao Tzu

Alfred Russell Wallace, the codiscoverer of the theory of natural selection, was stimulated in his thinking by what he considered to be evidence of a natural law regulating the introduction of species into areas (Wallace 1855: also known as the *Sarawak Law*). Wallace proposed a dual system to explain his observations. The first element of the duality was the extinction of resident inheritance systems in certain places and the second was their invariable replacement by others (Wallace 1855). Wallace inferred that new inheritance systems were evolving in some parts of the world where extinction was not occurring. Extinction became part of the evolutionary saga and would become a major point of distinction between Lamarckian and Darwinian theories.

The twentieth century's discovery of great upheavals having wracked our planet in the past, causing mass extinctions, followed by massive bursts of evolution producing species that never existed before are contrary to notions of a static "natural equilibrium." At least five perturbations to the biosphere have been of such great

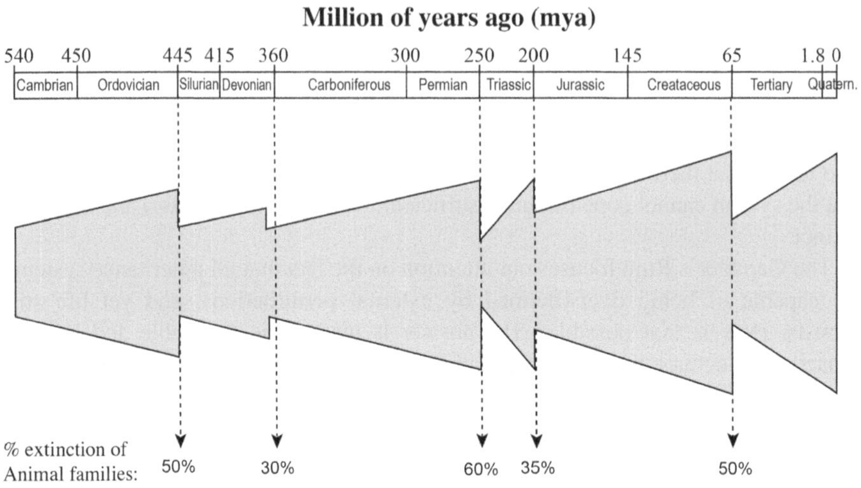

Fig. 10.4 Diagram showing the impact of the five *Great Mass Extinction* events on biodiversity, with time on the *x*-axis and diversity of animal families on the *y*-axis. After each extinction event, there is evolutionary renewal and re-diversification, no matter the severity of diversity lost. Redrawn and modified from Brooks et al. (2019)

magnitude that a huge portion of biodiversity was lost, unable to cope with the change. These are known as the *Great Mass Extinction* events (Fig. 10.4).

Traditional paleontological studies focus on the losers in evolution, giving the impression of a system on the brink of collapse, on the edge of blinking out, running out of evolutionary potential and dangerously close to succumbing to the Gambler's Ruin. In a Darwinian world, so long as some life survives, a vast amount of evolutionary potential is preserved. If the origin of mitochondria and the eukaryotic condition set the stage for us to emerge, so too did the fifth mass extinction, an encounter with an asteroid 65 million years ago that doomed the dinosaurs. In a Darwinian world, mass extinctions are better thought of as *Mass Evolutionary Re-sets* that allow the biosphere to produce new diversity following close encounters with the Gambler's Ruin.

Perhaps no body of research has shed more light on this reality than the work of Alycia Stigall (Stigall 2010, 2012a, b, 2014, 2017, 2019; Brame and Stigall 2014; Stigall et al. 2017, 2019; Lam et al. 2018). Like other paleontologists, Stigall has documented the impact of major external perturbations on the biosphere. But instead of focusing on the inheritance systems that went extinct (motivated initially by a desire to show that Lamarck was wrong, and extinction has actually occurred), she has focused on the survivors. Her work shows that when the conditions change, inheritance systems that cope well enough to survive do so in one or both of two ways. They either *move and find new places to live that have similar conditions* or they *move and buy enough time along the way to evolve new capacities to cope with different conditions.* A few species also survive in situ because they are so abundant and widespread that some populations escape the effects of the change. But most inheritance systems that do not move go extinct.

As Stigall's work so eloquently shows, the history of life is a saga of using preexisting capacities to try to cope with change, then fleeing when conditions change beyond the ability to cope, trying to find conditions with which it is possible to cope, or dying trying. It is the survival of the adequate. Life does what it needs to do to get by. Or it does not and goes extinct. When a crisis occurs, demanding a response, Necessity is not the mother of invention in evolution. Rather, Necessity is the mother of running away as fast as you can, hoping to find a livable place where you can cope with local conditions, survive long enough to participate in the next expansion event, and where you can afford to experiment with ecological innovations. Evolution is not heroic, not magical, but functional and effective. If the conditions change, cope using your capacity for ecological fitting; if you cannot cope, flee again using your capacity for ecological fitting; if you cannot flee, die. Extinction is a failure to have a coping mechanism already in place when conditions change. Once you have arrived in survivable new fitness space, the default switch to exploitation-biased behavior establishes the circumstance in which innovation is possible, setting the stage for survival when the next unanticipated perturbation occurs (Day et al. 2015; Keil et al. 2017; Edie et al. 2018; Kiesling et al. 2018; Grunert et al. 2019).

10.5 Summary

The biosphere has a dualistic nature—it is made up of inheritance systems all variously related to each other, and it is made up of many inheritance systems living together. Viewing the biosphere in its current state gives the impression that the most important thing to explain about the Entangled Bank is its current state: bats and birds and elephants and worms, and all of *this* biodiversity in its current form. We think this was not the central question Darwin was trying to communicate with his description of the Entangled Bank. Instead, he meant for the metaphor to be about things in motion, an image of a dynamical system, not a static object, with the central questions being about how such systems come to be and how they cope with constant change. The issue for evolutionary biology was never why are there are so many (Hutchinson 1959; Brown 1981) or so few (Felsenstein 1981; Brooks and Wiley 1986, 1988) inheritance systems, nor even how close relatives can coexist. We did not need a *paradigm of diversity*; we needed a *paradigm of diversification.*

Inheritance systems are *oscillators*, capable of exploring and exploiting, of generalizing and specializing to an extent possible based on their inherited characteristics and depending on the contingencies of the conditions of life in which they find themselves. These oscillators respond to perturbations by remaining relatively static, by going extinct, or by diversifying. This is the fundamental dynamic of evolvable, diversifying life.

Multiple inheritance systems living together respond to external perturbations by oscillating between phases of functional generalization in fitness space during geographic range expansion followed by phases of specialization during geographic isolation. If the perturbations are not strong enough to weaken or eliminate cohesion/connectivity in fitness space, each affected inheritance system will cope with the variable conditions by means of its particular (specific and phylogenetically conservative) capacities for ecological fitting. Environmental perturbations strong enough to alter cohesion/connectivity range from perturbations that impact a small number of co-occurring inheritance systems to phenomena that affect regional biotas to those that affect the entire biosphere. If the perturbations also reduce cohesion in fitness space, then exploration-biased behavior is favored. During periods of relative stability when cohesion is restored (connections to other parts of the inheritance are reduced), exploitation-biased behavior predominates, and inheritance systems tend to aggregate geographically where most of their activities focus on the immediate circumstances but where conditions also permit the possibility of emergent innovations. The key to persisting in a changing world is the capacity for life to bounce back and forth between the two states of relative exploration and exploitation in response to perturbations in the nature of the conditions.

Every episode of generalizing during geographic expansion is followed by an episode of specializing as inheritance systems become disconnected from previous associations. When this happens, the demographic tendency to switch from exploration to exploitation will be intensified as cohesive properties become the dominant foci of selection for each co-occurring inheritance system. From this emerge new geographically localized coexisting associations of specialized inheritance systems. Each inheritance system within the association is in a state of evolutionary reset, in which its fundamental fitness space is maximally sloppy, so the entire system has evolutionary potential in direct proportion to the number of coexisting, interacting, specialized inheritance systems.

Entangled banks are made up of a few who stayed behind and were able to cope with the new conditions and many who fled from elsewhere and found a safe haven. This is the Stockholm Paradigm, a "back to the future" Darwinian dynamic explaining how entangled banks come to be, elegant, simple, functional, devoid of magic yet astounding in its results.

We finish our saga in the next chapter, asking if entangled banks are greater than the sum of their parts and then putting evolution to work for humanity in a time of accelerating global climate change.

References

Ackerly DD (2003) Community assembly, niche conservatism, and adaptive evolution in changing environments. Int J Plant Sci 164(3 Suppl):S165–S184

Agosta SJ (2006) On ecological fitting, plant-insect associations, herbivore host shifts, and host plant selection. Oikos 114:556–565

Agosta SJ, Janzen DH (2005) Body size distributions of large Costa Rican dry forest moths and the underlying relationship between plant and pollinator morphology. Oikos 108:183–193

Agosta SJ, Klemens JA (2008) Ecological fitting by phenotypically flexible genotypes: implications for species associations, community assembly and evolution. Ecol Lett 11:1123–1134

Agosta SJ, Klemens JA (2009) Resource specialization in a phytophagous insect: no evidence for genetically based performance tradeoffs across hosts in the field or laboratory. J Evol Biol 22:907–912

Agosta SJ, Janz N, Brooks DR (2010) How generalists can be specialists: resolving the "parasite paradox" and implications for emerging disease. Fortschr Zool 27:151–162

Ando H, Setsuko S, Horikoshi K, Suzuki H, Umehara S, Inoue-Murayama M, Isagi Y (2013) Diet analysis by next-generation sequencing indicates the frequent consumption of introduced plants by the critically endangered red-headed wood pigeon (*Columba janthina nitens*) in oceanic island habitats. Ecol Evol 3:4057–4069

Araujo SBL, Braga MP, Brooks DR, Agosta SJ, Hoberg EP, von Hartenthal FW, Boeger WA (2015) Understanding host-switching by ecological fitting. PLoS One 10(10):e0139225. https://doi.org/10.1371/journal.pone.0139225

Audusseau H, Ryrholm, N, Stefanescu, C, Tharel S, Jansson C, Champeaux L, Shaw MR, Raper C, Lewis OT, Janz N, Schmucki, R (2020) Altered parasitism of a butterfly assemblage associated with a range expanding species. bioRxiv preprint. https://doi.org/10.1101/2020.02.13.947440

Banks JC, Palma RL, Paterson AM (2006) Cophylogenetic relationships between penguins and their chewing lice. J Evol Biol 19:156–166

Bernardi M, Gianolla P, Petti FM, Mietto P, Benton MJ (2018) Dinosaur diversification linked with the Carnian Pluvial Episode. Nat Commun 9:1499. https://doi.org/10.1038/s41467-018-03996-1

Bouchard P, Brooks DR (2004) Effect of vagility potential on dispersal and speciation in rainforest insects. J Evol Biol 17:994–1006

Braga MP, Razzolini E, Boeger WA (2014) Drivers of parasite sharing among Neotropical freshwater fishes. J Anim Ecol 84:487–497

Braga MP, Araujo SBL, Agosta S, Brooks DR, Hoberg EP, Nylin S, Janz N, Boeger WA (2018) Host use dynamics in a heterogeneous fitness landscape generates oscillations in host range and diversification. Evolution 72:1773–1783

Braga MP, Landis M, Nylin S, Janz N, Ronquist F (2020) Bayesian inference of ancestral host-parasite interactions under a phylogenetic model of host repertoire evolution. Syst Biol. https://doi.org/10.1093/sysbio/syaa019

Brame H-MR, Stigall AL (2014) Controls on niche stability in geologic time: congruent responses to biotic and abiotic environmental changes among Cincinnatian (Late Ordovician) marine invertebrates. Paleobiology 40:70–90

Brooks DR, Ferrao AL (2005) The historical biogeography of coevolution: emerging infectious diseases are evolutionary accidents waiting to happen. J Biogeogr 32:1291–1299

Brooks DR, Hoberg EP (2000) Triage for the biosphere: the need and rationale for taxonomic inventories and phyogenetic studies of parasites. Comp Parasitol 68:1–25

Brooks DR, McLennan DA (1993) Parascript: parasites and the language of evolution. Smithsonian University Press, Washington, DC

Brooks DR, McLennan DA (2002) The nature of diversity. University of Chicago Press, Chicago

Brooks DR, Wiley EO (1986) Evolution as entropy: toward a unified theory of biology, 1st edn. University of Chicago Press, Chicago

Brooks DR, Wiley EO (1988) Evolution as entropy: toward a unified theory of biology, 2nd edn. University of Chicago Press, Chicago

Brooks DR, McLennan DA, León Régagnon V, Hoberg EP (2006a) Phylogeny, ecological fitting and lung flukes: helping solve the problem of emerging infectious diseases. Rev Mex Biod 77:225–234

Brooks DR, León Régagnon V, McLennan DA, Zelmer D (2006b) Ecological fitting as a determinant of the community structure of platyhelminth parasites of anurans. Ecology 87(Supplement):S76–S85

Brooks DR, Hoberg EP, Boeger WA, Gardner SL, Galbreath KE, Herczeg D, Mejía-Madrid HH, Rácz SE, Dursahinhan AT (2014) Finding them before they find us: informatics, parasites and environments in accelerating climate change. Comp Parasitol 81:155–164

Brooks DR, Hoberg EP, Boeger WA (2015) In the eye of the Cyclops: the classic case of cospeciation and why paradigms are important. Comp Parasitol 83:1–8

Brooks DR, Hoberg EP, Boeger WA (2019) The Stockholm paradigm: climate change and emerging disease. University of Chicago Press, Chicago

Brown JH (1981) Two decades of homage to Santa Rosalia: toward a general theory of diversity. Am Zool 21:877–888

Burgess MD, Smith KW, Evans KL, Leech D, Pearce-Higgins JW, Branston CJ, Briggs K, Clark JR, du Feu CR, Lewthwaite K, Nager RG, Sheldon BC, Smith JA, Whytock RC, Willis SG, Phillimore AB (2018) Tritrophic phenological match–mismatch in space and time. Nat Ecol Evol 2:970–975

Cain SA (1938) The species-area curve. Am Midl Nat 19:573–581

Carbonaro FA, Langer FC, de Souza FG, Ghilardi RP (2018) Inferring ancestral range reconstruction based on trilobite records: a study-case on Metacryphaeus (Phacopida, Calmoniidae). Sci Rep 8:15179. https://doi.org/10.1038/s41598-018-33517-5

Carvalho JC, Cardoso P, Riga F, Kostas A, Triantis KA, Borges PAV (2015) Modeling directional spatio-temporal processes in island biogeography. Ecol Evol 5:4671. https://doi.org/10.1002/ece3.1632

Chagnon P-L, Magain N, Miadlikowska J, Lutzoni F (2019) Species diversification and phylogenetically constrained symbiont switching generated high modularity in the lichen genus *Peltigera*. J Ecol 107:1645. https://doi.org/10.1111/1365-2745.13207

Chan BKK, Xu G, Kim HK, Park J-H, Kim W (2018) Living with marginal coral communities: diversity and host-specificity in coral-associated barnacles in the northern coral distribution limit of the East China Sea. PLoS One 13:e0196309. https://doi.org/10.1371/journal.pone.0196309

D'Bastiani E, Campião KM, Boeger WA, Araújo SBL (2021) Influence of the ecological opportunity of interaction on the structure of host-parasite networks. Parasitology, in press. https://doi.org/10.1101/2020.01.13.904151

Darlington PJ Jr (1943) Carabidae of mountains and islands: data on the evolution of isolated faunas, and on atrophy of wings. Ecol Monogr 13:37–61

Darwin C (1872) The origin of species, 6th edn. Murray, London

Day MO, Ramezani J, Bowring SA, Sadler PM, Erwin DH, Abdala F, Rubidge BS (2015) When and how did the terrestrial mid-Permian mass extinction occur? Evidence from the tetrapod record of the Karoo Basin, South Africa. Proc R Soc B 282:20150834. https://doi.org/10.1098/rspb.2015.0834

de Vienne DM, Refreigier G, Lopez-Villavicencio M, Tellier A, Hood ME, Giraud T (2013) Cospeciation vs host-shift speciation: methods for testing, evidence from natural associations and relation to coevolution. New Phytol 198:347–385

Desdevises Y, Morand S, Jousson O, Legendre P (2002) Coevolution between *Lamellodiscus* (Monogenea: Diplectanidae) and Sparidae (Teleostei): the study of a complex host-parasite system. Evolution 56:2459–2471

Dick CW, Patterson BD (2007) Against all odds: explaining high host specificity in dispersal-prone parasites. Int J Parasitol 37:871–876

Dong F, Hung C-M, Li X-L, Gao J-Y, Zhang Q, Wu F, Lei F-M, Shou-Hsien Li S-H, Yang X-J (2017) Ice age unfrozen: severe effect of the last interglacial, not glacial, climate change on East Asian avifauna. BMC Evol Biol 17:244. https://doi.org/10.1186/s12862-017-1100-2

Eckstut ME, McMahan CD, Crother BI, Ancheta JM, McLennan DA, Brooks DR (2011) PACT in practice: comparative historical biogeographic patterns and species-area relationships of the Greater Antillian and windward Hawaiian Island terrestrial biotas. Glob Ecol Biogeogr 20:545–557

Economo EP, Sarnat EM, Janda M, Clouse R, Klimov PB, Fischer G, Blanchard BD, Ramirez LN, Andersen AN, Berman M, Guenard B, Lucky A, Rabeling C, Wilson EO, Knowles LL (2015)

Breaking out of biogeographical modules: range expansion and taxon cycles in the hyperdiverse ant genus *Pheidole* J. Biogeographica 42:2289–2301. https://doi.org/10.1111/jbi.12592
Edie SM, Huang S, Collins KS, Roy K, Jablonski D (2018) Loss of biodiversity dimensions through shifting climates and ancient mass extinctions. Integr Comp Biol 58:1179. https://doi.org/10.1093/icb/icy111
Ellis VA, Collins MD, Medeiros MCI, Sari EHR, Coffey ED, Dickerson RC, Lugarini C, Stratford JA, Henry DR, Merrill L, Matthews AE, Hanson AA, Roberts JR, Joyce M, Kunkel MR, Ricklefs RE (2015) Local host specialization, host-switching, and dispersal shape the regional distributions of avian haemosporidian parasites. PNAS 112:11294–11299. https://doi.org/10.1073/pnas.1515309112
Elton CS (1958) The ecology of invasions by animals and plants. Methuen, London
Erwin TL (1979) Thoughts on the evolutionary history of ground beetles: hypotheses generated from comparative faunal analyses of lowland forest sites in temperate and tropical regions. In: Erwin TL, Ball GE, Whitehead DR, Halpern AL (eds) Carabid beetles. Springer, Dordrecht, pp 539–592
Erwin TL (1981) Taxon pulses, vicariance, and dispersal: an evolutionary synthesis illustrated by carabid beetles. In: Nelson G, Rosen DE (eds) Vicariance biogeography: a critique. Columbia University Press, New York, pp 159–196
Erwin TL (1985) The taxon pulse: a general pattern of lineage radiation and extinction among carabid beetles. In: Ball GE (ed) Taxonomy, phylogeny, and zoogeography of beetles and ants. W. Junk, Dordrecht, pp 437–472
Estrada A, Meireles C, Morales-Castilla I, Poschlod P, Vieites D, Araújo MB, Early R (2015) Species' intrinsic traits inform their range limitations and vulnerability under environmental change. Global Ecol Biogeogr 24:849. https://doi.org/10.1111/geb.12306
Estrada-Peña A, Sprong H, Cabezas-Cruz A, de la Fuente J, Ramo A, Coipan EC (2016) Nested coevolutionary networks shape the ecological relationships of ticks, hosts, and the Lyme disease bacteria of the *Borrelia burgdorferi* (sl) complex. Parasit Vectors 9:517. https://doi.org/10.1186/s13071-016-1803-z
Felsenstein J (1981) Skepticism towards Santa Rosalia, or why are there so few kinds of animals? Evolution 35:124–138
Folinsbee K, Brooks DR (2007) Miocene hominoid biogeography: pulses of dispersal and differentiation. J Biogeogr 34:383–397
Forbes A, Devine S, Hippee A, Tvedte E, Ward A, Widmayer H, Wilson C (2017) Revisiting the particular role of host shifts in initiating insect speciation. Evolution 71:1126–1137
Grunert HR, Neil Brocklehurst N, Fröbisch J (2019) Diversity and disparity of *Therocephalia*: macroevolutionary patterns through two mass extinctions. Nat Sci Rep 9:5063. https://doi.org/10.1038/s41598-019-41628-w
Halas D, Zamparo D, Brooks DR (2005) A protocol for studying biotic diversification by taxon pulses. J Biogeogr 32:249–260
Heaney LR (2000) Dynamic disequilibrium: a long-term, large-scale perspective on the equilibrium model of island biogeography. Glob Ecol Biogeogr 9:59–74
Hoberg EP, Brooks DR (2008) A macroevolutionary mosaic: episodic host-switching, geographic colonization, and diversification in complex host-parasite systems. J Biogeogr 35:1533–1550
Hoberg EP, Brooks DR (2010) Beyond vicariance: integrating taxon pulses, ecological fitting, and oscillation in evolution and historical biogeography. In: Morand S, Krasnov BR (eds) The biogeography of host-parasite interactions. Oxford University Press, New York, pp 7–20
Hoberg EP, Brooks DR (2015) Evolution in action: climate change, biodiversity dynamics and emerging infectious disease. Philos Trans R Soc B 370:20130553
Hoberg EP, Zarlenga DS (2016) Evolution and biogeography of *Haemonchus contortus*: linking faunal dynamics in space and time. Adv Parasitol 93:1–30
Hoberg EP, Agosta SJ, Boeger WA, Brooks DR (2015) An integrated parasitology: revealing the elephant through tradition and invention. Trends Parasitol 3:128–133

Hoberg EP, Cook JA, Agosta SJ, Boeger WA, Galbreath KE, Laaksonen S, Kutz SJ, Brooks DR (2017) Arctic systems in the quaternary: ecological collision, faunal mosaics and the consequences of a wobbling climate. J Helminthol 91:409–421
Holland BS, Chiaverano LM, Howard CK (2017) Diminished fitness in an endemic Hawaiian snail in nonnative host plants. Ethol Ecol Evol 29:229–240
Horton DR, Kaur N, Cooper WR, Miliczky E, Badillo-Vargas IE, Esparza-Díaz G, Rashed A, Waters TD, Wohleb C, Johnson DL, Kawchuk L, Jensen AS (2019) Whence and whither the *Convolvulus* psyllid? An invasive plant leads to diet and range expansion by a native insect herbivore. Ann Entomol Soc Am 112:249–264
Hutchinson GE (1959) Homage to Santa Rosalia or why are there so many kinds of animals? Am Nat 93:145–159
Huxley JS (ed) (1940) The new systematics. Oxford University Press, Oxford
Huxley JS (ed) (1942) Evolution, the modern synthesis. Allen and Unwin, London
Janz N (2011) Ehrlich and Raven revisited: mechanisms underlying codiversification of plants and enemies. Annu Rev Ecol Evol Syst 42:71–89
Janz N, Nylin S (2008) The Oscillation Hypothesis of host-plant range and speciation. In: Tilmon KJ (ed) Specialization, speciation, and radiation: the evolutionary biology of herbivorous insects. University of California Press, Berkeley, pp 203–215
Janz N, Nylin S, Wahlberg N (2006) Diversity begets diversity: host expansions and the diversification of plant-feeding insects. BMC Evol Biol 6:4. https://doi.org/10.1186/1471-2148-6-4
Janzen DH (1980) When is it coevolution? Evolution 34:611–612
Jorge F, Perera A, Poulin R, Roca V, Carretero MA (2018) Getting there and around: host range oscillations during colonization of the Canary Islands by the parasitic nematode *Spauligodon*. Mol Ecol 27:533. https://doi.org/10.1111/mec.14458
Juarrero A (1999) Dynamics in action. MIT Press, Boston
Kaczvinsky C, Hardy NB (2019) Do major host shifts spark diversification in butterflies? Ecol Evol. https://doi.org/10.1002/ece3.6116
Keil P, Pereira HM, Cabral JS, Chase JM, May F, Martins IS, Winter M (2017) Spatial scaling of extinction rates: theory and data reveal nonlinearity and a major upscaling and downscaling challenge. Glob Ecol Biogeog 27:2–13. https://doi.org/10.1111/geb.12669
Kennedy JD, Borregaard MK, Marki PZ, Machac A, Fjeldsa J, Rahbek C (2018) Expansion in geographical and morphological space drives continued lineage diversification in a global passerine radiation. Proc R Soc B 285:20182181. https://doi.org/10.1098/rspb.2018.2181
Kiessling W, Schobben M, Ghaderi A, Hairapetian V, Leda L, Korn D (2018) Pre-mass extinction decline of latest Permian ammonoids. Geology 46:283–286
Kolbe JJ, Glor RE, Rodnguez Schettino L, Chamizo Lara A, Larson A, Losos JB (2004) Genetic variation increases during biological invasion by a Cuban lizard. Nature 431:177–181
Lam AR, Stigall AL, Matzke NJ (2018) Dispersal in the Ordovician: speciation patterns and paleobiogeographic analyses of brachiopods and trilobites. Palaeogeogr Palaeoclimatol Palaeoecol 489:147–165. https://doi.org/10.1016/j.palaeo.2017.10.006
Lieberman BS (2000) Paleobiogeography: using fossils to study global changes, plate tectonics, and evolution. Topics in geobiology, vol 16. Kluwer Academic, Plenum Press, New York, pp 1–208
Lieberman BS (2003a) Unifying theory and methodology in biogeography. Evol Biol 33:1–25
Lieberman BS (2003b) Paleobiogeography: the relevance of fossils to biogeography. Annu Rev Ecol Syst 34:51–69
Liebherr JK (1988) General patterns in West Indian insects, and graphical biogeographic analysis of some circum-Caribbean *Platynus* beetles (Carabidae). Syst Biol 37:385–409
Liebherr JK, Hajek AE (1990) A cladistic test of the taxon cycle and taxon pulse hypotheses. Cladistics 6:39–59
Lim BK (2008) Historical biogeography of New World emballonurid bats (tribe Diclidurini): taxon pulse diversification. J Biogeogr 35:1385–1401
Lomolino MV (2000) A call for a new paradigm of island biogeography. Glob Ecol Biogeogr 9:1–6

MacArthur RH, Wilson EO (1963) An equilibrium theory of insular zoogeography. Evolution 17:373–387

MacArthur RH, Wilson EO (1967) The theory of island biogeography. Princeton University Press, Princeton, NJ

Malcicka M, Agosta SJ, Harvey JA (2015) Multi level ecological fitting: indirect life cycles are not a barrier to host switching and invasion. Glob Chang Biol 21:3210–3218

Maynard Smith J (1974) The theory of games and the evolution of animal conflicts. J Theor Biol 47:209–221

Maynard Smith J, Price GR (1973) The logic of animal conflict. Nature 246:15–18

Mejía-Madrid HH (2012) Biogeographic hierarchical levels and parasite speciation. In: Stevens L (ed) Global advances in biogeography, ch 2: 23–48. InTech, Rijeka

Mellado A, Zamora R (2019) Ecological consequences of parasite host shifts under changing environments: more than a change of partner. J Ecol. https://doi.org/10.1111/1365-2745.13295

Messenger LA, Garcia L, Vanhove M, Huaranca C, Bustamante M, Torrico M, Torrico F, Miles MA, Llewellyn MS (2015) Ecological host fitting of *Trypanosoma cruzi* TcI in Bolivia: mosaic population structure, hybridization and a role for humans in Andean parasite dispersal. Mol Ecol 24:2406–2422

Moonjely S, Bidochka MJ (2019) Generalist and specialist *Metarhizium* insect pathogens retain ancestral ability to colonize plant roots. Fungal Ecol 41:209–217

Moser FN, van Rijssel JC, Mwaiko S, Meier JI, Ngatunga B, Seehausen O (2018) The onset of ecological diversification 50 years after colonization of a crater lake by haplochromine cichlid fishes. Proc R Soc B 285:20180171. https://doi.org/10.1098/rspb.2018.0171

Nurnberg S, Aberhan M (2015) Interdependence of specialization and biodiversity in Phanerozoic marine invertebrates. Nat Commun. https://doi.org/10.1038/ncomms7602comms7602

Nylin S, Janz N (2009) Butterfly host plant range: an example of plasticity as a promoter of speciation? Evol Ecol 23:137–146

Nylin S, Slove J, Janz N (2014) Host plant utilization, host range oscillations, and diversification in nymphalid butterflies: a phylogenetic investigation. Evolution 68:105–124

Nylin S, Agosta SJ, Bensch S, Boeger WA, Braga MP, Brooks DR, Forister ML, Hambäck PA, Hoberg EP, Nyman T, Schäpers A, Stigall AL, Wahlberg N, Wheat CW, Österling M, Janz N (2018) Embracing colonizations: a new paradigm for species association dynamics. Trends Ecol Evol 33:4–14. https://doi.org/10.1016/j.tree.2017.10.005

Pariselle A, Bukinga FM, Van Steenberge M, Vanhove MPM (2014) Ancyrocephalidae (Monogenea) of Lake Tanganyika: IV: Cichlidogyrus parasitizing species of Bathybatini (Teleostei, Cichlidae): reduced host-specificity in the Deepwater realm? Hydrobiologia 748:99–119

Patella L, Brooks DR, Boeger WA (2017) Phylogeny and ecology illuminate the evolution of associations under the Stockholm paradigm. Vie Et Milieu 44:91–102

Pérez-Ramos IM, Matías L, Gómez-Aparicio L, Godoy O (2019) Functional traits and phenotypic plasticity modulate species coexistence across contrasting climatic conditions. Nat Commun 10:2555. https://doi.org/10.1038/s41467-019-10453-0

Peterson DL, Cipollini D (2020) Larval performance of a major forest pest on novel hosts and the effect of stressors. Environ Entomol: nvz160. https://doi.org/10.1093/ee/nvz160

Pfennig DW, Rice AM, Martin RA (2006) Ecological opportunity and phenotypic plasticity interact to promote character displacement and species coexistence. Ecology 87:769–779

Price PW (1980) Evolutionary biology of parasites. Princeton University Press, Princeton, NJ

Rominger AJ, Fuentes MA, Marquet PA (2019) Nonequilibrium evolution of volatility in origination and extinction explains fat-tailed fluctuations in Phanerozoic biodiversity. Sci Adv 5: eaat0122. https://doi.org/10.1126/sciadv.aat0122

Rull V, Carnaval AC (eds) (2020) Neotropical diversification: patterns and processes. Springer, New York

Schindler DE, Armstrong JB, Reed TE (2015) The portfolio concept in ecology and evolution. Front Ecol Environ 13:257–263

Singer MC, Parmesan C (2019) Butterflies embrace maladaptation and raise fitness in colonizing novel host. Evol Appl. https://doi.org/10.1111/eva.12775

Spironello M, Brooks DR (2003) Dispersal and diversification in the evolution of *Inseliellium*, an archipelagic dipteran group. J Biogeogr 30:1563–1573

Spradling TA, Brant SV, Hafner MS, Dickerson CJ (2004) DNA data support a rapid radiation of pocket gopher genera (Rodentia: Geomyidae). J Mamm Evol 11:105–125

Stigall AL (2010) Invasive species and biodiversity crises: testing the link in the late Devonian. PLoS One 5:e15584

Stigall AL (2012a) Speciation collapse and invasive species dynamics during the late Devonian "mass extinction". Geol Soc Am Today 22:4–9

Stigall AL (2012b) Using ecological niche modelling to evaluate niche stability in deep time. J Biogeogr 39:772–781

Stigall AL (2014) When and how do species achieve niche stability over long time scales? Ecography 37:1–10

Stigall AL (2017) How is biodiversity produced? Examining speciation processes during the GOBE. Lethaia 11:1123. https://doi.org/10.1111/let.12232

Stigall AL (2019) The invasion hierarchy: ecological and evolutionary consequences of invasions in the fossil record. Annu Rev Ecol Evol Syst 50:355–380

Stigall AL, Bauer JE, Lam AR, Wright DF (2017) Biotic immigration events, speciation, and the accumulation of biodiversity in the fossil record. Glob Planet Chang 148:242–257

Stigall AL, Edwards CT, Freeman RL, Rasmussen CMØ (2019) Coordinated biotic and abiotic change during the Great Ordovician Biodiversification Event: Darriwilian assembly of early Paleozoic building blocks. Palaeogeogr Palaeoclimatol Palaeoecol 530:249–270

Thompson JN (1988) Coevolution and alternative hypotheses on insect/plant interactions. Ecology 69:893–895

Thompson JN (1994) The coevolutionary process. University of Chicago Press, Chicago

Thompson JN (1997) Evaluating the dynamics of coevolution among geographically structured populations. Ecology 78:1619–1623

Thompson JN (1999a) Specific hypotheses on the geographic mosaic of coevolution. Am Nat 153 (Suppl):1–14

Thompson JN (1999b) Coevolution and escalation: are ongoing coevolutionary meanderings important? Am Nat 153(Suppl):92–93

Thompson JN (1999c) The evolution of species interactions. Science 284:2116–2118

Thompson JN (2005) The geographic mosaic of coevolution. University of Chicago Press, Chicago

Van Valen L (1973) A new evolutionary law. Evol Theory 1:1–30

Wallace AR (1855) On the law which has regulated the introduction of new species. Ann Mag Nat Hist, 2nd Ser 16:184–196

Wijová M, Moravec F, Horák A, Lukeš J (2006) Evolutionary relationships of Spirurina (Nematoda: Chromadorea: Rhabditida) with special emphasis on Dracunculoid nematodes Inferred from SSU rRNA gene sequences. Int J Parasitol 36:1067–1075

Wilson EO (1959) Adaptive shift and dispersal in a tropical and fauna. Evolution 13:122–144

Wilson EO (1961) The nature of the taxon cycle in the Melanesian ant fauna. Am Midl Nat 95:169–193

Chapter 11
Putting Evolution to Work

Abstract Humanity is facing an existential crisis from global climate change. Evolutionary biology has a critical role to play in how we respond. The problem fundamentally involves two evolvable variables: *Ecosystems*, which we depend on for survival; and *Us*, who have the capacity to alter those ecosystems in a way that threatens our survival. Ecosystems are complex higher-order metabolic systems, closed-loop networks of interacting inheritance systems each with their own capacities for exploiting and exploring their surroundings. Collectively, this gives rise to an *Evolutionary Commons*, an emergent property of ecosystems that acts as a storehouse of potential that can be unleashed when the conditions change. Evolutionary potential is what makes ecosystems robust, relatively immune to the fate of any given inheritance system, without losing the ability to break apart and reform when disturbed in proportion to the capacities for ecological fitting of their constituent members. Putting evolution to work for humanity is about recognizing that the fundamental resource in the biosphere for coping with change is the potential stored in the evolutionary commons. For our own well-being, our economic policies and strategies should reflect this. To persist indefinitely, we must preserve the biosphere, and to do this we must preserve as many elements of the evolutionary process as possible. *Darwinism Then and Now* is fundamentally a story about coping with change by changing. Evolution is the only way the biosphere has coped with global climate change before, and the only way it will do it again. To sustain ourselves, we must exploit the biosphere without destroying the evolutionary potential to explore. We propose the Four Laws of Biotics to guide our efforts and discuss how they inform efforts in conservation biology, coping with emerging diseases, the circular economy, and the economics of well-being.

We think our metaphorical framework represents a helpful way for evolutionary biologists to begin sharing stories, achieving a unified framework encompassing the massive amount of information and insights gained in the past century and setting the stage for the ongoing evolution of the discipline. As we stated in the beginning, however, the goal of finding common ground across theoretical frameworks for

S. J. Agosta, D. R. Brooks, *The Major Metaphors of Evolution*, Evolutionary Biology – New Perspectives on Its Development 2,
https://doi.org/10.1007/978-3-030-52086-1_11

basic biology is only a means to an end for us. Half a century ago, Charles Elton wrote:

> We must make no mistake; we are seeing one of the greatest historical convulsions in the world's fauna and flora.—Elton (1958)

A fifth of the way through the twenty-first century, the living world faces accelerating global climate change at a scale and speed even Elton did not imagine. We know it exists, we measure its progress daily, we know what is causing it. If there was a simple solution to the issues associated with global climate change, we would have solved the problem by now. We understand this is a complex problem. And humans dislike, distrust, and are afraid of complexity. It is no surprise, therefore, that humanity has not converged on an effective sense of common purpose about coping with the problem. The challenges are not underappreciated; the responses, well intentioned as they are, have been inadequate. We believe humanity faces an existential crisis, and evolutionary biology has a critical role to play in helping us cope with and survive that crisis.

The evolutionary context of global climate change involves two variables: *Ecosystems*, upon which we depend for survival; and *Us*, who have the capacity to alter those ecosystems to such an extent that the continuation of humanity will be jeopardized.

11.1 Ecosystems: A Paradox

The planet is covered by geographically localized mix-and-match collections of inheritance systems produced by multiple Stockholm Paradigm cycles. They are full of life, but they are not alive. That is why we give them titles and locations (*African Savannah* is like the *Duchy of Cornwall*) but do not name them (we only name inheritance systems and not all of them). Phylogenetic conservatism in the functional traits of the associated inheritance systems accounts for the substantial similarities in how those inheritance systems interact with each other in different associations. The flatworm parasites inhabiting frogs in the deciduous hardwood forests of Central Europe and eastern North America, temperate grasslands of North America, and dry and wet tropical forests of Mexico and Costa Rica, for example, all have mosaic histories of geographic and host associations. They nonetheless exhibit virtually identical trophic structures emerging from phylogenetically conservative microhabitat preferences and transmission dynamics on the part of the parasites and affinities for aquatic habitats and feeding habits on the part of the hosts (Brooks et al. 2006a, b; for an example based on freshwater stream fish assemblages, see Bower and Winemiller 2019).

If each organism in every one of these entangled banks degrades its necessary resources in order to survive, then every inheritance system should ultimately deplete its surroundings and leave or go extinct. And yet, they persist. How can that be—what makes ecosystems greater than the sum of their parts? Part of the

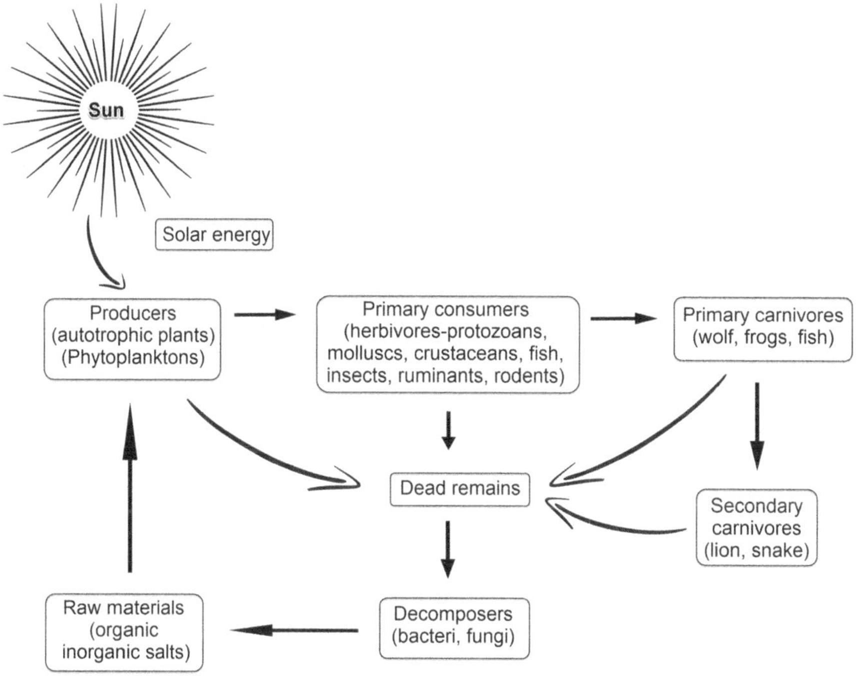

Fig. 11.1 Diagram of a closed-loop ecosystem. Energy comes from outside the system, is harnessed by *producers* and converted to biomass, which is harvested and degraded by *consumers*, and then harvested and degraded further by *decomposers* which transform it back into abiotic resources that can be recycled in the system

answer is that they use matter and energy coming from outside the system. But the fate of that matter and energy, once inside the system, is a critical part of explaining their persistence. Organisms degrade their surroundings with respect to their functional requirements, but in doing so alter their surroundings and convert resources into biomass. The altered surroundings and new biomass may serve as suitable functional resources for organisms representing other inheritance system. In this way, inheritance systems are linked in complex interactions in which the primary degraders of the abiotic world—the primary producers of the living realm—transform the abiotic world partially into waste and partially into biomass. The biomass embarks on a journey through a complex web of biomass transformers, which we call consumers. At each step of the way, the increasingly abstract representation of the original abiotic surroundings is degraded a little more, and eventually the terminal transformers of biomass, which we call decomposers, transform biomass back into abiotic resources. This allows the interactions among otherwise independent inheritance systems to form closed metabolic loops (Fig. 11.1).

Each inheritance system ensures its own survival by providing the means for other inheritance systems in the larger system to survive. When a living system dies or goes extinct, it ceases to have meaning to itself but because it becomes part of the nature of the conditions it may continue to have meaning to other living systems. Those that die provide fitness space for decomposers of equal value as those that reproduce and die from the perspective of those decomposers. These associations of metabolic transformations, exchanges, and recycling among inheritance systems are quintessential examples of division of labor and cooperation described by Maynard Smith and Szathmáry (1995), analogous to the *egalitarian transition* (Queller 1997) that produced eukaryotes, with mitochondria and chloroplasts entrained on a common metabolic economy. It is also compatible with David Sloan Wilson's notion of selection operating among inheritance systems (Wilson 1980; Wilson and Dugatkin 1997).

Today, the notion that ecosystems are complex higher-order metabolic systems comprised of closed-loop networks of individual organisms and their trophic interactions is standard fare (see Braakman and Smith 2013 for an in-depth review). It was not always so (Hagen 1992; Willis 1997). The botanist Arthur G. Tansley (1935) is credited with coining the term *ecosystem*. At the time, there was much support for the idea that ecological associations of multiple inheritance systems (communities) were "super-organisms," following the perspective of another botanist, Frederic Clements (Clements 1905, 1916). Tansley saw the problem with this analogy, recognizing that communities are complex systems that emerge from the interactions among their constituent parts. What is more, he had a notion of the duality of the nature of the organism and nature of the conditions, and that while the organism may have primacy, both factors were inextricably entwined.

> Though the organisms may claim our primary interest, when we are trying to think fundamentally we cannot separate them from their special environment, with which they form one physical system. It is the systems so formed which, from the point of view of the ecologist, are the basic units of nature on the face of the earth. Our natural human prejudices force us to consider the organisms (in the sense of the biologist) as the most important parts of these systems, but certainly the inorganic "factors" are also parts-there could be no systems without them, and there is constant interchange of the most various kinds within each system, not only between the organisms but between the organic and the inorganic. These ecosystems, as we may call them, are of the most various kinds and sizes.—Arthur G. Tansley (1935)

At the same time that Clements was promoting the idea of communities as superorganisms, Lotka (1913, 1925) was busy characterizing metabolism as the physical basis for understanding living systems. Together, Lotka's metabolic view and Tansley's complex systems view set the stage for our modern conception of ecosystems. Lindeman (1942) is credited with another major advance by emphasizing trophic interactions as the processes that bind inheritance systems together into ecosystems. He compartmentalized organisms into discrete trophic levels, with the now-familiar succession of energy transfer from producer to consumer to decomposer. Eugene Odum (1953, 1957, 1964, 1968) and his brother Howard (Odum 1983) and others (e.g., Evans 1956; Whittaker 1962) advanced the view of ecosystems as energy-transfer systems and argued that they are the fundamental unit of ecology. Eugene Odum put special emphasis on the notion of ecosystem stability arising from the maintenance of a static equilibrium through ecosystem-level homeostasis. This paralleled a similar set of ideas from community ecology, which became a core element of evolutionary ecology encompassed by MacArthur and Wilson's (1967) theory of island biogeography. This includes the ideas that communities (1) are geographic places containing a set number of niches, (2) are structured by competition for occupancy of those niches, and (3) can contain only a certain number of species equal to the number of available niches (*species packing*; MacArthur and Levins 1967; MacArthur 1969; May and MacArthur 1972; Roughgarden and Feldman 1975; Werner 1977; Hairston 1980; Case 1981). Like Odum's conception of ecosystems, the overall notion was of stasis, with communities existing in a balanced equilibrium (e.g., Levin 1975). Too many species coming into the system and there will be competition until some are eliminated; too few species and others will rush in from outside to fill the empty niches.

But the surroundings are not structured into niches and biological systems never exist in static equilibrium. They are in constant motion and always changeable according to the interaction between the nature of the organism and the nature of the conditions. The ways in which energy and matter are put to work within ecosystems (the fluxes) are at least as important as measuring overall inputs and outputs (the flows) (Waring 1989). Beginning near the end of the twentieth century, Robert Ulanowicz (1980, 1986, 1997) showed that stable ecosystems can emerge and maintain themselves without attaining a static equilibrium. This set the stage for ecosystems to be open-ended with respect to how many species could coexist—as long as the metabolic loop remains closed, the ecosystem functions.

The flow of energy and biomass (structural information) from one organism to another produces the metabolic networks of trophic interactions that characterize ecosystems. Ecosystems are powered ultimately by organisms that use external energy sources (chemoautotrophs and photoautotrophs) in lieu of inherited systems to power their internal production. All other inheritance systems comprising ecosystems derive their useful energy and matter from other organisms (Fig. 11.2). One living system (Q_i) represents inherited useful energy and matter ($\psi_\mu{}^i$) for another living system (Q_j). Q_j expends energy ($\psi_\alpha{}^e$, d_eS) taking all or part of Q_i from the surroundings represented by the keystone icon in the figure (a filter: Odum 1983). Some of that internalized energy comprises fluxes (ψ_μ) and flows ($\psi_\alpha{}^e$) perpetuating Q_j and the rest comprises fluxes ($\psi_\mu{}^i$) and flows ($\psi_\alpha{}^s$) associated with Q_j serving as inherited useful information and matter for an organism at a different trophic level (level k). Just as Q_i represents higher quality energy (lower entropy) than Q_j, Q_j represents higher quality energy (lower entropy) than whatever system uses Q_j at level k. The rates of entropy dissipation are consequently slower at higher levels of biological organization above the organism.

Extending Lotka's ideas, Ulanowicz characterized the fundamental ecosystem function as follows:

$$A = C - (E + H + S)$$

where A = ascendancy (power, potential, information, organization); C = capacity (H_{max}); ($E + H + S$) = what actually happens (H_{obs}); E = export (moving materials

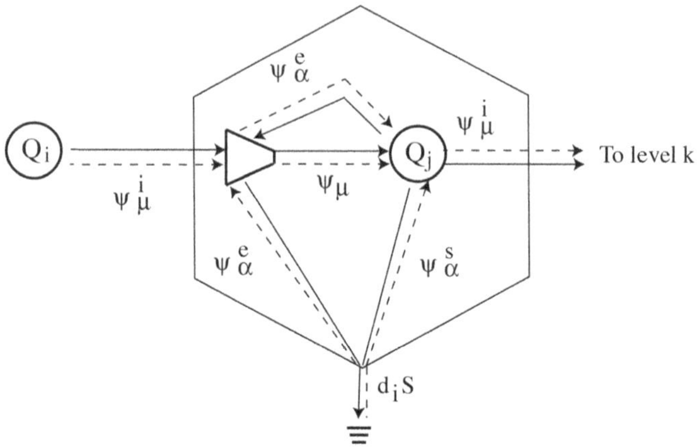

Fig. 11.2 Cartoon diagram of ecosystems as metabolic systems in which internal fluxes constrain the flows. The diagram depicts a single example of trophic interaction where one living system (Q_i) represents inherited useful energy and matter ($\psi_\mu{}^i$) for another living system (Q_j). See text for further details and definition of terms. Redrawn and modified from Maurer and Brooks (1991)

from one place to another inside the system, part of d_iS); H = redundancy (of E, also part of d_iS); and S = entropy production of the system as a whole (dS). Like organisms, ecosystems are maintained through time by the exploitation of "entropy gradients" or "resource gradients" in the surroundings (Collier 1986, 1988, 1990, 2000; Wicken 1987; Ulanowicz 1980, 1986, 1988, 1997; Matsuno 1989, 1995, 1996, 1998, 2000; Maurer and Brooks 1991; Hirata 1993; Depew and Weber 1995) degrading those necessary resources as a result (Gladyshev 1996; Ulanowicz 1997; Brooks and McLennan 2000). In doing so, ecosystems are persistent, stable, and resilient. Organisms use metabolism to "buy circular time" to stay alive, and so do their emergent ecosystems. As new inheritance systems are added, the structural complexity of the ecosystem changes. Ecosystem order is a direct consequence of the flow of energy through biological systems, mediated by fluxes in the form of metabolic circularity that increase the complexity of the system for a finite period of time in the face of constant energy flows. Stable and resilient ecosystems are characterized by a high degree of circularity in their trophic interactions. Circular time in ecosystems is generated by metabolic closure among inheritance systems. Ecosystems have no coherent space–time of their own because linear time is generated by each of the constituent inheritance systems independently of their interactions within the ecosystems.

On a day-to-day basis, ecosystems function by coping with the conditions at hand and with conflicts within the system based on the ecological fitting capacities of their constituent inheritance systems. As a result, ecosystems are centers of exploitation for their constituent inheritance systems who nonetheless retain their capacity to explore. This gives rise to a kind of *evolutionary commons*, a storehouse of potential capacity (*evolvability*: Sniegowski and Murphy 2006; Daniels et al. 2008; Vasas et al. 2015; Crother and Murray 2018, 2019; Hansen et al. 2019) that is unleashed when external perturbations break the ecosystem apart and basic survival becomes the preoccupation of each individual inheritance system (it is every man for himself). But because the capacity for ecological fitting is always present, new ecosystems can form readily, driven again by the primacy of exploitation over exploration when conditions warrant it. Ecosystems as metabolic systems are robust, stable, and resilient, not fragile, volatile, driven by stochastic external phenomena. They do not exist on the edge of chaos, but comfortably within a *window of vitality* (Ulanowicz 1997). The stability, robustness, and resilience of ecosystems is proportional to how good they are as metabolic systems.

During periods of environmental stability, ecosystems are as predictable as living organisms—they exploit their surroundings as much as their capacities will allow. Though tied together in a common metabolic economy, their constituent parts are opportunistic due entirely to the Darwinian nature of their individual inheritance systems. When there are external perturbations that catalyze a shift from exploitation-biased to exploration-biased behavior, therefore, there will be many individual responses depending on the nature of the organism for each inheritance system. As a result of shared metabolic closure among inheritance systems that retain their own capacities, ecosystems manifest four states associated with changing conditions: (1) an *establishment* state when the system is converting from exploration-biased to exploitation-biased behavior; (2) a *successional* state when the system is at or near its maximum growth rate (entropy production), is maximally exploitative, and has maximum evolutionary potential; (3) a *climax* state when there is high production but no net growth of the system, only the steady-state maintenance of maximum potential; and (4) a *disassembly* state when perturbations to the system cause it to convert from exploitation-biased to exploration-biased behavior, with a growing number of disconnections from the original system, coupled with exploration by the constituent inheritance systems of potential for assembling new ecosystems.

11.1.1 Debunking the Butterfly Effect

One of the hallmarks of complex systems is that higher levels of organization are largely insensitive to perturbations at lower levels of organization (Kitano 2004, 2007; Page 2011). Multicellular organisms are generally immune to the fate of any given cell, ant colonies to the fate of any given ant, nations to the fate of any one person, and so on. This degree of insensitivity of higher levels to lower levels is part of what makes complex systems *robust* (Kitano 2004, 2007), allowing them to maintain functionality in the face of both internal and external perturbations to the system. As systems become more complex, with more emergent levels of organization, their robustness can increase if additional functional layers emerge, each providing a buffer against perturbations to the subsystems. Complex systems are multilevel (hierarchical or heterarchical, depending on your nomenclatural preferences), hyper-diverse, hyper-connected, functionally redundant, and modular

(Kitano 2004, 2007; Page 2011). They do not fall apart when a few pieces conflict, break, change their trophic allegiances, or leave. If they did, they would not last long enough to become very complex in the first place.

And yet, there is a common perception that the biosphere is a fragile thing, that ecosystems are so delicately balanced and fine-tuned that the extinction of just one species can cause them to collapse. Colloquially, this is often referred to as the *butterfly effect*, the idea that small changes in one part of a system can have large effects on the whole system, potentially causing its demise. The concept is attributed to the American meteorologist Edward Lorenz. In 1972, he gave a presentation about predicting the weather at the 139th meeting of the American Association for the Advancement of Science titled "Does the flap of a butterfly's wings in Brazil set off a tornado in Texas?" (Lorenz 1972). The title stemmed from his earlier work in chaos theory showing how nonlinear systems, like weather, were sensitive to initial conditions (Lorenz 1963). Small perturbations, like a single butterfly flapping its wings, at the beginning of a weather event could theoretically lead to large unintended consequences for the whole system, like a violent tornado.

While the initial context was weather prediction, the butterfly effect became a popular metaphor for environmentalists, a symbol for the delicate "balance of nature," representing a dire warning about the loss of *any* biodiversity, through our own actions or otherwise. It urged us to act by attempting to control nature and maintain the *status quo*. *Save this particular organism, because if you do not, the entire biosphere will be broken beyond repair.*

In a Darwinian world, the butterfly effect is truly a poor metaphor for the nature of the biosphere. At best, it misleads and distracts from what should be the focus, the vast evolutionary potential stored in ecosystems, ready to be unleashed when conditions change. At worst, it directs critical resources for dealing with environmental problems to the wrong places, often in some well-intentioned but misguided effort to save, reintroduce, or even resurrect some particular species. What ultimately needs to be conserved to maintain the biosphere is *function not form*, *process not product* (e.g., Agosta 2002). If the history of life teaches us one thing, it is that the biosphere is extraordinarily responsive to changes in its constituent inheritance systems. Each mass extinction has been a mass evolutionary reset—large blips for the biosphere, loss of standing diversity, ecosystem disassembly, and renewal.

The issue is not whether the butterfly effect could cause ecosystems to collapse, but that it is highly improbable. Complex systems become more robust and insensitive to perturbations as more complexity emerges. For evolvable life, that emergent complexity is embodied in inheritance systems, whose conservative nature produces not only diversified realized capacities but also growing potential. Life is the most complex system in the universe that we know, and the collection of ecosystems called the biosphere is its highest level of organization. If the actions of a single butterfly, or loss of single inheritance system, were enough to cause critical disruption, ecosystems would not exist. At the same time, complex systems can absorb perturbations only to a point; they all have a vulnerability threshold, beyond which they undergo abrupt phase transitions including collapse (Page 2011; Clements and Ozgul 2018). It is at this point that the butterfly effect becomes a true threat, when the system is so close to a threshold that even the beat of a single butterfly's wings can tip it. While some ecosystems may be close to, or even at, tipping points, we do not believe the biosphere as a whole approaches that state. All the evidence is to the contrary (e.g., D'Alelio et al. 2019). In response to climate change and other large-scale perturbations, the biosphere is coping by exploring new opportunities through ecological fitting, possibly to our own detriment, but by no means collapsing. It is changing in accordance with the predictions of the Stockholm Paradigm, beginning the process of evolutionary renewal. Ecological fitting allows the potential to be expressed in the form of altered trophic connections, buffering ecosystems against sudden collapse. It is the inheritance systems that are persistent and their trophic interactions that are flexible, not the reverse. When a species goes extinct, its fitness space does not persist as some ghostly hole in the biosphere, waiting to suck a new species into it. The old fitness space will disappear as the ecological fitting capacities of other species co-opt it piecemeal into their own fitness spaces. This produces new trophic connections, buffering the ecosystem's integrity despite the impact of species loss. Equally, ecological fitting provides the foundation for new ecosystems—new entangled banks—to emerge after large-scale perturbations lead to disassembly of old ecosystems.

11.2 Us: A New View of "The Commons"

The *Tragedy of the Commons* has become an umbrella concept for the inherent challenges of managing the resources necessary for human survival (Hardin 1968, 1974; Ostrom et al. 1999). Put simply, it is the idea that individuals using shared public resources (the "commons") only on their own behalf may produce negative outcomes for their society at large (the "tragedy"). As the human population gets larger, the problem grows more pervasive, threatening to become intractable. Garrett Hardin (1968) famously called this situation a "no technical solution" problem because technology alone cannot solve it and can actually make it worse. Technology may allow us to procrastinate, but in the end what is needed is a change in *behavior* (Hardin 1968, 1974), including more education and cooperation to better manage the commons (Ostrom et al. 1999). Humanity is faced increasingly with demands to change its fundamental behavior with respect to how it interacts with the environment in a variety of ways (see contributions in Vasbinder 2019).

So long as the commons refers to the accumulation, growth, management, or distribution of material goods, one could argue about the relevance of evolutionary biology to the discussion. But if humanity is facing an existential crisis, as we believe, then the focus of discussion moves to the issue of survival. Survival is the primary arena of Darwinian evolution, so evolutionary principles become critical to any discussion of humanity's future. The basis of survival is, as we have emphasized, evolutionary potential. In order to keep the actual flowing, life needs to generate a lot of potential, and does so. That potential is stored in all extant organisms and their interactions up to ecosystems, so biodiversity is a bank account of potential solutions to future change. The *Tragedy of the Evolutionary Commons* is loss of evolutionary potential.

11.2.1 The Myth of Control: Why Domestication Is Not the Answer

It is a general feature of complex systems that external perturbations of great magnitude may cause the system to respond in ways that could well be contrary to what might be expected given only the external force. This should be understandable to all readers; human history is littered with people who died as a result of unintended consequences stemming from attempts to control the world. Whether fighting wars, building cities, or setting aside nature reserves, we have ample evidence that efforts expended trying to force the biosphere into compliance with human aspirations can backfire horrifically. This is why humans fear complex systems that seem to have their own agendas. And that fear of the unknown potential future drives our desire to control and predict. But biological systems cannot be tamed in this way. This is why attempts by physicists and chemists to explain life using both bottom-up and top-down control principles have failed to make inroads into the main body of professional biology.

A broad coalition of humans acts as if the biosphere is static and change needs to be resisted by all the technological capabilities we can muster. Engineers and preservationists (environmentalists who believe we should freeze existing biodiversity into an unchanging state) often think they fundamentally disagree. Engineers have been quite willing to say, "Tell us what you want, and we will make it that way." Without realizing it, preservationists have an analogous perspective, "Give us the material resources to stop further changes, and we will do it." In North America, the preservationist perspective stems from the nineteenth-century transcendentalist movement, based in part on the idea that in the past humans knew more because they were better integrated with nature, and the loss of that integration has actually reduced our knowledge. Neither group has been able to engineer sustainable stasis because we, and the biosphere within which we reside, are complex evolutionary systems. Efforts by preservationists and engineers to impose stasis therefore may alter the evolutionary timeline and trajectory, but they do not stop it. And each alteration makes it more difficult for us to know what will happen next, limiting our ability to plan effectively for a safe future.

Ecosystems are powerful and resilient, capable of coping with perturbations and capable of reforming even when perturbations break them apart. They are not, however, capable of being controlled. If we wish to actively "preserve" any part of the biosphere, we should focus on preserving the evolutionary process in general (Brooks and McLennan 2002) rather than trying to manage particular places or inheritance systems or ecosystems (Jørgensen et al. 2019). The biosphere is not a collection of variable yet static entities; it is an evolutionary system with an indefinite ability to survive even massive external perturbations. This leads to a simple-sounding policy—save as many places as possible, link them together as much as possible, and then let evolution take over. This is the way the biosphere has responded to previous mass extinction events, with remarkable success each time. We may lose some species for which we have formed attachments, but what remains will be capable of sustaining a transitional biosphere that will evolve into a new (meta)stable one—remember that the most recent mass extinction set the stage for us to evolve.

Given that both ecosystems and the inheritance systems that comprise them are complex systems, we should not be surprised that 15,000 years of human efforts to control some of them through domestication have resulted only in a reduction of diversity and potential. Domestication aims to homogenize and eliminate as much potential as possible. But because the foci of domestication are Darwinian inheritance systems, the more finely focused selection is, the more unanticipated outcomes there will be—finely focused selection maximizes exploitation, making fitness space sloppier. This is why domestication increases yield but also leads to disease susceptibility and the need for more investment in care; this is why biological control does not work, and why coevolutionary arms races devolve into Red Queen dynamics where the pathogens have all the advantages. Efforts at engineering or controlling biodiversity are analogous to the enclosure of public commons by rich and powerful people in seventeenth-century Europe to increase their emerging corporate profits, at the expense of the poor people who previously used the commons to produce food for their families (Blom 2017). Such efforts always risk reducing evolutionary potential.

The biosphere is an evolutionary system with an indefinite capacity to evolve; at the same time, the dynamics of the biosphere do not favor any particular species. If we wish to play an active role in our own survival during this time of evolutionary renewal, we need to accommodate ourselves to the evolutionary commons. The "civilization bet," the idea that our technological advancement will offset the costs indefinitely, is incomplete. Technological advances may allow us to persist through global climate change and other existential threats, *but only if we put evolution to work for us*. Without that, attempts to control are a fast track to the Gambler's Ruin.

11.2.2 The Laws of Biotics

Darwinism then and now is a theory of *coping with change by changing*. Evolution is intimate, brutally short-sighted, relentless, and all conflict resolutions produce unanticipated consequences. Evolution promises that: (1) the only sense of progress is persistence in time (survival); (2) the great majority of the traits species use for persistence and coping with the conditions of life are specific and conservative elements of inheritance; (3) if you persist long enough, you may come up with better solutions for coping with the conditions of life, but they will always be contingent and temporary because the conditions always change; (4) there will always be unanticipated consequences no matter how good a solution appears to be; (5) the consequences of overgrowth can be deferred but never escaped; (6) the longer the deferred penalties for overgrowth persist without resolution, the greater the chances of extinction; and (7) evolution recycles but *conserves nothing*.

These insights lead to some general principles humanity should abide by in our exploitation and exploration of the biosphere, embodied in what we call the *Four Laws of Biotics* (with a large tip of the hat to Isaac Asimov):

First Law: Humanity may not harm the biosphere, or by inaction, allow the biosphere to come to harm. The most fundamental form of "harm" is restricting the biosphere's ability to cope with changes, including those produced by any species.

Second Law: Humanity may not injure any portion of the biosphere or, through inaction, allow any portion of the biosphere to come to harm, except when required to do so in order to prevent greater harm to the biosphere itself. We can use some portions of the biosphere, and protect ourselves from others, so long as the ability of the biosphere to cope and persist is not endangered.

Third Law: Humanity may exploit any portion of the biosphere, except where such exploitation would conflict with the first two laws. We can use our understanding of the biosphere to determine the maximum amount of allowable exploitation.

Fourth Law: Humanity must protect its existence as long as such protection does not conflict with the first three laws. We cannot destroy the biosphere in an effort to preserve ourselves. We must live within our means or go extinct.

11.3 Changing from "Conservation and Restoration" to "Encouraging the Exploration of Evolutionary Potential"

The love for all living creatures is the most noble attribute of man.—Charles Darwin (1872)

Conservation biology has become center stage for discussions about humanity and a changing climate. It brings together a galaxy of concerned people, each with different motivations and backgrounds. Evolution is the only natural process by which the biosphere has recovered from all previous episodes of climate change. And yet, major policy decisions in conservation biology do not fully integrate the evolutionary nature of the biosphere. Many policies in conservation biology revolve around a perceived need to preserve existing diversity where it is and as it is; or in a more extreme version, to try to return the biosphere to a state presumed to exist at some point before widespread human impacts (e.g., Lundgren et al. 2020). We believe these are inappropriate and unreachable goals, given the evolutionary nature of the biosphere. In doing so, we reject the notion that humans have transcended their evolutionary legacy and exist outside the rest of the biosphere; this is what makes the four Laws of Biotics a human imperative.

Natural philosophy flourished in eighteenth century Europe led by the likes of Jean-Jacques Rousseau. He inspired a tradition of looking to nature for philosophical insights. One overarching concept—that nature maintains itself in an unchanging state—stretches as far back as Herodotus in western philosophy. Rousseau felt that the evident order and predictability in biological systems was a manifestation of the essential goodness of the natural state. Anything disrupting this "Balance of Nature" must, therefore, be evil (Neiman 2002). Romanticism was eclipsed by the modernism and naturalism of natural sciences during the late nineteenth century, but it resurfaced in the second half of the twentieth century as the notion that "healthy" ecosystems existed in a stable equilibrium (e.g., Levin 1975). The concept was a core principle of ecological research and management of natural resources until being abandoned in the latter quarter of the twentieth century for lack of evidence.

As an evolved system, the notion of a stable "natural equilibrium" with some kind of "fixed carrying capacity" in the biosphere is antithetical. Evolution implies that *change* is the fundamental mechanism that allows biological systems to survive and persist. Any notion of stasis is contrary to this. It is also contrary to the great mass extinctions the biosphere has experienced in the past, followed by periods of rapid evolutionary renewal, producing biodiversity that never existed before. During periods of isolation, when the conditions are relatively stable and inheritance systems are localized in disconnected fitness space, new biodiversity emerges easily. At the same time, many inheritance systems persist when the conditions change, making new connections in fitness space, because of their capacities for ecological fitting. Still, some go extinct when conditions change because they lack the capacity to either stay and cope with the new conditions or take advantage of opportunities to explore new connections in fitness space (e.g., move to a new geographic area).

Within biology, there is general acceptance that any appearance of stasis in nature is an illusion. The biosphere is dynamic, never static, always changing. But too many involved in conservation biology, from researchers and policy makers to concerned citizens, appear to have missed this message (Brooks et al. 2019). In fact, there are elements of conservation biology that serve as the last remnants of Rousseauian natural philosophy. This is perhaps best exemplified by the Gaia hypothesis, which posits a mythical "balance of nature" based on the assumption that the biosphere is a set of coordinated sub-systems acting to maintain system-level homeostasis. Such notions reinforce the idea of a timeless unchanging biosphere that without humans trying to destroy it, would remain "in balance". This in turn reinforces the notion that at some point in the past, the biosphere existed in a "better" state; therefore, we should attempt to control nature and impose stasis to make sure things don't get worse. For the Romanticist, returning nature to "the way it was" becomes an imperative for humanity, motivated by a "nostalgia for a past that never existed" (Brooks et al. 2019).

A call to not impede evolution is the same as a call to impede extinction. The biosphere is in constant motion, renewing life. Inheritance produces phylogenetically conservative and specific features of form and function that maintain continuity in that change and from which immense potential emerges. Biological stability is thus long-term persistence as a result of coping with constant environmental change.

This has fooled many people into thinking that the proper "conditions of existence" is a state in which nothing changes. Worse, it has fooled some of those people into nostalgia for a past that never existed, the belief that things were "better in the past," often a past in which humans did not exist.

11.3.1 What Lessons About Survival Can We Learn by Studying What Is Happening Today?

We should be paying closer attention to inheritance systems that show anticipatory behavior, the propensity for exploration. The easiest way for an organism to appear to be anticipatory is for tomorrow to be like today. When that is not true, the second easiest is for tomorrow to embody conditions the organism can cope with by virtue of its inherited capacities—ecological fitting. The least likely way for an organism to appear to be anticipatory is to evolve an appropriate new capacity at precisely the time changes in the nature of the conditions demand it. This is the foundational neo-Darwinian position, articulated so long ago by Kellogg, and it is simply wrong. Inheritance systems that have gone extinct in the past, and are going extinct today are evidence that the right solution does not show up at the right time.

Ecological fitting in sloppy fitness space is the general explanation for how biological systems respond initially to change (Agosta and Klemens 2008). The capacity for ecological fitting is pervasive; capacity is always greater than opportunity, but opportunity is the catalyst. Perturbations caused by climate change unleash capacity by increasing opportunities for ecological fitting. While we have planning conferences about what might happen next, the rest of the biosphere is already on the move, exploring what is possible. This is both good and bad news for us. The good news is that life will persist, despite our actions, and this persistence may include our own continued survival. The bad news is that things will keep changing in ways that are unpredictable and possibly harmful to us. To illustrate both aspects, we briefly describe three examples that we think highlight the fundamental role of ecological fitting in fueling life's propensity to cope when conditions change.

11.3.1.1 Terraforming Ascension Island

Ascension Island in the middle of the southern Atlantic Ocean underwent a remarkable transformation in fewer than 100 years (Duffey 1964; Wilkinson 2004). When Darwin arrived at the island during the return voyage of the H.M.S Beagle in 1836 he commented on how sparsely vegetated it was with a lack of trees and abundance of rock. Less than a century later, in 1925, the marine biologist Alistair Hardy visited the island and had the opposite reaction (Wilkinson 2004). Upon encountering the island's highest peak, he commented that "it alone supports a rich vegetation and is known as Green Mountain" (Hardy 1967). What took place in the 89 years between

Darwin's and Hardy's visit illustrates the power of ecological fitting on a grand scale (Wilkinson 2004). It turns out that 7 years after Darwin's visit, the botanist Joseph Hooker visited the island at the request of the British Admiralty to develop an action plan for making it a sustainable habitat for humans. He made four major recommendations aimed at increasing productivity and water availability on the island (Duffey 1964): (1) plant trees on the mountain; (2) encourage more vegetation on the steep sides of the valleys; (3) plant drought-tolerant trees and shrubs in the lowlands; and (4) introduce a mix of tropical and European crops to the "mountain gardens". In the years that followed, consignments of plants were routinely sent to the island to the point that Hardy eventually wrote:

> Tall eucalyptus trees now lined the road, flowering shrubs, conifers and palms of many kind appeared, and sheep grazed on the slopes of grass in between patches of almost dense jungle.—Alistair Hardy (1967)

In fewer than 100 years, Ascension Island had been terraformed into a forested-agricultural ecosystem comprised primarily of introduced plants. Although none of the species had evolved together and many were from an entirely different region of the world (temperate Europe), they were nonetheless able to assemble into a functioning "tropical" ecosystem through what Wilkinson (2004) recognized as widespread ecological fitting.

11.3.1.2 Life in the Plastisphere

It is by now common knowledge that the world's oceans are filled with our plastic debris. Plastics are accumulating everywhere in the oceans from the coasts, to the surface, to the water column, and down to the deepest depths at 10,898 m in the Mariana Trench (Zettler et al. 2013; Chiba et al. 2018). Due to the ingestion of microplastics by zooplankton, which can be transferred to higher trophic levels, and the risks of accidental ingestion and entanglement to larger animals posed by floating debris, plastic pollution has emerged as a major threat to the marine environment (Chiba et al. 2018). But there is another side to this story that illustrates how readily life can explore new possibilities. In the approximately 3.5 billion years since its origin on this planet, evolvable life has never encountered plastics. So, what happens when this foreign material suddenly invades the oceans in the last 60 years? In addition to the potential threats to biodiversity discussed above, the incessant drive for life to explore new opportunities is on full display. Zettler et al. (2013) used scanning electron microscopy and genomic sequencing to study microbial communities found attached to plastic debris in the open oceans and made the remarkable discovery of an emerging new ecosystem that they called the *plastisphere*. The evidence suggests that some bacteria are actually metabolizing the plastics, while other microbes are playing the roles of autotrophs, predators, and symbionts. The plastisphere communities are distinct from the surrounding surface water microbial communities, reflecting the novelty of plastic as a "habitat." Zettler et al. (2013) argued that the plastisphere has been assembled over vast areas of open ocean, from the large pool of surface water species, through a sorting process that includes colonization by species that share "a propensity to adhere to surfaces" and a "sticky nature...that allowed them to adhere to plastic." Or in other words, in response to something novel in life's surroundings, the familiar "ecological fitting in sloppy fitness space" occurred.

11.3.1.3 Invasion of the Giant Liver Fluke

The fact that some parasites and pathogens have complex life cycles requiring more than one host species to complete development has been used to suggest that they are unlikely to become invasive species (e.g., Dobson and May 1986; Henttonen et al. 2001; Torchin et al. 2003; Taraschewski 2006). But the evidence suggests otherwise (Brooks et al. 2006a, b; Hoberg 2010; Lymbery et al. 2014; Malcicka et al. 2015). Consider the case of the giant liver fluke *Fascioloides magna* (Platyhelminths: Trematoda). It is native to North America but has been accidentally introduced to Europe with imported game animals at least twice and is currently spreading in the region along the River Danube (Malcicka et al. 2015). The species has a complex multi-host life cycle: newly hatched larvae use freshwater snails as intermediate hosts and mature flukes develop in and use ungulates as definitive hosts. In North America, *F. magna* has a narrow intermediate host range which includes five congeneric species of snails in the genus *Lymnaea*, subfamily Lymaeinae. In Europe, it has switched to two new snail species in the same subfamily but different genus (*Galba*) as its intermediate host. The range of definitive hosts in North America is also narrow and includes four species of ungulates where development can be completed. In Europe, it has switched to three new ungulate species as definitive hosts. Despite its complex multi-host life cycle, the giant liver fluke *F. magna* has managed to leave its ancestral hosts behind to become a successful invader in Europe by ecological fitting with multiple new intermediate and definitive hosts (Malcicka et al. 2015).

The Romanticist influences in the extended hardened synthesis feed the mythical narrative that *Necessity is the Mother of Invention*, but that is not true. *Necessity is the Mother of Coping if you can, Fleeing if you cannot cope, and Dying if you cannot flee*. And yet, what we know of the history of life on this planet supports an optimistic view of the capacity of the biosphere to cope with even massive perturbations. Ecological fitting provides an effective means of spreading your bets—diversifying your portfolio—when conditions change through natural processes or anthropogenic forcing. The more limited an inheritance system's capacity to oscillate between generalizing and specializing, the more likely it will go extinct. Those that survive have indefinite built-in evolutionary capabilities that increase the chances of surviving ecological perturbations and also increase the chances that new, robust ecosystems will emerge rapidly. The biosphere has always shown itself to be capable of novel recovery and to be indifferent to the fate of any particular inheritance system, no matter how self-important. There is strong evidence that the current biosphere is beginning to cope with climate change in the way it always has, without asking our permission and without waiting to see if we will participate, and sometimes in spite of our efforts. Accommodating ourselves to the evolutionary renewal that is underway does not preclude us from taking proactive measures in our own behalf.

11.3.2 Being Proactive About Emerging Infectious Disease

Forewarned, forearmed.—Robert Greene (1591)

One of the most direct manifestations of the Stockholm Paradigm in the context of today's accelerating global climate change is the crisis of Emerging Infectious Diseases. During biotic expansions, inheritance systems mix together and pathogens capable of infecting hosts to which they have never been exposed have a chance to enlarge their host range. Newcomer hosts from adjacent areas will often be closely related to the residents, increasing the ease of host range expansion by ecological fitting. Residents that were not in contact with the pathogen before the biotic expansion event may find themselves susceptible. The mixing process itself may provide opportunities for host range expansion by the stepping-stone mechanism (Braga et al. 2014, 2018; Araujo et al. 2015) that recently allowed SARS-CoV-2 to move from bats to humans (Brooks et al. 2020; Morens and Fauci 2020; Zhou et al. 2020). This is how evolution produces persistent pathogen–host associations, lots of them. And because evolution is about coping with and resolving conflict, pathogens that can coexist with available hosts, even if they do not get along that well, will persist.

The current episode of global climate change increases connectivity in fitness space, inviting all life to explore new opportunities. It is in the nature of the organism to accept those invitations with as much enthusiasm as possible, given their inherited capacities and the opportunities at hand. Humans are no exception, and our explorations have shown us that our fitness space is enormous and is enormously full of pathogens. Failing to link our exploration of fitness space to its fundamental Darwinian principle, however, has allowed pathogens to find us much faster than we have found them. Human population expansion and density, and globalized trade and travel are threat multipliers that intensify the climate changed-catalyzed perturbations underlying the EID crisis. This is a global security concern and an existential threat to human civilization (Vasbinder 2019; Brooks et al. 2019). And it is going to get much worse in the near future, although outbreaks are virtually a daily occurrence already—humans, livestock, crops, wild species upon which we depend directly and indirectly will all be visited by new and returning pathogens. Even if we are not hit by one or more global pandemics, the socioeconomic costs of waiting until each small crisis emerges before reacting will inevitably overwhelm health services, as happened in the past 5 years with Ebola, Zika, African Swine Fever, and SARS-CoV-2. This will lead to a "death by a thousand cuts" scenario (Brooks et al. 2019), a steady drip of high probability/low impact pathogens that accumulate (pathogen pollution) in places that are significant to human survival, no one likely to do major damage but each one costing something. It is the cumulative effect of many small costs that can bring a system down. As an analogy, HIV is a virus with so many antigenic sites that a person's immune system goes into hyperdrive trying to produce antibodies to all of them. But there are too many, and eventually, the person's immune system fails leaving them with acquired immunodeficiency syndrome (AIDS). Now think of the public and agricultural health care system as society's immune response.

There are many, not a few, evolutionary "accidents waiting to happen" out there (Mollentze and Streicher 2020), requiring only the catalyst of ecological alteration resulting from climate change, species introductions, and the intrusion of humans into areas they have never inhabited before (Brooks et al. 2019). Those accidents already cost the world at least $1 trillion per year in treatment costs and production losses (Brooks et al. 2019) before the pandemic driven by SARS-CoV-2. Emerging infectious diseases are a national security concern for every country. As climate and environment changes, humans, plants, and animals increasingly migrate, allowing disease-causing organisms of all kinds to find new areas and new hosts to live in. They link wildlands to agricultural landscapes, infecting our crops and livestock, and find their way into the food supply chain. They travel with us through globalized trade and travel. This leads to frequent emergence of new diseases, making them an existential threat to all of humanity. And surprisingly, technological, heavily urbanized humanity may be at particular risk.

City life brings with it many technological advantages and those living in cities may feel relatively immune to the impacts of global climate change, including emerging infectious disease. But despite their wealth and technology, cities are especially vulnerable to disease (Brooks et al. 2019). In discussing the pandemic driven by SARS-CoV-2, Brooks et al. (2020) highlighted several reasons for this including (1) cities are warm—large metropolitan centers can be 3–5 °C warmer than their surroundings, warm enough to create their own micro-climates in which pathogens and their vectors and reservoirs can survive winter months that previously would have eliminated them; (2) cities rely on a constant flow of water, food, and other materials from the outside increasing opportunities for the introduction of new disease-causing agents; (3) cities support a large number of urbanized species, such as hedgehogs, rats, mice, foxes, pigeons, squirrels, that can be reservoirs of diseases affecting humans and their pets. Cities are attractive ecosystems for many species, making the community dynamics among people, pathogens and other species far more complicated than we thought; (4) cities have people living at high densities, increasing the probability of disease exposure and transmission in proportion to population size; and most importantly (5) all cities include at least one group who are poor, under-educated, and mal-nourished, all factors that bring about stress and poor physical health thereby increasing susceptibility to disease. While these may be the most vulnerable groups for introducing disease into cities, once a disease is established, it will not stay confined to the poor.

It is not only a possibility that, given the state of the planet and our influence on it, there will be a continuing parade of emerging and persisting pathogens, it is a certainty. This all sounds very dire with respect to humanity. We cannot stop pathogens from coming to us as a result of opportunities provided by climate change, globalization, and human population density without violating the First Law of Biotics. We can, however, be better prepared to cope with them when they arrive. The traits that characterize transmission dynamics and microhabitat preference for pathogens are both highly specific and phylogenetically conservative. The capacities that allow them to persist in one place give us insights into how they will behave in novel conditions. This means we can anticipate the arrival of disease-causing organisms and their behavior once they have arrived; and that means we can mitigate their impacts on society. In many areas of health practice, we have internalized the message that crisis response is not only more expensive than prevention, it is unsustainably more expensive. Early changes in diet and exercise patterns are always preferable to and cheaper than open-heart surgery.

The *DAMA Protocol* (Document, Assess, Monitor, Act) (Brooks et al. 2014, 2015, 2019) serves as the umbrella framework for achieving this goal, extending human and material resources devoted to coping with the wave of emerging diseases and buying time for the development of new vaccinations, medications, and control measures. DAMA integrates activities ranging from the local, boots on the ground contributions of citizen scientists led by field biologists to the most sophisticated technologies of bioinformatics, molecular biology, and satellite surveillance. For a given area, the first step is to *document* the existence of organisms that actually or potentially cause disease. It is impossible to cope with emerging diseases and the organisms that cause them without knowing their identity, location, and potential for infection. Most, if not all, disease-causing organisms reside in at least one species of *reservoir host* that is not diseased. Often times, these reservoirs are already known or suspected for a given area, allowing the attention to be focused on a relatively small subset of local biodiversity. The primary interface for disease transmission between reservoir hosts and humans (and their domesticated crops, livestock, and pets) is the intersection of habitats where both groups occur, for example where agriculture meets wildlands. This knowledge allows us to focus on specific places, thereby saving time and money. Furthermore, the modes of transmission of disease-causing organisms from host to host are highly specific, including through food, water, and contact with infected surfaces. Many are transmitted by vectors, such as mosquitoes and ticks. All pathogens encountered in documentation are *assessed* to determine their relative importance, beginning with *phylogenetic triage*. By using phylogenetic information (evolutionary history), the potential for an organism to cause disease in humans and their domesticated species in one area can be assessed quickly by asking if it causes disease in another area, in a closely related species, or if it is a close relative of another organism known to cause disease. The assessment procedure allows us to focus on species of particular interest. In addition, we can use what we know of climate change, migration, and trade connections to anticipate the likely arrivals and risk pathways of disease-causing organisms not encountered in documentation activities. Assessments produce recommendations about *monitoring* pathogens of significance, and actions to mitigate their impact or making their arrivals less certain. To monitor the potential for emerging disease, we are looking for changes at the interface between humans and disease-causing organisms in a given area. These include changes in geographic distributions, abundance, host range, transmission dynamics, variation, emergence of rare variants, etc. The results of monitoring activities must be translated into effective *action* as rapidly as possible when necessary. One way to achieve this is through Action Councils including representatives of all relevant public agencies—public health, veterinary health, plant protection, disaster preparedness, tourism, etc. Such councils would be well-suited for making recommendations determining effective courses of action and coordinating their implementation.

11.3.3 A Specter Returns

External perturbations that disrupt ecosystem structure produce what we called "every man for himself" behavior in which inheritance systems flee the disintegrating ecosystem (accelerating the disintegration process). As they flee, circular (metabolic) interactions within the ecosystem as a whole and between the ecosystem and the fleeing constituent inheritance systems(s) become attenuated, linearized. This leads to a loss of stability, resilience, and potential. We have emphasized our belief that humans are not, and cannot be, outside of the evolutionary context that connects them to the rest of the biosphere. This does not mean that humans are incapable of trying. If humans trying to live outside their evolutionary context, either through ignorance or hubris, are like any other inheritance system leaving its ecosystem, they risk creating a tragedy of their own evolutionary commons, creating an unstable existence in which the butterfly effect becomes a real possibility.

In 1918, the global Spanish Influenza pandemic killed an estimated 10% of the human population. Occurring at the end of the First World War, this was a human tragedy of global proportions. And yet, the impact on the world's economy was minimal. A century later, the human population is almost five times larger and population density due to urbanization is almost five times greater than in 1918. As well, the world is linked by a global trade, travel, and communications system fed by a small number of monocultures of domestic plants and animals. Fueled by the industrial revolution humanity seems to have built itself a highly technological niche—perhaps even its own geological period called the Anthropocene—largely outside of the rest of the biosphere. For some, the goal of finally transcending humanity's biological origins—and thus protecting ourselves from the contingent natural evils of nature discussed by Neiman (2002)—was in sight. SARS-CoV-2, causal agent of COVID-19, has killed only a tiny fraction of the number of humans compared with the Spanish Influenza, and yet the world's public health services were caught unawares and unprepared, creating a crisis of confidence that froze government officials into inaction and led to a global economic catastrophe. Had the Chinese government paid attention to three scientific reports from more than a decade previously (Li et al. 2005; Poon et al. 2005; Cheng et al. 2019) and reiterated only 2 years before the outbreak (Cyranoski 2017), and closed or closely regulated game meat and live animals being sold in open-air markets, there would have been no pandemic, no panic, and no global economic meltdown (Brooks et al. 2020). The lesson of COVID-19, it seems, is that Anthropocene humanity—not the biosphere—is in a precarious "butterfly effect" state.

When real ecosystems disintegrate, their surviving members are capable of forming new ecosystems if they are able to form new, circularized metabolic associations. Could this be true for Anthropocene humanity?

11.3.3.1 Circular Economies

The four Laws of Biotics prescribe a Darwinian philosophy to follow if the goal is to best ensure the survival of a biosphere that includes humans. The ultimate solution to the environmental problems that confront us—too many people crowded together, coupled with too high an incentive for short-term gain—is not technical. Technology can help us, buying time until the next crisis, but can also make things worse by compounding the population problem and by extension all other environmental problems (Hardin 1968). The true innovation needs to be in the way we behave, which stems fundamentally from our value systems. What we place value on is what determines our actions. This includes the economic and resource management systems we create and the constellation of cultural norms, rules, and regulations we use to enforce them. If the majority of humanity continues to place the largest value on short-term gains, then the only logical outcome for technological human is ecological ruin (Hardin 1968, 1974).

One exciting emerging concept for a more sustainable approach is the *Circular Economy* (Ellen MacArthur Foundation 2012; Preston 2012; CIRAIG 2015; Stahel 2016; Geissdoerfer et al. 2017; Sairiatli 2017; Skawinska and Zalewski 2018). This idea stems from the argument that by placing too high a value on short-term gains, the current global economic system produces an economy of *linear consumption*. From this flows a pattern of "make-and-dispose," creating enormous amounts of waste coupled with little incentive to recycle, repurpose, and reuse. Given the inherent limitations of this behavior, finding an alternative is seen as an essential step toward sustainability. The Circular Economy is an alternative with origins in the study of biological systems:

> The notion of a circular economy has its roots in industrial ecology, a theory first developed by environmental academics in the 1970s and still used today. It involves remodelling industrial systems along lines of ecosystems, recognizing the efficiency of resource cycling in the natural environment.—Felix Preston (2012)

The Circular Economy is a case of *biomimicry*—designing and producing materials and systems patterned from nature—on a grand scale, as can be seen by comparing Figs. 11.1 and 11.2 with Fig. 11.3.

The Circular Economy seeks to "close the loops" in the economic machine, placing far more value and incentive on recycling, repurposing, and reusing resources for maximum long-term use within the system as opposed to maximum

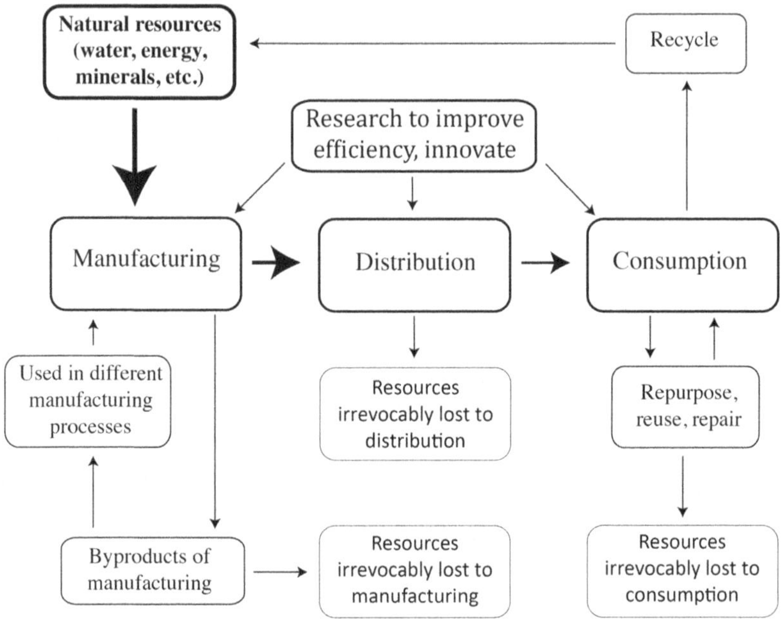

Fig. 11.3 Diagram of a circular economy. Note the similarity to ecosystems as closed-loop metabolic energy-transfer systems depicted in Figs. 11.1 and 11.2

short-term gain by linear input–output. Just like evolution, rather than "make-and-dispose," which is expensive and wasteful, it is "make-and-reuse," which is cheap and persistent. The vision is for the economy to be more like:

> . . .a regenerative system in which resource input and waste, emission, and energy leakage are minimised by slowing, closing, and narrowing material and energy loops. This can be achieved through long-lasting design, maintenance, repair, reuse, remanufacturing, refurbishing, and recycling.—Martin Geissdoerfer et al. (2017)

If we ask what kind of economic system best mimics evolution, the answer would be whichever kind leads to long-term persistence (survival of the adequate), not maximizes short-term gain (survival of the fittest). *Go slow and live within your means.* Effective exploitation of the surroundings by life is slow and sustainable because metabolism is circular and energy use is incremental (mediated by metabolic enzymes and reactions). This allows harvested energy to be stored (as biomass) and used as many times as possible within the system. And as complexity increases to include interactions between organisms, it allows some energy to be passed on to different trophic levels (slowing its degradation even further) to ultimately be recycled by decomposers within the system. Metabolic systems have thus evolved to be the quintessential *recyclers*, and this is the key to how they continue to "buy" the time to be alive. Yet, humans are doing just the opposite. A linear economy of short-term gain is fast, wasteful, and inherently unsustainable because, without a focus on the long term, there is little value placed on reuse and replenishment—it is use it and lose it. Just as metabolism buys organisms and their ecosystems the circular time to "slow down and live," sequestering energy and evolutionary potential, a more circular economy can allow technological humans to live more sustainably, getting more out of harvested resources, and buying the most time to find new solutions to environmental problems.

Inheritance systems persist evolutionarily because they can oscillate between exploiting and exploring. This allows ecosystems to persist during times of environmental stability, then disassemble and reassemble easily in response to perturbations. For most of humanity's history, we coped with climate change and related threats with which we could not cope by fleeing into new fitness space. In the past 7000 years, however, we have become increasingly sedentary to the point that today more than half of humans live in urban settings. This has reduced our ability (and desire) to flee from changing conditions. Circular economies increase the efficiency of exploiting local conditions. How can we keep this from becoming a trap during times of accelerating climate change? Hillary Brown (Brown et al. 2019) has proposed linking circular economies into cooperating networks. The "realized fitness space" for each community would thus include the realized fitness space of all linked communities, and their joint "fundamental fitness space" would be large enough to allow innovations to arise in one place and be shared more widely, leading to oscillations between exploiting and exploring within the network of cooperating circular economies. This would allow smaller villages and cities to have the benefits of a single larger urban area, while retaining the benefits of more rural life.

Charles Elton concluded in 1958 that human civilization was about to experience severe socioeconomic disruptions as a result of global climate change. He believed the major manifestations of those changes for humanity would be conflict and migration. More than 60 years later, it is clear that Elton was prescient. Cooperation always reduces conflict, of course, and both circular economies and networks of circular economies require it. The system envisioned by Hillary Brown could also serve to mitigate Elton's second concern, migration. Global climate change is expected to create a large number of climate refugees. Currently, most people think of this in terms of people being displaced by climate change or violence from their own countries. But as climate change impacts become more global and severe, many nations will face issues of internal migration, especially from large coastal cities inundated by rising sea levels. Networks of cooperating small cities and villages situated inland could serve as pre-positioned safe havens for internal migrants and potential centers for societal renewal.

As exciting as these ideas are, circularizing economies is not enough for a Darwinian rethink of human sustainability in a time of global climate change. We must link the need to live through extreme changes within our means environmentally—through large-scale biomimicry—with policies that incentivize such behavior.

11.3.3.2 The Economics of Well-Being

The extended hardened synthesis has inherited a long-standing set of assumptions about the relationship between human economics and the living world at large (sometimes called the natural economy). Most believe that these assumptions stem from Darwin, but we think that is not the case. We begin at a place familiar to most students of biology, London at the end of the eighteenth century. The Reverend Thomas Robert Malthus had observed two things about the city in the first 40 years of the Industrial Revolution: there were too many people and there was not enough food. Malthus published his ideas about what this meant in *An Essay on the Principle of Population* (Malthus 1797). He believed that the reason for this mismatch between food availability and the number of people wanting food was an inherent difference in the reproductive rate between ourselves and the organisms we eat. He proposed that human populations grew exponentially while food derived from crops and livestock grew only arithmetically. Malthus extended this argument to say that charity, especially food charity, given to the poor would be self-defeating since it would simply be turned into ever more humans vying for proportionately ever less food.

It is popularly believed that Darwin based his theory of natural selection on Malthus's notions, but our reading of the *Origin* suggests that Darwin used Malthus's ideas as an analogy or as auxiliary supporting evidence for his theory. In any event, Darwin viewed the reproductive overrun as a property of the nature of the organism for all inheritance systems—human, crops, livestock—and something that would be corrected by the nature of the conditions whenever it got out of control (which would be often). From this perspective, we could reinterpret Malthus's observations as indicating that in the second half of the eighteenth century, so many people had left the countryside and moved to London that the *density* of people exceeded the food supply chain for the city. The problem was not poor people having too many children, it was too many people coming together in one place so rapidly that available food supplies were depleted. Blom (2017) has shown that at the time, the situation in the countrysides of Europe was desperate due to a small number of rich and powerful people enclosing public lands for commercial production, reducing the amount of land available for local people to feed themselves. The lack of income from food production in the countryside and the (often forlorn) hope of new and better-paying jobs in London led to a rapid influx of people dependent on external food sources. There was certainly a struggle for survival among the poor in London, but it was not a conflict that could lead to a positive resolution. It could only lead to continuing conflict and replacement, winners and losers. It is not surprising, therefore, that Malthus's incorrect core proposition was seized upon by Spencerians and incorporated into neo-Darwinism, the hardened synthesis and the extended hardened synthesis. It was also incorporated in various economic theories of the late nineteenth century and beyond.

Regardless of the outcome of debates about Darwin's precise view of Malthus's ideas that our assertions may engender, it is true that the contemporary global economy focuses on short-term gain encapsulated by the central metric used to compare national economies, the Gross Domestic Product (GDP). The GDP is a measure of a nation's overall economic performance, putting a monetary market value on all the goods and services produced in a short time period. It is all about who is making the most profit at the moment, a *survival of the fittest* mentality. But when it is only about maximum exploitation, occupying the most space, and having the most stuff, everyone and everything else tends to suffer. Inequalities for many emerge more rapidly than increased profits for a few and this is unsustainable, especially when the conditions change. Growth-based economic policies are theories of conflict and replacement, of linear production, and all have led to increasing inequality wherever they have been adopted throughout the world (for another prescient thinker, see Polanyi 1944).

World governments and other organizations increasingly recognize that the GDP and other one-dimensional indicators based only on material profits may be poor measures of a nation's ability to survive. This is because they do not reflect *how well people are living*. If not material wealth, what should be the central indicator of how well a society is performing? One idea to emerged recently is what is being called the *Economics of Well-being* (Stiglitz et al. 2008; Gurría 2019; Jacobs 2019; Llena-Nozal et al. 2019; Eker and Ilmola-Sheppard 2020; Ilmola-Sheppard et al. 2020)

> Even though economic welfare is one of the key prerequisites of citizens' well-being, it has been recognized that there is a need for a more comprehensive approach to measuring well-being to inform policymakers and the general public, as well as to support efficient policymaking.—Leena Ilmola-Sheppard et al. (2020)

If "healthfulness," "happiness," and "hopefulness" are all a part of "wealth," then how well people are living—their material living conditions, non-material "quality of life," and outlook for the future—is a critical part of gauging how well a society is functioning. Morale matters. Societies of mentally and physically healthy people are more likely to create systems and policies that target a sustainable future, not just short-term gain.

Again, what interests us most about this newly emerging idea in economics is that it is fundamentally Darwinian. It is more compatible with the Darwinian emphasis on survival and persistence than with neo-Darwinism's emphasis on maximizing growth and fitness, conflict, and replacement. Most economic systems are based on maximizing the growth of some analog of fitness and pursuing the most direct line to that maximum value without regard to the consequences of our actions. This is why current economic systems function poorly in today's small, fast, crowded world facing accelerating global climate change. The need for the general well-being of the system (to function) puts a damper on any seeking of maximum absolute fitness. Fit is what is required, and as high fitness as is possible within the constraints of the functional needs of the larger system. The economics of Darwinism is the economics of well-being. As much as you can have without destroying your own future. This holds true for humanity in particular, and for the biosphere in general, upon which we depend for our existence.

11.4 Summary

Between stasis and oblivion, there is evolution

We have an ingrained, evolutionary affinity for nature—a deep love that E.O. Wilson (1984) called *biophilia.* As with any love, it is easy to let our emotions get the best of us. *Bring this place back to the way it was in the past because we loved it that way. Save all the big cute furry creatures because we love them.* But this is not the way forward if we hope to best ensure the persistence of the biosphere in a way that favors our own survival. Instead, we must be more disciplined in our decision-making—we must listen to the lessons that evolution teaches us. Fortunately, these lessons are simple and provide a clear message. *Preserve as much of the evolutionary process, and thereby potential, as possible. Do not allow a Tragedy of the Evolutionary Commons.* Set aside large enough tracts of connected heterogeneous habitat for the whole process to occur by itself. Then slowly and carefully intrude human footprints in accordance with the four Laws of Biotics—a steady-state ecosystem can take a fair amount of perturbation because life is not on the edge of chaos (survival of the fittest) but nestled securely in a window of vitality (survival of the fit).

Darwinian sustainability is about how long you can persist, not about how much space you occupy or how many resources you accumulate. It is not a matter of *he who dies with the most toys wins*, but a matter of *he who dies last wins*. It is how you cope with change that matters. And how has life been coping so far? In its 3.5 billion years on this planet, it has been doing just fine. It has continued to survive. It does so through the relentless application of what we might call Biological Assumption Zero: organisms will do what they can, where they can, when they can to survive and reproduce, within the constraints of history (inherited capacity) and opportunity.

Policies that limit evolution will either fail immediately or achieve short-term gains at the expense of long-term losses and unintended consequences. Attempts to impose stasis on the biosphere can only limit its ability to respond to change, thereby reducing its overall chances at survival. To survive indefinitely, we must preserve the evolutionary commons (evolvability)—losing it is a sure path to the Gambler's Ruin. Fortunately, all life on this planet is evolvable life, and so we get evolution for free. In placing humans back into an evolvable biosphere, we reveal the risk space for our own survival—we become part of the solution rather than part of the problem.

Evolution teaches that everything old eventually passes away. We all experience this personally and naturally we resent it. But evolution also teaches that the passing away of the old makes way for the new. We also experience this personally, most directly by the birth of new children, but also in other ways including the discovery of new knowledge.

We believe objective knowledge is embedded dualistically within subjective possibilities, just as realized information is embedded within potential information. If information grows in the universe, so does objective knowledge. The rate at which objective knowledge emerges is faster than the rate at which we discover it, so we can take some comfort in believing that there is always more out there to be discovered, including possible solutions to today's problems. Knowledge, insight, and wisdom are out there, but there is no free lunch in the universe, so it is up to us to do the work to find the path to the continued survival of our species.

No time like the present—let's work together and get going.

References

Agosta SJ (2002) Habitat use, diet and roost selection by the Big Brown Bat (*Eptesicus fuscus*) in North America: a case for conserving an abundant species. Mammal Rev 32:179–198

Agosta S, Klemens JA (2008) Ecological fitting by phenotypically flexible genotypes: implications for species associations, community assembly and evolution. Ecol Lett 11:1123–1134

Araujo SBL, Braga MP, Brooks DR, Agosta SJ, Hoberg EP, von Hartenthal FW, Boeger WA (2015) Understanding host-switching by ecological fitting. PLoS One 10:e0139225. https://doi.org/10.1371/journal.pone.0139225

Blom P (2017) Nature's mutiny. Carl Hansen Verlag, Munich

Bower LM, Winemiller KO (2019) Intercontinental trends in functional and phylogenetic structure of stream fish assemblages. Ecol Evol 9(2). https://doi.org/10.1002/ece3.5823

Braakman R, Smith E (2013) The compositional and evolutionary logic of metabolism. Phys Biol 10:011001. https://doi.org/10.1088/1478-3975/10/1/011001

Braga MP, Razzolini E, Boeger WA (2014) Drivers of parasite sharing among Neotropical freshwater fishes. J Anim Ecol 84:487–497

Braga MP, Araujo SBL, Agosta SJ, Brooks DR, Hoberg EP, Nylin S, Janz N, Boeger WA (2018) Host use dynamics in a heterogeneous fitness landscape generates oscillations in host range and diversification. Evolution 72–79:1773–1783. https://doi.org/10.1111/evo.13557

Brooks DR, McLennan DA (2000) The nature of the organism and the emergence of selection processes and biological signals. In: Taborsky E (ed) Semiotics, evolution, energy. Shaker Verlag, Aachen, pp 185–218

Brooks DR, McLennan DA (2002) The nature of diversity: an evolutionary voyage of discovery. University of Chicago Press, Chicago, IL

Brooks DR, León Régagnon VA, McLennan DA, Zelmer D (2006a) Ecological fitting as a determinant of the community structure of platyhelminth parasites of anurans. Ecology 87 (Supplement):S76–S85

Brooks DR, McLennan DA, León Régagnon VA, Hoberg EP (2006b) Phylogeny, ecological fitting and lung flukes: helping solve the problem of emerging infectious diseases. Rev Mex Biod 77:225–234

Brooks DR, Hoberg EP, Boeger WA, Gardner SL, Galbreath KE, Herczeg D, Mejía-Madrid HH, Rácz SE, Dursahinhan AT (2014) Finding them before they find us: informatics, parasites and environments in accelerating climate change. Comp Parasitol 81:155–164

Brooks DR, Hoberg EP, Boeger WA (2015) In the eye of the cyclops: the classic case of cospeciation and why paradigms are important. Comp Parasitol 83:1–8

Brooks DR, Hoberg EP, Boeger WA (2019) The Stockholm paradigm: climate change and emerging disease. University of Chicago Press, Chicago

Brooks DR, Hoberg EP, Boeger WA, Gardner SL, Araujo SBL, Bajer K, Botero-Cañola S, Byrd B, Földvári G, Cook JA, Dunnum J, Dursahinhan A, Garamszegi L, Herczeg D, Juarrero A, Jakab F, Kemenesi G, Kurucz K, León-Règagnon V, Mejía-Madrid HH, Molnár O, Nisbett RA, Preiser W, Drive F, Stuart M, Szathmary E, Trivellone V (2020) Before the pandemic ends: making sure it never happens again. WCSA J 1:1–10

Brown H, Armstrong P, Csuvar A, Gossler J, Herrera G, Ibrahim T, Kasz K, Leone J, Kurucz K, Molohides L, Murai M, Rajczi A, Stryker K, Tesser D, Turani J, Vajda V, Vorosmarty C,

Kerekes S (2019) Resiliency and regeneration in the Pannonian region of Hungary: towards a circular economy for Koszeg and beyond. City University of New York, Institute of Advanced Studies, New York, Kozeg, 114 pp
Case TJ (1981) Niche packing and coevolution in competition communities. Proc Natl Acad Sci U S A 78:5021–5025
Cheng VCC, Lau SKP, Woo PCY, Yuen KY (2019) Severe acute respiratory syndrome coronavirus as an agent of emerging and reemerging infection. Clin Microbiol Rev 20:660–694
Chiba S, Saito H, Fletcher R, Yogi T, Kayo M, Miyagi S, Ogido M, Fujikura K (2018) Human footprint in the abyss: 30 year records of deep-sea plastic debris. Mar Policy 96:204–212
CIRAIG (2015) Circular economy: a critical literature review of concepts. International Reference Centre for the Life Cycle of Products, Processes and Services, Ecole Polytechnique, Paris
Clements FE (1905) Research methods in ecology. University Publishing Company, Lincoln, NE [now University of Nebraska Press]
Clements FE (1916) Plant succession: an analysis of the development of vegetation. Publication no. 242. Carnegie Institution of Washington, Washington, DC
Clements FE, Ozgul A (2018) Indicators of transitions in biological systems. Ecol Lett 21:905–919
Collier J (1986) Entropy in evolution. Biol Philos 1:5–24
Collier J (1988) The dynamics of biological order. In: Weber B, Depew DJ, Smith JD (eds) Entropy, information and evolution: new perspectives on physical and biological evolution. MIT Press, Cambridge, MA, pp 227–242
Collier J (1990) Intrinsic information. In: Hanson PP (ed) Information, language and cognition. Oxford University Press, Oxford, pp 390–409
Collier J (2000) Autonomy and process closure as the basis for functionality. Ann N Y Acad Sci 901:280–291
Crother BI, Murray CM (2018) Linking a biological mechanism to evolvability. J Phylogenetics Evol Biol 6:192. https://doi.org/10.4172/2329-9002.1000192
Crother BI, Murray CM (2019) Early usage and meaning of evolvability. Ecol Evol 9 (7):3784–3793. https://doi.org/10.1002/ece3.5002
Cyranoski D (2017) SARS outbreak linked to Chinese bat cave. Nature 522:15–16
D'Alelio D, Mele BH, Libralato S, d'Alcala MR, Jordan F (2019) Rewiring and indirect effects underpin modularity reshuffling in a marine food web under environmental shirts. Ecol Evol 2019:11631–11646. https://doi.org/10.1002/ece3.5641
Daniels BC, Chen Y-J, Sethna JP, Gutenkunst RN, Myers CR (2008) Sloppiness, robustness, and evolvability in systems biology. Curr Opin Biotechnol 19:389–395
Darwin C (1872) The origin of species, 6th edn. John Murray, London
Depew D, Weber B (1995) Darwinism evolving. Bradford Books, Cambridge, MA
Dobson AP, May RM (1986) Patterns of invasions by pathogens and parasites. In: Mooney HA, Drake JA (eds) Ecology and biological invasions of North America and Hawaii. Springer, Berlin, pp 58–76
Duffey E (1964) The terrestrial ecology of Ascension Island. J Appl Ecol 1:219–251
Eker S, Ilmola-Sheppard L (2020) Systems thinking to understand national well-being from a human capital perspective. Sustainability 12:1931. https://doi.org/10.3390/su12051931
Ellen MacArthur Foundation (2012) Towards the circular economy: economic and business rationale for an accelerated transition. Report of the Ellen MacArthur Foundation
Elton CS (1958) The ecology of invasions by animals and plants. Methuen, London
Evans FE (1956) Ecosystem as the basic unit in ecology. Science 123:1127–1128
Geissdoerfer M, Savaget P, Bocken NMP, Hultink EJ (2017) The circular economy—a new sustainability paradigm? J Clean Prod 143:757–768. https://doi.org/10.1016/j.jclepro.2016.12.048
Gladyshev GP (1996) Thermodynamic direction of biological evolution: model and reality. Izv Akad Nauk Ser Biol 4:389–397
Greene R (1591–1592) A notable discovery of coosnage, 1591. In: Harrison GB (ed) The second part of conny-catching, 1592. 1923 reprint. John Lane, London

Gurría A (2019) The economy of well-being. Speech to OECD Brussels, 8 July 2019
Hagen JB (1992) An entangled bank: the origins of ecosystem ecology. Rutgers University Press, New Brunswick, NJ
Hairston NG (1980) Species packing in the salamander genus *Desmognathus*: what are the interspecific interactions involved? Am Nat 115:354–366
Hansen TF, Solvin TM, Pavlicev M (2019) Predicting evolutionary potential: a numerical test of evolvability measures. Evolution 73:689–703
Hardin G (1968) The Tragedy of the Commons. Science 162:1243–1248
Hardin G (1974) Living on a lifeboat. Bioscience 24:561–568
Hardy A (1967) Great waters. Collins, London
Henttonen H, Fuglei E, Gower CN, Haukisalmi V, Ims RA, Niemimaa J, Yoccoz NG (2001) *Echinococcus multilocularis* on Svalbard: introduction of an intermediate host has enabled the local life-cycle. Parasitology 123:547–552
Hirata H (1993) Information of organization in ecological systems: nutrient > energy > carbon. J Theor Biol 162:187–194
Hoberg E (2010) Invasive processes, mosaics and the structure of helminth parasite faunas. Rev Sci Tech 29:255–272
Ilmola-Sheppard L, Strelkovskii N, Rovenskaya E, Abramzon S, Bar R (2020) A systems description of the national well-being system. Version 1.0. Working paper, International Institute for Applied Systems Analysis
Jacobs M (2019) Beyond growth: towards a new economic approach. Report of the Secretary-General's Advisory Group on a new growth narrative OECD New Approaches to Economic Challenges Group Conference Paris, 17 Sept 2019
Jørgensen PS, Folke C, Carroll SP (2019) Evolution in the Anthropocene: informing governance and policy. Annu Rev Ecol Evol Syst 50:231–23.20
Kitano H (2004) Biological robustness. Nat Rev Genet 5:826–837
Kitano H (2007) Towards a theory of biological robustness. Mol Syst Biol 3:137
Levin SA (ed) (1975) Ecosystem analysis and prediction. Proceedings of a Conference on Ecosystems, Alta, Utah, July 1974. Society for Industrial and Applied Mathematics Institute for Mathematics and Society, Philadelphia, PA
Li W, Shi Z, Yu M, Ren W, Smith C, Epstein JH, Wang H, Crameri G, Hu Z, Zhang H, Zhang J, McEachern J, Field H, Daszak P, Eaton BT, Zhang S, Wang LF (2005) Bats are natural reservoirs of SARS-like coronaviruses. Science 310:676–679
Lindeman RL (1942) The trophic-dynamics aspect of ecology. Ecology 23:399–418
Llena-Nozal A, Martin N, Murtin F (2019) The economy of well-being: creating opportunities for people's well-being and economic growth. OECD Statistics Working Papers 2019/02
Lorenz EN (1963) Deterministic non-periodic flow. J Atmos Sci 20:130–141
Lorenz EN (1972) Predictability: does the flap of a butterfly's wings in Brazil set off a tornado in Texas? Presentation given at the 139th meeting for the American Association for the Advancement of Science, Washington, DC
Lotka AJ (1913) Evolution from the standpoint of physics, the principle of the persistence of stable forms. Sci Am Supplement 75:345–6, 354, 379
Lotka AJ (1925) Elements of physical biology. Williams and Wilkins, Baltimore, MD
Lundgren EJ, Ramp D, Rowan J, Middleton O, Schowanek SD, Sanisidro O, Carroll SP, Davis M, Sandom CJ, Svenning JC, Wallach AD (2020) Introduced herbivores restore Late Pleistocene ecological functions. Proc Natl Acad Sci U S A. https://doi.org/10.1073/pnas.1915769117
Lymbery AJ, Morine M, Kanani HG, Beatty SJ, Morgan DI (2014) Co-invaders: the effects of alien parasites on native hosts. Int J Parasitol Parasites Wildl 3:171–177
MacArthur R (1969) Species packing, and what competition minimizes. Proc Natl Acad Sci 64:1369–1371
MacArthur RH, Levins R (1967) The limiting similarity, convergence and divergence of coexisting species. Am Nat 101:377–385

MacArthur RH, Wilson EO (1967) The theory of island biogeography. Princeton University Press, Princeton, NJ
Malcicka M, Agosta SJ, Harvey JA (2015) Multi level ecological fitting: indirect life cycles are not a barrier to host switching and invasion. Glob Chang Biol 21:3210–3218
Malthus TR (1797) An essay on the principle of population, 1st edn. Allen and Unwin, London
Matsuno K (1989) Protobiology: physical basis of biology. CRC Press, Boca Raton, FL
Matsuno K (1995) Consumer power as the major evolutionary force. J Theor Biol 173:137–145
Matsuno K (1996) How many trophic levels are there? J Theor Biol 180:105–109
Matsuno K (1998) Competence of natural languages for describing the physical origin of life. In: van de Vijver G, Salthe SN, Delpos M (eds) Evolutionary systems: biological and epistemological perspectives on selection and self-organization. Kluwer Academic, Dordrecht, pp 295–306
Matsuno K (2000) Material contextualization in time. In: Taborsky E (ed) Semiotics, evolution, energy. Shaker Verlag, Aachen, pp 219–230
Maurer BA, Brooks DR (1991) Energy flow and entropy production in biological systems. J Ideas 2:48–53
May RM, MacArthur RH (1972) Niche overlap as a function of environmental variability. Proc Natl Acad Sci U S A 69:1109–1113
Maynard Smith J, Szathmáry E (1995) The major transitions in evolution. Oxford University Press, Oxford
Mollentze N, Streicher DG (2020) Viral zoonotic risk among taxonomic orders of mammalian and avian reservoir hosts. Proc Natl Acad Sci U S A 117(17):9423–9430. https://doi.org/10.1073/pnas.1919176117
Morens DM, Fauci AS (2020) Emerging pandemic diseases: how we got to COVID-19. Cell. https://doi.org/10.1016/j.cell.2020.08
Neiman S (2002) Evil in modern thought: an alternative history of philosophy. Princeton University Press, Princeton, NJ
Odum EP (1953) Fundamentals of ecology, 1st edn. W.B. Saunders, Philadelphia
Odum EP (1957) The ecosystem approach in the teaching of ecology illustrated with sample class data. Ecology 38:531–535
Odum EP (1964) The new ecology. Bioscience 14:14–16
Odum EP (1968) Energy flow in ecosystems: a historical review. Am Zool 8:11–18
Odum HT (1983) Systems ecology. Wiley, New York, NY
Ostrom E, Burger J, Field CB, Norgaard RB, Policansky D (1999) Revisiting the commons: local lessons, global challenges. Science 284:278–282
Page SE (2011) Diversity and complexity. Princeton University Press, Princeton, NJ
Polanyi K (1944) The great transformation. Farrar & Rinehart, New York
Poon LLM, Chu DKW, Chan KH, Wong OK, Ellis TM, Leung YHC, Lau SKP, Woo PCY, Suen KY, Yuen KY, Guan Y, Peiris JSM (2005) Identification of a Novel Coronavirus in Bats. J Virol 79:2001–2009
Preston F (2012) A global redesign? shaping the circular economy. Chatham House Energy, Environment and Resource Governance Briefing Paper. Royal Institute for International Affairs, London
Queller DC (1997) Cooperators since life began. Q Rev Biol 72:184–188
Roughgarden J, Feldman M (1975) Species packing and predation pressure. Ecology 56:489–492
Sariatli F (2017) Linear economy versus circular economy: a comparative and analyzer study for optimization of economy for sustainability. Visegrad J Bioecon Sustain Dev 6:31–34
Skawinska E, Zalewski RI (2018) Circular economy as a management model in the paradigm of sustainable development. Management 22:217–233
Sniegowski PD, Murphy HA (2006) Evolvability. Curr Biol 16:831–834
Stahel WR (2016) Circular economy. Nature 531:435–438
Stiglitz J, Sen A, Fitoussi J-P (2008) Report by the Commission on the measurement of economic performance and social progress, Paris
Tansley AG (1935) The use and abuse of vegetational concepts and terms. Ecology 16:284–307
Taraschewski H (2006) Hosts and parasites as aliens. J Helminthol 80:99–128

Torchin ME, Lafferty KD, Dobson AP, McKenzie VJ, Kuris AM (2003) Introduced species and their missing parasites. Science 421:628–630
Ulanowicz RE (1980) An hypothesis on the development of natural communities. J Theor Biol 85:223–245
Ulanowicz RE (1986) Growth & development: ecosystems phenomenology. Springer, New York
Ulanowicz RE (1988) On the importance of higher-level models in ecology. Ecol Model 43:45–56
Ulanowicz RE (1997) Ecology: the ascendent perspective. Columbia University Press, New York
Vasas V, Fernando C, Szilagyi A, Zachar I, Santos M, Szathmary E (2015) Primordial evolvability: impasses and challenges. J Theor Biol 381:29–38
Vasbinder JW (ed) (2019) Disrupted balance—society at risk. World Scientific, Singapore
Waring RH (1989) Ecosystems: fluxes of matter and energy. In: Cherrett JM (ed) Ecological concepts: the contribution of ecology to an understanding of the natural world. Blackwell Scientific, Oxford, pp 17–41
Werner EE (1977) Species packing and niche complementarity in three sunfishes. Am Nat 111:553–578
Whittaker RH (1962) Classification of natural communities. Bot Rev 28:1–239
Wicken JS (1987) Evolution, thermodynamics and information: extending the Darwinian paradigm. Oxford University Press, Oxford
Wilkinson DM (2004) The parable of Green Mountain: Ascension Island, ecosystem construction and ecological fitting. J Biogeogr 31:1–4
Willis AJ (1997) The ecosystem: an evolving concept viewed historically. Funct Ecol 11:268–271
Wilson DS (1980) The natural selection of populations and communities. Benjamin/Cummings, New York
Wilson EO (1984) Biophilia. Harvard University Press, Cambridge, MA
Wilson DS, Dugatkin LA (1997) Group selection and assortative interactions. Am Nat 149:336–351
Zettler ER, Mincer TJ, Amaral-Zettler LA (2013) Life in the "Plastisphere": microbial communities on plastic marine debris. Environ Sci Technol 47:7137–7146
Zhou P, Yang X, Wang X, Hu B, Zhang L, Zhang W, Si H, Zhu Y, Li B, Huang C, Chen H, Chen J, Luo Y, Guo H, Jiang R, Liu M, Chen Y, Shen X, Wang X, Zheng X, Zhao K, Chen Q, Deng F, Lin-Lin Liu L, Yan B, Zhan F, Wang Y, Xiao G, Shi Z (2020) A pneumonia outbreak associated with a new coronavirus of probable bat origin. Nature 579:270–273. https://doi.org/10.1038/s41586-020-2012-7

The manufacturer's authorised representative in the EU is Springer Nature Customer Service Centre GmbH, Europaplatz 3, 69115 Heidelberg, Germany. If you have any concerns regarding our products, please contact ProductSafety@springernature.com

Printed and bound by CPI Group (UK) Ltd, Croydon, CR0 4YY

30/09/2026

02225696-0014